# *Handbook of Materials Behavior Models*

VOLUME III Multiphysics Behaviors

Includes Three-Volume Index

# *Handbook of Materials Behavior Models*

VOLUME III Multiphysics Behaviors
Includes Three-Volume Index

EDITOR

JEAN LEMAITRE

*Université Paris 6*
*LMT-Cachan*
*Cachan Cedex*
*France*

ACADEMIC PRESS

A Harcourt Science and Technology Company

San Diego San Francisco New York Boston
London Sydney Tokyo

This book is printed on acid-free paper. ♾



ACADEMIC PRESS
A Division of Harcourt, Inc.
525 B Street, Suite 1900, San Diego, CA 92101-4495, USA
http://www.academicpress.com

Academic Press
Harcourt Place, 32 Jamestown Road, London, NW1 7BY, UK
http://www.academicpress.com

Library of Congress Catalog Number: 2001089698

Set International Standard Book Number: 0-12-443341-3
Volume 1 International Standard Book Number: 0-12-443342-1
Volume 2 International Standard Book Number: 0-12-443343-X
Volume 3 International Standard Book Number: 0-12-443344-8

Printed in the United States of America
01 02 03 04 05 MB 9 8 7 6 5 4 3 2 1

# CONTENTS

## CHAPTER 1
# Background on mechanics of materials



## CHAPTER 2
# Elasticity, viscoelasticity



## CHAPTER 3
# Yield limit

## CHAPTER 4
## *Plasticity*

## CHAPTER 5
## *Viscoplasticity*

## CHAPTER 6
## *Continuous damage*

## CHAPTER 7
## *Cracking and fracture*

## CHAPTER 9
## *Multiphysics coupled behavior*

## CHAPTER 10
## *Composite media, biomaterials*

## CHAPTER 11
## *Geomaterials*

# FOREWORD

We know that there is an abundance of models for particular materials and for specific types of mechanical responses. Indeed, both the developers of models and their users sometimes criticize this situation, for different reasons. The presence of different models that attempt to describe the same material and response is due not only to the personal style of their inventors, but also to a desirable element of competition that drives the progress in the field.

Given this situation, the selection of the proper constitutive model from all the available ones can be difficult for users or even materials modelers when they are not experts in the field. This Handbook is the first attempt to organize a wide range of models and to provide assistance in model selection and actual application. End-users will find here either potential models relevent for their application and ready to be used for the problem at hand, or an entrance to the specific technical literature for more details.

Recognizing the breadth of the field as well as the unavoidable personal touch of each approach, Jean Lemaitre has chosen to include in this Handbook the writings of as many as 130 authors. Drawing on his wide experience developing and using constitutive models for many materials, he has addressed his worldwide network of colleagues, all experts in their pertinent subject, to accomplish this difficult task. Yet, even though the Handbook covers an unprecedented range of materials and types of behavior, it is only a sample of currently available models, and other choices would have been possible. Indeed, more choices will become possible as the development of novel and improved material models continues.

*Erik van der Giessen*
*Koiter Institute Delft*
*Delft University of Technology*
*The Netherlands*

# INTRODUCTION

**Why a Handbook of models?** Handbooks are often compilations of characteristic numbers related to well-established laws or formulae that are ready to apply. In this case of the behavior of materials, no unique law exists for any phenomenon, especially in the range of nonlinear phenomena. This is why we use the term *model* instead of *law.* During the past thirty years many models have been proposed, each of them having its own domain of validity. This proliferation is partly due to advances in computers. It is now possible to numerically simulate the "in-service life" of structures subjected to plasticity, fatigue, crack propagation, shock waves and aging for safety and economy purposes. The time has come to try to classify, compare, and validate these models to help users to select the most appropriate model for their applications.

**How is the Handbook organized?** All solid materials are considered, including metals, alloys, ceramics, polymers, composites, concrete, wood, rubber, geomaterials such as rocks, soils, sand, clay, and biomaterials. But the Handbook is organized first by phenomena because most engineering mesomodels apply to different materials.

- In the first volume: "Deformation of Materials," the first chapter is an attempt to give general methodologies in the "art" of modeling with special emphasis, on domains of validity in order to help in the choice of models, in the selection of the appropriate materials for each specific application, and in the consideration of the so-called "size effect" in engineering structures. Chapter 2 to 5 deal, respectively, with elasticity and viscoelasticity, yield limit, plasticity, and viscoplasticity.
- The second volume is devoted to "Failure of Materials": continuous damage in Chapter 6, cracking and fracture in Chapter 7, friction and wear in Chapter 8.
- In the third volume "Multiphysics Behaviors" are assembled. The different possible couplings are described in Chapter 9. Chapters 10 and 11 are devoted to special classes of materials: composites and

geomaterials, respectively, because they each corresponds to a particular modeling typed and moreover to a self-organized community of people.
- In each chapter the different sections written by different authors describe one model with its domain of validity, its background, its formulation, the identification of material parameters for as many materials as possible, some advice on implementation or use of the model, and some references. The order of the sections follows as much as possible from physical and micromechanical oriented models to more phenomenological and engineering oriented ones.

**How to use the Handbook?**

- Search by phenomena: This is the normal order of the Handbook described in the "Contents".
- Search by model name: Unfortunately, not all models have a name, and some of them have several. Look in the list of contributors, where the names of all authors are given.
- Search by type of application: Each chapter begins with a **chapter introduction** in which a few words are written on each section.

If you do not find exactly what you are looking for, please remember that the best model is the simplest which gives you what you need and nothing more! In case of any difficulty, get in touch with the author(s), whose address is given after the title of each section.

**Some personal comments.** This Handbook has been initiated by the editor of "Academic Press" who gave me much freedom to organize the book. It took me two years to prepare the contents, to obtain the agreement of more than 100 authors, to ask for manuscripts, to ask again and again (and again for some of them!) to review and to obtain the final material. It was an exciting experience for which all actors must be thanked: the editors Z. Ruder, G. Franklin, and M. Filion, all the authors who are still my friends, my colleagues and friends from the LMT–Cachan who often advised me on subjects and authors and particularly Erik van der Giessen, who helped me in the selection of the subjects, who corrected the chapter introductions, and who agreed to write the foreword, Catherine Genin who was so kind and so efficient with letters, fax, e-mail, telephone, disks and manuscripts and answered so many questions in order to obtain the materials in due time. I must also mention Annie, my wife, who accepted 117 articles on the table at home!

*Merci à tous,*
*Jean Lemaitre*
*Septembre 2000*

# CONTRIBUTORS

*Numbers in parentheses indicate the section of authors' contributions.*

**ELIAS C. AIFANTIS** (4.12), Aristotle University of Thessaloniki, Thessaloniki, 54006 Greece, and Michigan Technological University, Houghton, Michigan

**HOLM ALTENBACH** (3.6), Fachbereich Ingenieurwissenschaften, Martin-Luther-Universität Halle-Wittenberg, D-06099 Halle (Saale), Germany

**P. ARCHAMBAULT** (9.12), Laboratoire de Science et Génie des Matériaux et de Métallurgie, UMR 7584 CNRS/INPL, Ecole des Mines de Nancy, Parc de Saurupt, 54042 Nancy Cedex, France

**ELLEN M. ARRUDA** (5.12), Department of Mechanical Engineering, University of Michigan, Ann Arbor, Michigan

**C. BATHIAS** (6.11), Laboratoire de Mécanique de la Rupture, CNAM/ITMA, 2 rue Conté, 75003 Paris, France

**ZDENEK P. BAZANT** (1.3), Northwestern University, Evanston, Illinois, USA

**AHMED BENALLAL** (4.11, 9.2), Laboratoire de Mécanique et Technologie, ENS de Cachan/CNRS/Université Paris 6, 61 avenue du Président Wilson, 94235 Cachan, France

**ALBRECHT BERTRAM** (5.2), Otto-von-Guericke-University Magdeburg, Universitätsplatz 2, 39106 Magdeburg, Germany

**YVES BERTHIER** (8.2), Laboratoire de Mécanique des Contacts, UMR CNRS-INSA de Lyon 5514, Batiment 113, 20, Avenue Albert Einstein, 69621 Villeurbanne Cedex, France

**B. J. BESSON** (7.8), École des Mines de Paris, Centre des Matériaux, UMR CNRS 7533, BP 87, 91003 Evry Cedex, France

**J. F. BESSELING** (5.5), j.f.besseling@wbmt.tudelft.nl

**M. BERVEILLER** (4.2), Laboratoire de Physique et Mécanique des Matériaux, Ile du Saulcy, 57045 Metz Cedex, France

**RENÉ BILLARDON** (9.16), ENS de Cachan/CNRS/Université Paris 6, 61 avenue du Président Wilson, 94235 Cachan Cedex, France

**SOL R. BODNER** (5.7), Technion Israel Institute of Technology, Haifa 32000, Israel

**MARY C. BOYCE** (5.12), Department of Mechanical Engineering, Center for Materials Science and Engineering, Massachusetts Institute of Technology, Cambridge, Massachusetts, USA

**YVES BRECHET** (1.2), L.T.P.C.M. BP75, Institut National Polytechnique de Grenoble, 38402 St Martin d'Heres Cedex, France

**DAVID BROEK** (7.14), 263 Dogwood Lane, Westerville, Ohio, USA

**HUY DUONG BUI** (7.3, 8.5), Laboratoire de Mécanique des Solides, École Polytechnique, 91128 Palaiseau, France Electricité de France, R&D, Clamart, France

**ALAIN BURR** (10.7), Laboratoire de Physico-Chimie Structurale et Macro-moléculaire, UMR 7615, ESPCI, 10 rue Vauquelin, 75231 Paris Cedex 05, France

**ESTEBAN P. BUSSO** (9.7), Department of Mechanical Engineering, Imperial College, University of London, London, SW7 2BX, United Kingdom

**GEORGES CAILLETAUD** (5.3), Centre des Matériaux de l'École des Mines de Paris, UMR CNRS 7633, BP 87, F91003 Evry Cedex, France

**VALTER CARVELLI** (10.8), Department of Structural Engineering, Technical University (Politecnico) of Milan, Piazza Leonardo Da Vinci 32, 20133 Milano, Italy

**SERGE CESCOTTO** (4.3, 9.3), Département MSM, Université de Liege, 1, chemin des Chevreuils bât.B52/3, 4000 Liege, Belgique

**J. L. CHABOCHE** (5.8), O.N.E.R.A., DMSE, BP 72, 92322 Châtillon Cedex, France and LASMIS, Troyes University of Technology, BP 2060, 10010 Troyes Cedex, France

**A. H. C. CHAN** (11.10), School of Engineering, University of Birmingham, United Kingdom

**C. L. CHOW** (6.4), Department of Mechanical Engineering, University of Michigan-Dearborn

**TRAIAN CICONE** (8.4), Dept. of Machine Elements and Tribology, Polytechnic University of Bucharest, Romania

**N. D. CRISTESCU** (2.3), 231 Aerospace Building, University of Florida, Gainesville, Florida

**OLIVIER COUSSY** (11.7), Laboratoire Central des Ponts et Chaussées, Paris, France

**STEPHEN C. COWIN** (10.10), New York Center for Biomedical Engineering, School of Engineering, The City College, New York

**CHRISTIAN CUNAT** (2.6, 9.17), LEMTA, UMR CNRS 7563, ENSEM INPL 2, avenue de la Forêt-de-Haye, 54500 Vandoeuvre-lès-Nancy, France

**PATRICK DANGLA** (11.7), Laboratoire Central des Ponts et Chaussées, Paris, France

**FÉLIX DARVE** (11.2), L'INP Grenoble, L3S-BP 53 38041 Grenoble, France

**YANNIS F. DAFALIAS** (4.8, 11.11), Civil and Environmental Engineering, The George Washington University, Washington, D.C.

**S. DENIS** (9.12), Laboratoire de Science et Génie des Matériaux et de Métallurgie, UMR 7584 CNRS/INPL, Ecole des Mines de Nancy, Parc de Saurupt, 54042 Nancy Cedex, France

**CHANDRA S. DESAI** (9.10), Department of Civil Engineering and Engineering Mechanics, The University of Arizona, Tucson, Arizona, USA

**RODRIGUE DESMORAT** (6.14, 6.15), Université Paris 6-LMS, 8, Rue du Capitaine Scott, F-75015 Paris, France

**MARTA DRAGON-LOUISET** (8.5), Laboratoire de Mécanique des Solides, Ecole Polytechnique, 91128 Palaiseau, France

**DANIEL C. DRUCKER** (3.2, 3.7), Department of Aerospace Engineering, Mechanics Engineering Service, University of Florida, 231 Aerospace Building, Gainesville, Florida 32611

**GEORGE J. DVORAK** (10.5), Rensselaer Polytechnic Institute, Troy, New York

**L. DUCHÊNE** (4.3), Département MSM, Université de Liege, 1, chemin des chevreuils bât.B52/3, 4000 Liege, Belgique

**FERNAND ELLYIN** (5.10), Department of Mechanical Engineering, University of Alberta, Edmonton, AB, Canada

**F. D. FISCHER** (9.13), Montanuniversit. at Leoben, Franz-Josef-Strasse 18, A-8700 Leoben, Austria

**DOMINIQUE FRANÇOIS** (7.5), École Centrale de Paris, Châtenay-Malabry, F92 295, France

**JEAN FRÊNE** (8.4), Laboratoire de Mécanique des Solides, Université de Poitiers, France

**E. GAUTIER** (9.12), Laboratoire de Science et Génie des Matériaux et de Métallurgie, UMR 7584 CNRS/INPL, École des Mines de Nancy, Parc de Saurupt, 54042 Nancy Cedex, France

**A. GODINAS** (4.3), Département MSM, Université de Liege, 1, chemin des Chevreuils bât.B52/3, 4000 Liege, Belgium

**DIETMAR GROSS** (7.6), Institute of Mechanics, TU Darmstadt, Hochschulstrasse 1, D 64289 Darmstadt

**ANNE MARIE HABRAKEN** (4.3, 9.3), Département MSM, Université de Liege, 1, chemin des Chevreuils bât.B52/3, 4000 Liege, Belgique

**JEAN-MARC HAUDIN** (4.9), CEMEF – BP 207, 06904 Sophia Antipolis, France

**D. R. HAYHURST** (9.4), Department of Mechanical Engineering, UMIST, PO Box 88, Manchester M60 1QD, United Kingdom

**FRANÇOIS HILD** (7.4, 10.7), LMT-Cachan, 61 avenue du Président Wilson, F-94235 Cachan Cedex, France

**LAURENT HIRSINGER** (9.16), ENS de Cachan/CNRS/Université Paris 6, 61 avenue du Président Wilson, 94235 Cachan Cedex, France

**K. HO** (5.6), Yeungnam University, Korea

**GERHARD A. HOLZAPFEL** (10.11), Institute for Structural Analysis, Computational Biomechanics, Graz University of Technology, 8010 Graz, Austria

**KOZO IKEGAMI** (2.5), Tokyo Denki University, Kanda-Nishikicho 2-2, Chiyodaku, Tokyo 101-8457, Japan

**TATSUO INOUE** (9.11), Department of Energy Conversion Science, Graduate School of Energy Science, Kyoto University, Yoshida-Honmachi, Sakyo-ku, Kyoto, Japan

**HIROMASA ISHIKAWA** (4.7), Hokkaido University, N13, W8, Kita-ku, Sapporo 060-8628, Japan

**HENRIK MYHRE JENSEN** (7.7), Department of Solid Mechanics, 404, Technical University of Denmark, DK-2800 Lyngby, Denmark

**KOJI KATO** (8.7), Tohoku University, Aramaki-Aza-Aoba 01, Sendai 980-8579, Japan

**JANUSZ R. KLEPACZKO** (4.10), Metz University, Laboratory of Physics and Mechanics of Materials, Ile du Saulcy, 57045 Metz, France

**W. G. KNAUSS** (7.13), California Institute of Technology, Pasadena, California

**DUSAN KRAJCINOVIC** (6.3), Arizona State University, Tempe, Arizona

**E. KREMPL** (5.6), Mechanics of Materials Laboratory, Rensselaer Polytechnic Institute, Troy, New York

**TADEUSZ KURTYKA** (3.5), CERN – European Organization for Nuclear Research, CH-1211 Geneve 23, Switzerland

**PIERRE LADEVÈZE** (10.6), LMT-Cachan, ENS de Cachan/CNRS/Université Paris 6, 61 avenue du Président Wilson, 94235 Cachan Cedex, France

**FREDERICK A. LECKIE** (10.7), Department of Mechanical and Environmental Engineering, University of California, Santa Barbara, California

**J-B. LEBLOND** (7.3, 9.14), Laboratoire de Modélisation en Mécanique, Université de Pierre et Marie Curie, Paris, France

**JEAN LEMAITRE** (1.1, 2.1, 3.1, 4.1, 5.1, 6.1, 6.2, 6.14, 6.15, 7.1, 8.1, 9.1, 10.1, 11.1), Université Paris 6, LMT-Cachan, 61, avenue du Président Wilson, F-94235 Cachan Cedex, France

**GIULIO MAIER** (10.8), Department of Structural Engineering, Technical University (Politecnico) of Milan, Piazza Leonardo Da Vinci 32, 20133 Milano, Italy

**DIDIER MARQUIS** (4.4), Laboratoire de Mécanique et Technologie, Ecole Normale Supérieure de Cachan, 61 avenue du Président Wilson, 94230 Cachan, France

**MAJID T. MANZARI** (11.11), Department of Mechanics, National Technical University of Athens, 15773, Hellas, and Civil and Environmental Engineering, University of California, Davis, California

**JACKY MAZARS** (6.13, 7.2), LMT-Cachan, Ecole Normale Superieure de Cachan, 61, avenue du Président Wilson, 94235 Cachan, France and L35-Institut National Polytechniquede Grenoble, F38041 Grenoble Cedex 9, France

**FREDERIC MESLIN** (10.9), LMT-Cachan, ENS de Cachan, Université Paris 6, 61 avenue du Président Wilson, 94235 Cachan Cedex, France

**ALAIN MOLINARI** (5.4), Laboratoire de Physique et Mécanique des Matériaux, École Nationale d'Ingénieurs, Université de Metz, Ile du Saulcy, 57045 Metz-Cedex, France

**BERNARD MONASSE** (4.9), CEMEF – BP 207, 06904 Sophia Antipolis, France

**HAEL MUGHRABI** (7.11), Universität Erlangen-Nürnberg, Institut für Werkstoffwissenschaften, Martensstr. 5, D-91058 Erlangen, Germany

**N. STALIN-MULLER** (7.3), Laboratoire de Mécanique des Solides, École Polytechnique, 91128 Palaiseau, France

**Z. MRÓZ** (4.5), Institute of Fundamental Technological Research, Warsaw, Poland

**SUMIO MURAKAMI** (6.7), Nagoya University, Furo-cho, Chikusa-ku, Nagoya, 464-8603 Japan

**ROBERTO NOVA** (11.9), Milan University of Technology (Politecnico), Department of Structural Engineering, Milan, Italy

**A. NEEDLEMAN** (6.5), Brown University, Division of Engineering, Providence, Rhode Island and Department of Mechanical Engineering, Solid Mechanics, Technical University of Denmark, 2800 Lyngby, Denmark

**SIA NEMAT-NASSER** (5.11, 11.5), Center of Excellence for Advanced Materials, Department of Mechanical and Aerospace Engineering, University of California, San Diego, California

**R. W. OGDEN** (2.2, 2.4), Department of Mathematics, University of Glasgow, Glasgow G12 8QW, United Kingdom

**NOBUTADA OHNO** (4.6), Department of Mechanical Engineering, Nagoya University, Chikusa-ku, Nagoya 464-8603, Japan

**J. URGEN OLSCHEWSKI** (5.2), BAM-V.2, Unter den Eichen 87, 12200 Berlin, Germany

**FLORENCE OSSART** (9.16), ENS de Cachan/CNRS/Université Paris 6, 61 avenue du Président Wilson, 94235 Cachan Cedex, France

**M. PASTOR** (11.10), Centro de Estudios y Experimentación de Obras Públicas and ETS de Ingenieros de Caminos, Madrid, Spain

**PIOTR PERZYNA** (9.5), Institute of Fundamental Technological Research, Polish Academy of Sciences, Świętokrzyska 21, 00-049 Warsaw, Poland

**GILLES PIJAUDIER-CABOT** (6.13), Laboratoire de Génie Civil de Nantes Saint-Nazaire, École Centrale de Nantes, BP 92101, F-44321 Nantes Cedex 03, France

**A. PINEAU** (7.8), École des Mines de Paris, Centre des Matériaux, UMR CNRS 7533, BP 87, 91003 Evry Cedex, France

**ARNAUD POITOU** (10.9), LMT-Cachan, ENS de Cachan, Université Paris 6, 61 avenue du Président Wilson, 94235 Cachan Cedex, France

**RACHID RAHOUADJ** (2.6, 9.17), LEMTA, UMR CNRS 7563, ENSEM INPL 2, Avenue de la Forêt-de-Haye, 54500 Vandoeuvre-lès-Nancy, France

**MICHEL RAOUS** (8.6), Laboratoire de Mécanique et d'Acoustique, 31, chemin Joseph Aiguier, 13402 Marseille Cedex 20, France

**GILLES ROUSSELIER** (6.6), EDF/R&D Division, Les Renardières, 77818 Moret-sur-Loing Cedex, France

**J. W. RUDNICKI** (11.6), Department of Civil Engineering, Northwestern University, Evanston, Illinois, USA

**KATSUHIKO SASAKI** (4.7), Hokkaido University, N13, W8, Kita-ku, Sapporo 060-8628, Japan

**A. R. SAVKOOR** (8.3), Vehicle Research Laboratory, Delft University of Technology, Delft, The Netherlands

**R. A. SCHAPERY** (2.7), Department of Aerospace Engineering and Engineering Mechanics, The University of Texas, Austin, Texas

**ROBERT SCHIRRER** (6.12), Institut Charles Sadron, 6 rue Boussingault, F-67083 Strasbourg, France

**SABINE M. SCHLOGL** (9.8), Koiter Institute Delft, Delft University of Technology, The Netherlands

**B. A. SCHREFLER** (11.8), Department of Structural and Transportation Engineering, University of Padua, Italy

**L. SIMONI** (11.8), Department of Structural and Transportation Engineering, University of Padua, Italy

**PETROS SOFRONIS** (9.9), Department of Theoretical and Applied Mechanics, University of Illinois at Urbana-Champaign, 104 South Wright Street, Urbana, Illinois

**DARRELL SOCIE** (6.8), Department of Mechanical Engineering, University of Illinois, Urbana, Illinois

**CLAUDE STOLZ** (8.5), Laboratoire de Mécanique des Solides, École Polytechnique, 91128 Palaiseau, France

**PIERRE M. SUQUET** (10.3), LMA/CNRS, 31 Chemin Joseph Aiguier, 13402, Marseille, Cedex 20, France

**ALBERTO TALIERCIO** (10.8), Department of Structural Engineering, Technical University (Politecnico) of Milan, Piazza Leonardo Da Vinci 32, 20133 Milano, Italy

**EIICHI TANAKA** (5.9), Department of Mechano-Informatics and Systems, Graduate School of Engineering, Nagoya University, Furo-cho, Chikusa-ku, Nagoya 464-8603, Japan

**MASATAKA TOKUDA** (9.15), Department of Mechanical Engineering, Mie University, Kamihama 1515 Tsu 514-8507, Japan

**V. TVÉRGAÁRD** (6.5), Department of Mechanical Engineering, Solid Mechanics, Technical University of Denmark, 2800 Lyngby, Denmark

**K. DANG VAN** (6.9), Laboratoire de Mechanique des Solid, École Polytechnique, 91128 Palaiseau, France

**ERIK VAN DER GIESSEN** (9.8, 10.2), University of Groningen, Applied Physics, Micromechanics of Materials, Nyenborgh 4, 9747 AG Groningen, The Netherlands

**P. VAN HOUTTE** (3.3), Department MTM, Katholieke Universiteit Leuven, B-3000 Leuven, Belgium

**J. G. M. VAN MIER** (11.12), Delft University of Technology, Faculty of Civil Engineering and Geo-Sciences, Delft, The Netherlands

**IOANNIS VARDOULAKIS** (11.4), National Technical University of Athens, Greece

**GEORGE Z. VOYIADJIS** (9.4), Department of Civil and Environmental Engineering, Louisiana State University, Baton Rouge, Louisiana

**MICHAEL WAUTERS** (9.3), MSM-1, Chemin des Chevreuils B52/3 4000 Lïege, Belgium

**YONG WEI** (6.4), Department of Mechanical Engineering, University of Michigan-Dearborn, USA

**J. R. WILLIS** (10.4), Department of Mathematical Sciences, University of Bath, Bath BA2 7AY, United Kingdom

**A. TOSHIMITSU YOKOBORI, JR.** (7.9), Fracture Research Institute, Graduate School of Engineering, Tohoku University, Aoba 01 Aramaki, Aoba-ku Sendai-shi 980-8579, Japan

**TAKEO YOKOBORI** (7.10), School of Science and Engineering, Teikyo University, Utsunomiya, Toyosatodai 320-2551, Japan

**XING ZHANG** (7.12), Division 508, Department of Flight Vehicle Design and Applied Mechanics, Beijing University of Aeronautics and Astronautics, Beijing 100083, China

**JUN ZHAO** (7.12), Division 508, Department of Flight Vehicle Design and Applied Mechanics, Beijing University of Aeronautics and Astronautics, Beijing 100083, China

**Y. ZHU** (9.3), ANSYS Inc., Houston, Texas

**O. C. ZIENKIEWICZ** (11.10), Department of Civil Engineering, University of Wales at Swansea, United Kingdom

**MICHA ŻYCZKOWSKI** (3.4), Cracow University of Technology, ul. Warszawska 24, PL-31155 Kraków, Poland

CHAPTER 9

# *Multiphysics Coupled Behavior*

SECTION 9.1

# Introduction to Coupled Behaviors

Jean Lemaitre
*Université Paris 6, LMT-Cachan, 61 avenue du Président Wilson, 94235 Cachan Cedex, France*

All previous chapters, more or less, have dealt with one physical phenomenon even if several micromechanisms are involved, but in practice sometimes two or more phenomena may occur simultaneously with *interactions* between them. It is this *coupling* which is the subject of Chapter 9. For example, the cooling of a steel from high temperature may induce a phase transformation, which induces heterogeneous volume change, which induces internal stresses, which induce plastic strains, which induce damage, which may induce cracks, etc. Unfortunately, this is what can happen in welded structures: the damage changes the elastic strains, which change the stresses, which change the phase transformation.

The *thermodynamics of irreversible* processes is intensively used to build models of coupled phenomena because if *internal variables* for each phenomenon may be *qualitatively* identified, the coupling is contained directly in the *state and dissipative potentials*, provided their *quantitative identification* can be performed.

The state potential $\psi$ is a function of all the states variables $v_i$,

$$\psi = \psi(v_1, v_2, v_3, \ldots, v_i)$$

from which the associated variables (or thermodynamical forces $A_i$) are derived as

$$A_i = \frac{\partial \psi}{\partial v_i}$$

There is *state coupling* between the phenomena *i* and *j* if

$$\frac{\partial^2 \varphi}{\partial v_i \partial v_j} \neq 0$$

*Handbook of Materials Behavior Models.* ISBN 0-12-443341-3.

and uncoupling if

$$\frac{\partial^2 \varphi}{\partial v_i \partial v_j} = 0$$

The dissipative potential is a function of the associate variables

$$\varphi = \varphi(A_1, A_2, A_3, \ldots, A_i)$$

from which the kinetic evolution laws are derived according to

$$\dot{v}_i = \frac{D\varphi}{DA_i}$$

There is *kinetic coupling* between phenomena *i* and *j* if

$$\frac{D^2\varphi}{DA_i DA_j} \neq 0$$

The difficulty is, of course, to *choose* the proper variables $v_i$ to *choose* the form of the function $\psi$, and to *choose* the form of the function $\varphi$. The various models differ essentially in the choices made for $v_i$, $\psi$, and $\varphi$.

*Damage* induces an elastic and plastic *softening* effect which must be taken into account for precise calculation of metal forming or the limit state of ductility and fatigue. See Section 9.2 for isotropic damage, Sections 9.3 and 9.4 for anisotropic damage, and Sections 9.5 and 9.6 for the additional effect of temperature.

Coupling *oxidation* with viscoplasticity is described in Section 9.7, and interaction with *hydrogen* may be found in Sections 9.8 and 9.9. Section 9.10 is a general description of *disturbed state* in a hierarchical framework.

Metallurgical couplings such as *phase transformations* inducing plasticity in metallic materials are important in many industrial processes such as quenching, welding, casting, surface treatments. Basic aspects suitable for numerical simulations may be found in Sections 9.11, 9.12, 9.13, and 9.14.

*Shape memory* properties of some alloys are the result of another coupling between martensite and austenite exchanges (see Section 9.15).

Modeling the coupling between *elasticity* and *magnetism* needs a good representation of hysteresis behavior (see Section 9.16).

*Aging* is also an important phenomenon to take into consideration for long time range structures because it modifies the strength of materials, in particular *polymers* (see Section 9.17).

SECTION 9.2

# Elastoplasticity and Viscoplasticity Coupled with Damage

AHMED BENALLAL
*Laboratoire de Mécanique et Technologie, ENS de Cachan/CNRS/Université Paris 6, 61 avenue du Président Wilson, 94235 Cachan, France*

Contents



## 9.2.1 APPLICATION

The model to be described herein was developed mainly for metals and metallic alloys at room or high temperatures. It can model their behaviors, progressive degradation, and final rupture along monotonic or cyclic loadings. It can also be used to analyze ductile rupture, creep rupture, and low-cycle fatigue.

## 9.2.2 BACKGROUND

Within continuum damage mechanics and in the case of isotropic damage, a scalar variable $D$ is introduced which measures the average effects of the

*Handbook of Materials Behavior Models.* ISBN 0-12-443341-3.

degradation properties of the material on its mechanical response. In the context of continuum thermodynamics, this is an internal variable which completes the set of variables describing the behavior of the material. In the case of elastic-plastic or elastic-viscoplastic materials, the other main variables are the plastic strain, the accumulated plastic strain $p$ describing isotropic hardening, and the internal strain $\alpha$ dealing with kinematic hardening effects. The coupling between the deformation behavior and the degradation of the material is usually undertaken by the effective stress concept $\tilde{\sigma} = \sigma/1 - D$ associated with the strain equivalence principle. The total strain $\varepsilon$ is partitioned into an elastic part and an inelastic part.

The free energy potential is of the form

$$\Psi = \frac{1}{2\rho}(1 - D)E_{ijkl}(\varepsilon_{kl} - \varepsilon^p_{kl})(\varepsilon_{kl} - \varepsilon^p_{kl}) + \frac{1}{2}C\alpha_{ij}\alpha_{ij} + g(p)$$

with $E_{ijkl}$ being the matrix of elastic constants. In the case of isotropy we have

$$E_{ijkl} = \left(K - \frac{2G}{3}\right)\delta_{ij}\delta_{kl} + G(\delta_{ik}\delta_{jl} + \delta_{il}\delta_{kj})$$

where $K$ and $G$ are the bulk and shear moduli, respectively.

This leads to the state laws

$$\tilde{\sigma}_{ij} = \rho\frac{\partial\Psi}{\partial\varepsilon_{ij}} = \left(K - \frac{2G}{3}\right)(\varepsilon_{kk} - \varepsilon^p_{kk})\delta_{ij} + 2G(\varepsilon_{ij} - \varepsilon^p_{ij})$$

$$R = -\rho\frac{\partial\Psi}{\partial p} = -g'(p)$$

$$X_{ij} = -\rho\frac{\partial\Psi}{\partial\alpha_{ij}} = -C\alpha_{ij}$$

$$Y = -\rho\frac{\partial\Psi}{\partial D} = W_e$$

$R$, **X**, and $Y$ are the forces associated with the internal variables $p$, $\alpha$, and $D$, respectively. $R$ is the size of the elastic domain, **X** is the back stress, and $Y$ is the elastic strain energy. A very common form for $g(p)$ is

$$g(p) = Q[1 - \exp(-bp)]$$

where $Q$ and $b$ are material-dependent parameters.

## 9.2.3 DESCRIPTION OF THE MODEL

Beside these ingredients, the model is based on a yield domain defined by a yield function:

$$f(\tilde{\sigma}, \mathrm{X}, R) = J_2(\tilde{\sigma} - \mathrm{X}) - R - \sigma_y$$

where $J_2(\tilde{\sigma} - \mathrm{X}) = \sqrt{\frac{3}{2}(\tilde{s}_{ij} - X_{ij})(\tilde{s}_{ij} - X_{ij})}$ is the second invariant of the active stress $\tilde{\sigma} - \mathrm{X}$. $S$ is the stress deviator, and $\sigma_y$ the initial yield stress in uniaxial tension. The model is also based on an inelastic potential

$$F(\tilde{\sigma}, \mathrm{X}, R) = J_2(\tilde{\sigma} - \mathrm{X}) + \frac{X_{ij}X_{ij}}{4\gamma} + \frac{Y^2}{2S} - R - \sigma_y$$

involving the material-dependent parameters $S$ and $\gamma$ and leading to the evolution laws for the internal variables

$$\dot{\varepsilon}^p_{ij} = \dot{\lambda}\frac{\partial F}{\partial \sigma_{ij}} = \frac{3}{2(1-D)}\dot{p}\frac{s_{ij} - X_{ij}}{J_2(\tilde{\sigma} - \mathrm{X})}$$

$$\dot{p} = \dot{\lambda}\frac{\partial F}{\partial R} = \dot{\lambda}$$

$$\dot{\alpha}_{ij} = \dot{\lambda}\frac{\partial F}{\partial X_{ij}} = \dot{\varepsilon}^p_{ij} - \gamma\dot{p}X_{ij}$$

$$\dot{D} = \dot{\lambda}\frac{\partial F}{\partial Y} = \frac{Y}{S}\dot{p}$$

The inelastic multiplier $\dot{\lambda}$ is obtained by the consistency condition $\dot{f} = 0$ for rate-independent elastic-plastic material. For rate-dependent elastic-visco-plastic materials, it is given by the viscosity law

$$\dot{\lambda} = \frac{1}{\eta}\langle\Phi(f)\rangle$$

where $\eta$ is a material parameter and $\Phi$ a positive increasing function of $f$. Currently used forms for $\Phi$ are the power law and the exponential formulae, given respectively by

$$\Phi(f) = \left(\frac{f}{K}\right)^n, \quad \Phi(f) = \exp\left(\frac{f}{K} - 1\right)$$

Here $K$ and $n$ are also material-dependent parameters.

The model is completed by

- a local initiation criterion often defined by a damage threshold in terms of a critical accumulated plastic strain $p_D$, i.e., such that $D = 0$ if $p \leq p_D$;

- a rupture criterion often defined by critical value $D_c$ for the damage parameter.

## 9.2.4 IDENTIFICATION

The identification procedure is dependent on the rate-dependent or rate-independent character of the material. For both types of materials, elastic and hardening properties are first obtained during the first stages of tests, when no damage is present. These are generally obtained by cyclic tests under strain control. For proportional loadings, tension and compression tests are usually sufficient.

For rate-independent materials, the unloading branches of the stress-strain loops define the elastic domain. Its center represents exactly the back stress $X$ (in the loading direction), and its size gives the amount of istropic hardening through $2(R + \sigma_y)$. Experimental measures of the position of the center and its size and their plots versus the plastic strain or the cumulated plastic strain allow us to obtain the hardening parameters $C$, $\gamma$, and the function $g(p)$.

For rate-dependent materials, the unloading branches of the stress-strain response also include the viscous stress. Therefore, before carrying out the previous procedure, one must first identify viscous properties. This is usually done by carrying out tests at different strain rates or more simply by ending the cyclic tests after saturation of the hardening by a relaxation period.

Finally, obtention of the degradation properties is obtained by measuring the evolution of Young's modulus after damage has started. The decrease of this modulus gives the damage parameter and its plot versus the plastic or cumulated plastic strain and allows us to obtain the parameter $S$. For a general procedure identification, see Reference [3].

## 9.2.5 HOW TO USE THE MODEL

The model has been implemented both in its rate-independent and rate-dependent forms in ABAQUS through the UMAT subroutine. Generalized integration schemes have been used together with Newton–Raphson procedures. The corresponding consistent tangent operators were defined. For the rate-independent case, the onset of localization into planar bands (or loss of ellipticity) is signaled in order to avoid mesh dependency. The loss of the ellipticity criterion is implemented. Also, gradient and nonlocal procedures are being studied.

## REFERENCES

1. Lemaitre, J. (1994). A Course on Damage Mechanics, Springer Verlag.
2. Lemaitre, J., and Chaboche, J. L. (1989). Mechanics of Solid Materials, Cambridge.
3. Benallal, A. Thermoviscoplasticité et endommagement des structures.

SECTION 9.3

# A Fully Coupled Anisotropic Elastoplastic Damage Model

SERGE CESCOTTO[1], WAUTERS MICHAËL[2], ANNE-MARIE HABRAKEN[1], and Y. ZHU[3]

[1] *University of Liège, Liège, Belgium*
[2] *MSM-1, chemin des Chevreuils B52/3 4000 Liège, Belgium*
[3] *ANSYS Inc., Houston, Texas*

Contents



## 9.3.1 VALIDITY

This model predicts the damage growth and fracture appearance in ductile materials. Initially proposed by Zhu and Cescotto [3], it has been developed in the case of sheets, especially for deep drawing processes [2].



*Handbook of Materials Behavior Models*. ISBN 0-12-443341-3.

Important characteristics of this macroscopic model are its easy parameter identification and the anisotropic evolution of the damage and plastic surfaces computed from the energy equivalence assumption. Deep drawing simulations, described in Zhu and Cescotto [3], confirm its validity.

## 9.3.2 BACKGROUND

### 9.3.2.1 Basic Concepts

This damage model is included in the continuum theory of damage. The damage in the material is represented by a variable $\underline{D}$ corresponding to an average material degradation affecting stiffness, strength, and anisotropy. It reflects various types of damage at the microscale level, such as nucleation, growth and coalescence of voids, and microcracks. In the present model, $\underline{D}$ is a vector of three components, the damage in each orthotropic direction of the sheet:

$$\underline{D} = \langle D_1 \quad D_2 \quad D_3 \rangle^{\mathrm{T}}$$

The well-known concept of effective stress is used:

$$\underline{\bar{\sigma}} = \underline{\underline{M}}(\underline{D})\underline{\sigma}$$

with the "damage effect" tensor $\underline{\underline{M}}(\underline{D})$ of fourth order defined by its diagonal:

$$\underline{\underline{M}}(\underline{D}) = \mathrm{diag}\left[\frac{1}{1-D_1} \; \frac{1}{1-D_2} \; \frac{1}{1-D_3} \; \frac{1}{\sqrt{(1-D_2)(1-D_3)}} \right.$$
$$\left. \times \frac{1}{\sqrt{(1-D_1)(1-D_3)}} \; \frac{1}{\sqrt{(1-D_1)(1-D_2)}}\right]$$

The principle of energy equivalence is taken into account. It states that the complementary elastic energy stored in the damaged material has the same form as the one for a fictitious undamaged material except that the true stress tensor is replaced by the effective stress tensor. This principle is able to take into account not only the apparent Young moduli decrease but also the Poisson's coefficients decrease.

$$W_e(\underline{\sigma}, \underline{D}) = W_e(\underline{\bar{\sigma}}, \underline{D})$$

## 9.3.2.2 Thermodynamic Frame

The Helmholtz free energy takes the following form:

$$\rho\psi(\underline{\varepsilon}, T, \underline{D}, \alpha, \beta) = W_e(\underline{\varepsilon_e}, T, \underline{D}) + \psi_p(T, \alpha) + \psi_d(T, \beta)$$

where $W_e(\underline{\varepsilon_e}, T, \underline{D})$ is the elastic strain energy, $T$ is the temperature, $\psi_p(T, \alpha)$ is the free energy due to the plastic hardening, $\underline{\varepsilon}$ is the total strain tensor, $\psi_d(T, \beta)$ is the free energy due to the damage "hardening," $\underline{\varepsilon_e}$ is the elastic part of the strain tensor, $\alpha$ is an internal variable representing the cumulated plastic strain, and $\beta$ is an internal variable representing the cumulated damage.

Since the thermodynamic force $\underline{\sigma}$ (Cauchy or true stress tensor) is associated with the elastic strain $\underline{\varepsilon_e}$, a thermodynamic force $\underline{Y}$ can be associated with the damage tensor $\underline{D}$:

$$\underline{Y} = \rho\frac{\partial\psi}{\partial\underline{D}} = \frac{\partial W_e(\underline{\varepsilon_e}, T, \underline{D})}{\partial\underline{D}} = -\frac{\partial W_e(\underline{\sigma}, T, \underline{D})}{\partial\underline{D}}$$

$\underline{Y}$ is called the damage energy release rate; the negative sign of $\underline{Y}$ corresponds to the energy restitution due to damage growth. The forces associated with the cumulated plastic strain $\alpha$ and cumulated damage $\beta$ are, respectively, $R$ and $B$, called the plastic hardening threshold and the damage strengthening threshold.

With the hypothesis of uncoupling between mechanical plastic and damage dissipations, the second law of thermodynamics yields for an isothermal process:

$$\underline{\sigma} : \underline{\dot{\varepsilon}_p} - R\dot{\alpha} \geq 0 \qquad -\underline{Y}\,\underline{\dot{D}} - B\dot{\beta} \geq 0$$

This induces the existence of a plastic dissipative potential and a damage dissipative potential, chosen in this associated theory frame as the plastic yield criterion and the damage evolution criterion.

$$F_p(\underline{\sigma}, \underline{D}, R) = 0 \qquad F_d(\underline{Y}, B) = 0$$

By introducing the Lagrange multipliers $\dot{\lambda}_p$ and $\dot{\lambda}_d$, we define the $\Phi$ function:

$$\Phi = \underline{\sigma} : \underline{\dot{\varepsilon}_p} - R\dot{\alpha} - \underline{Y}\,\underline{\dot{D}} - B\dot{\beta} - \dot{\lambda}_p F_p - \dot{\lambda}_d F_d$$

If the criteria $F_P$ and $F_d$ are satisfied, the current values of $\underline{\sigma}$, $R$, $\underline{Y}$, $B$, and $\underline{D}$ will make the $\Phi$ function have a stationary value. We can build the following evolution laws:

$$\frac{\partial \Phi}{\partial \underline{\sigma}} = 0 \;\Rightarrow\; \dot{\underline{\varepsilon}}_p = \dot{\lambda}_p \frac{\partial F_p}{\partial \underline{\sigma}} \qquad \frac{\partial \Phi}{\partial \underline{Y}} = 0 \;\Rightarrow\; \dot{\underline{D}} = -\dot{\lambda}_d \frac{\partial F_d}{\partial \underline{Y}}$$

$$\frac{\partial \Phi}{\partial R} = 0 \;\Rightarrow\; \dot{\alpha} = -\dot{\lambda}_p \frac{\partial F_p}{\partial R} \qquad \frac{\partial \Phi}{\partial B} = 0 \;\Rightarrow\; \dot{\beta} = -\dot{\lambda}_d \frac{\partial F_d}{\partial B}$$

## 9.3.3 MODEL DESCRIPTION

### 9.3.3.1 Anisotropic Elasticity and Damage

When the material is damaged, the constitutive elastic law is given hereafter as

$$\underline{\sigma} = \overline{\underline{\underline{C}}}_e \underline{\varepsilon}_e$$

with $\overline{\underline{\underline{C}}}_e$ the elastic stiffness matrix of the damaged material. Using the principle of energy equivalence, the following relation can be written:

$$\overline{\underline{\underline{C}}}_e = \underline{\underline{M}}^{-1}(\underline{D}) \underline{\underline{C}}_e \underline{\underline{M}}^{-1}(\underline{D})$$

In the case of orthotropic materials, the damaged elastic tensor $\overline{\underline{\underline{C}}}_e^{-1}$ is recalled in Reference [2].

### 9.3.3.2 Anisotropic Plastic Yield Surface

In this model, the plastic yield surface is chosen as the Hill's one:

$$F_p(\underline{\sigma}, \underline{D}, R) = F_p(\bar{\underline{\sigma}}, R) = \overline{\sigma_{eq}} - R_0 - R(\alpha) = 0$$

with $R_0$ the initial elastic stress threshold and $\overline{\sigma_{eq}}$ the effective anisotropic equivalent stress:

$$\overline{\sigma_{eq}} = \{\tfrac{1}{2}(\bar{\underline{\sigma}} - \bar{\underline{\gamma}})^{\mathrm{T}} \underline{\underline{H}}(\bar{\underline{\sigma}} - \bar{\underline{\gamma}})\}^{1/2} = \{\tfrac{1}{2}(\underline{\sigma} - \underline{\gamma})^{\mathrm{T}} \bar{\underline{\underline{H}}}(\underline{\sigma} - \underline{\gamma})\}^{1/2}$$

where $\underline{\gamma}$ is the back-stress tensor and $\underline{\underline{H}}$ the plastic characteristic Hill tensor for the fictitious undamaged material:

$$\underline{\underline{H}} = \begin{bmatrix} G+H & -H & -G & 0 & 0 & 0 \\ -H & H+F & -F & 0 & 0 & 0 \\ -G & -F & F+G & 0 & 0 & 0 \\ 0 & 0 & 0 & N & 0 & 0 \\ 0 & 0 & 0 & 0 & L & 0 \\ 0 & 0 & 0 & 0 & 0 & M \end{bmatrix}$$

$F$, $G$, $H$, $L$, $M$, and $N$ are parameters characterizing the current state of plastic anisotropy. For a strain-hardening material, the uniaxial stress in one direction varies with an increase of plastic strain, and therefore the anisotropic parameters should also vary, since they are a function of the current yield stress. To determine them for the current state, we consider that the plastic work should be the same in each direction. For instance, in the case of a linear workhardening material, we have (Fig. 9.3.1):

$$W_{\text{pl}} = \frac{1}{2}\varepsilon_{eq}^{pl}(\sigma_F + \sigma_y) = \frac{1}{2Et}(\sigma_F^2 - \sigma_y^2) = \frac{1}{2Et_i}(\sigma_{Fi}^2 - \sigma_{yi}^2)$$

with $i = 1$ to 6 (three tensile curves and three shear curves) and $Et_i$ is the slope of the stress–plastic strain curve $i$.

We build the following ratios in terms of the new equivalent stress $\sigma_F$:

$$a_i = \left(\frac{\sigma_F}{\sigma_{Fi}}\right)^2 = \frac{\sigma_F^2}{\left(\frac{Et_i}{E}\right)(\sigma_F^2 - \sigma_y^2) + \sigma_{yi}^2} \qquad \begin{matrix} H = a_1 + a_2 - a_3 & N = a_4 \\ G = a_1 - a_2 + a_3 & L = a_5 \\ F = -a_1 + a_2 + a_3 & M = a_6 \end{matrix}$$

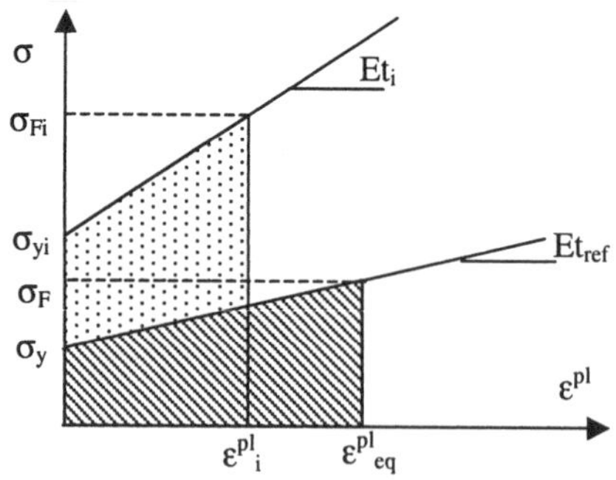


FIGURE 9.3.1.

In the present model, the rolling direction (direction 1) is taken as the reference one, so $a_1 = 1$. Using the evolution laws, we can develop the following expressions:

$$\dot{\underline{\varepsilon}}_p = \dot{\lambda}_p \frac{\partial F_p}{\partial \underline{\sigma}} = \dot{\lambda}_p \frac{\underline{\underline{M}}\,\underline{\underline{H}}\,\underline{\underline{M}}(\underline{\sigma} - \underline{\gamma})}{2\overline{\sigma}_{eq}} \quad \text{plastic flow rule}$$

$$\dot{R} = (1 - m)\dot{\lambda}_p \frac{\partial R}{\partial \alpha} \quad \text{isotropic hardening rule}$$

$$\dot{\underline{\gamma}} = m\dot{\lambda}_p \frac{(\underline{\sigma} - \underline{\gamma})}{\overline{\sigma}_{eq}} \frac{\partial R}{\partial \alpha} + \underline{\underline{\Omega}}\,\underline{\gamma}' - \underline{\gamma}'\underline{\underline{\Omega}} \quad \text{kinematic hardening rule}$$

$$F_p \leq 0 \qquad \dot{\lambda}_p \geq 0 \qquad \dot{\lambda}_p F_p = 0 \quad \text{plastic loading–unloading rule}$$

where $m$ is the ratio of kinematic to isotropic hardening and $\underline{\gamma}'$ is the deviatoric tensor of $\underline{\gamma}$.

#### 9.3.3.3 Damage Evolution Law and Damage Surface

By analogy to the plasticity, a damage criterion, chosen as a quadratic homogeneous function of the damage energy release rate $\underline{Y}$, is proposed [1]:

$$F_d = Y_{eq} - B_0 - B(\beta) = 0$$

with the equivalent damage energy release rate $Y_{eq}$ defined thanks to the damage charateristic tensor $\underline{\underline{J}}$:

$$Y_{eq} = \left\{ \frac{1}{2} \underline{Y}^t \underline{\underline{J}}\,\underline{Y} \right\}^{1/2}$$

A suitable tensor $\underline{\underline{J}}$, simple enough to be applied and able to describe the damage growth, has been proposed by Zhu and Cescotto [3]:

$$\underline{\underline{J}} = 2 \begin{bmatrix} J_1 & \sqrt{J_1 J_2} & \sqrt{J_1 J_3} \\ \sqrt{J_1 J_2} & J_2 & \sqrt{J_2 J_3} \\ \sqrt{J_1 J_3} & \sqrt{J_2 J_3} & J_3 \end{bmatrix}$$

In the case of damage hardening materials, the equivalent damage energy release rate increases with an increase of the total damage growth. As for the $\underline{\underline{H}}$ matrix components, the anisotropic parameters should also vary. Again, we suppose that for a current state of damage, the damage work done in each direction should be the same. In the case of a linear damage hardening characterized by its slope $Dt$, we have:

$$J_i = \left(\frac{Y_{eq}}{Y_i}\right)^2 = \frac{Y_{eq}^2}{\left(\frac{Dt_i}{Dt_{eq}}\right)(Y_{eq}^2 - Y_{0eq}^2) + Y_{0i}^2}$$

with $i = 1$ to 3 (the three principal directions of an orthotropic material). In this model, the reference direction is the rolling direction and $J_1 = 1$.

Similarly to the plastic flow, the anisotropic damage evolution laws are characterized as follows:

$$\dot{\underline{D}} = -\dot{\lambda}_d \frac{\partial F_d}{\partial \underline{Y}} = -\dot{\lambda}_d \frac{J\underline{Y}}{2Y_{eq}} \qquad \dot{\beta} = -\dot{\lambda}_d \frac{\partial F_d}{\partial B} = \dot{\lambda}_d \quad \text{damage evolution rules}$$

$$\dot{B} = \dot{\beta}\frac{\partial B}{\partial \beta} = \dot{\lambda} \quad \text{damage hardening rule}$$

$$F_d \leq 0 \quad \dot{\lambda}_d \geq 0 \quad \dot{\lambda}_d F_d = 0 \quad \text{damage loading–unloading rule}$$

## 9.3.4 CALIBRATION OF THE MODEL

All the parameters of this model can be identified only with tensile tests. These tests are characterized by $\alpha$, the angle between the rolling direction of the sheet and the axial direction of the sample (Fig. 9.3.2).

### 9.3.4.1 Elastic Parameters

Tensile tests are done in the domain of small displacements for the directions $\alpha = 0°$, 45°, and 90°. They allow us to compute the Young's moduli and the Poisson's coefficients. For sheets, tensile tests in the thickness direction are

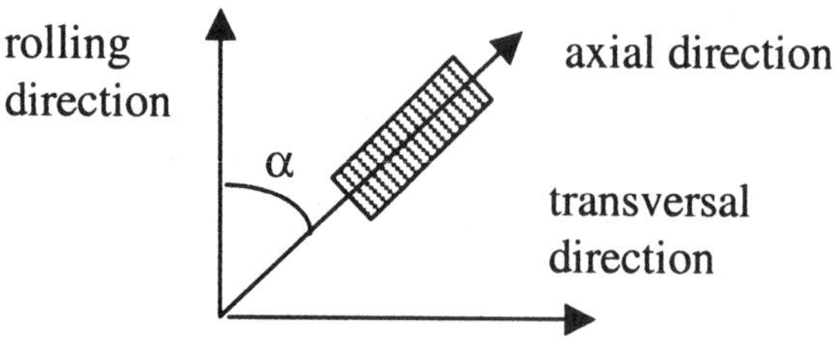


FIGURE 9.3.2.

not possible; we assume the following relationships or use values deduced from texture measures:

$$E_3 = \frac{(E_1 + E_2)}{2} \quad \text{and} \quad G_{13} = G_{23} = G_{12}$$

### 9.3.4.2 Plastic Parameters

We need to know the stress-strain curves in the three orthotropic directions and in the three shear planes. Uniaxial tensile tests in the domain of large displacements are done. We use a statistical method which consists of tests in the directions $\alpha = 0°, 15°, 30°, 45°, 60°, 75°$, and $90°$. For a given plastic work level, we minimize a functional of the corresponding stresses in these seven directions, leading to the Hill's parameters $F$, $G$, $H$, and $N$ ($N = M = L$ for sheets) [4]. It is then possible to deduce the stress-strain curves in the thickness direction and in the shear plane 1–2. The shear curves in the planes 1–3 and 2–3 are supposed to be equal to the one in the plane 1–2. The model is able to use multilinear stress-strain curves, with the number of points chosen by the user.

Obviously, the model needs to be furnished with effective stress-strain curves. Once we know the damage parameters of the material, these six effective stress-strain curves can be computed analytically from the true stress-strain curves.

$$\underline{\bar{\sigma}} = \underline{\underline{M}}(\underline{D}).\underline{\sigma} \quad \text{and} \quad \underline{\bar{\varepsilon}_e} = \underline{\underline{M}}^{-1}(\underline{D}).\underline{\varepsilon_e}$$

### 9.3.4.3 Damage Parameters

From the theory, we have the following relationship for a uniaxial test in direction $i$:

$$Y_i = \frac{\sigma_i^2}{E_i(1 - D_i)^3}$$

We need the three damage curves, characterized by the initial value of $Y_i$ and the slope $Dt_i$ of the damage curve (hypothesis of a linear behavior). If we suppose that the damaging phenomenon begins with the entry in plasticity, we have:

$$Y_{i0} = \frac{\sigma_{iy}^2}{E_i(1 - D_i)^3} = \frac{\sigma_{iy}^2}{E_i}$$

To find the effective curves, we compute the damage values $\langle D_1\, D_2\, D_3 \rangle$ associated with a given stress value. It corresponds to the resolution of the system hereafter:

$$Y_i = \frac{\sigma_i^2}{E_i(1 - D_i)^3} = Dt_i D_i + Y_{i0}$$

This yields a function $\sigma_i(D_i)$ presenting a maximum for a precise value of $D_i$. If we write the expression of $\sigma_{max}$ according to $D_i$, we obtain the following relation:

$$\sigma_{i\,max} = \sqrt{\frac{27E_i}{256Dt_i^3}}(Dt_i + Y_{i0})^2$$

Physically, $\sigma_{i\,max}$ should have the same value as the maximum stress on the real stress-strain curve. Therefore, if we know this value, knowing $Y_{i0}$ and $E_i$, it is possible to determine $Dt_i$.

This leads to the conclusion that, for this model, *no particular damage test is necessary* to find the damage parameters. This fact is simply linked to the strong hypothesis of a linear damage curve.

## 9.3.5 HOW TO USE THE MODEL

The model has been implemented in the nonlinear finite element code LAGAMINE developed by the MSM team; the coupled integration scheme is described in Zhu and Cescotto [3]. We only introduce the six effective stress-strain curves, the Poisson's coefficients, and the damage parameters. The predictions of the model can be illustrated by Figures 9.3.3 and 9.3.4, which describe the effect on a uniaxial tensile loading.

Figure 9.3.3 shows the negative influence of the damage on the plasticity. It compares the plastic surface obtained by using the true stresses (with damage) and the one which is defined in the fictitious case of no damaging phenomenon (using effective stresses). As it can be observed, the elastic zone is reduced by the damage.

Figure 9.3.4 illustrates the shape evolution of the yield locus during the deformation process. It can be seen that the anisotropic behavior of the material varies with the deformation. This anisotropy variation is intensified by the damaging phenomenon. See Table 9.3.1.

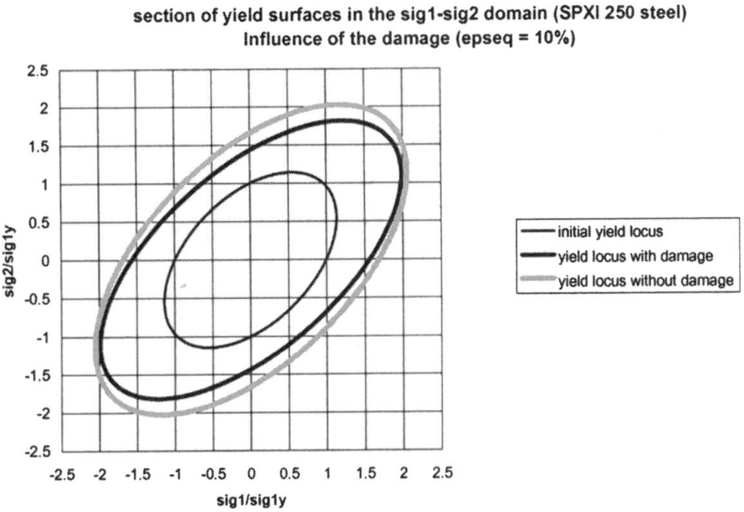


FIGURE 9.3.3.

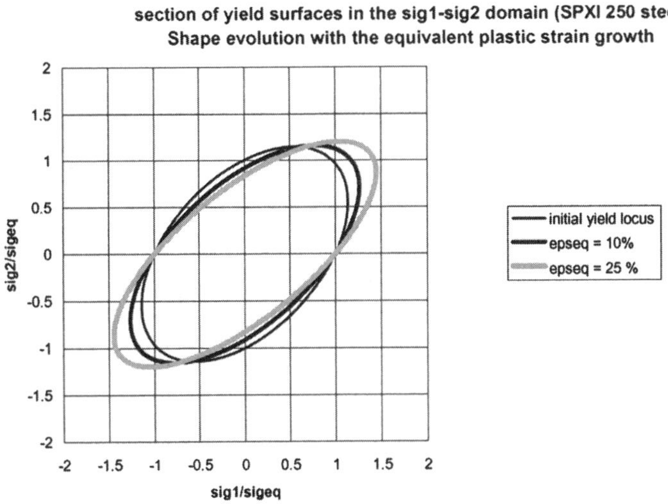


FIGURE 9.3.4.

TABLE 9.3.1 **Table of Parameters** ($E_i$, $G_{12}$, $\sigma_i$, $Y_{i0}$, **and** $Dt_i$ **are given in N/mm$^2$;** $\varepsilon_i$ **is given in %)**

| High-strength steel | | | | | | Deep drawing steel | | | | | |
|---|---|---|---|---|---|---|---|---|---|---|---|
| $E_1$ | $E_2$ | $G_{12}$ | $\nu_{12}$ | $\nu_{13}$ | $\nu_{23}$ | $E_1$ | $E_2$ | $G_{12}$ | $\nu_{12}$ | $\nu_{13}$ | $\nu_{23}$ |
| 203407 | 209274 | 77268 | 0.2884 | 0.3135 | 0.3188 | 204129 | 185016 | 194573 | 0.3371 | 0.3028 | 0.3092 |
| $\sigma_{y1}$ | | $\sigma_{y2}$ | $\sigma_{y3}$ | | $\sigma_{y12}$ | $\sigma_{y1}$ | | $\sigma_{y2}$ | $\sigma_{y3}$ | | $\sigma_{y12}$ |
| 249.3 | | 250.1 | 245.2 | | 142 | 112.8 | | 109.6 | 112.2 | | 69.4 |
| $\sigma_{max1}$ | $\varepsilon_1$ | $\sigma_{max2}$ | $\varepsilon_2$ | $\sigma_{max3}$ | $\varepsilon_3$ | $\sigma_{max1}$ | $\varepsilon_1$ | $\sigma_{max2}$ | $\varepsilon_2$ | $\sigma_{max3}$ | $\varepsilon_3$ |
| 509 | 35 | 521 | 30 | 540 | 30 | 420.5 | 41.4 | 409.5 | 37.7 | 573.6 | 32.8 |
| $Y_{10}$ | $Y_{20}$ | $Y_{30}$ | $Dt_1$ | $Dt_2$ | $Dt_3$ | $Y_{10}$ | $Y_{20}$ | $Y_{30}$ | $Dt_1$ | $Dt_2$ | $Dt_3$ |
| 0.3055 | 0.2989 | 0.2914 | 10.8 | 11.05 | 12.17 | 0.0623 | 0.0649 | 0.0647 | 7.96 | 8.33 | 15.77 |

# REFERENCES

1. Cordebois, J.P., and Sidoroff, F. (1979). Damage induced elastic anisotropy, EUROMECH 115, Villard de Lans.
2. Habraken, A.M., et al. (1997). Calibration and validation of an anisotropic elasto-plastic damage model for sheet metal forming, Mc NU97, Voyiadjis, ed., "Damage in Engineering Materials", *Studies in Applied Mechanics*, Elsevier.
3. Zhu, Y., and Cescotto, S. (1995). A fully coupled elasto-visco-plastic damage theory for anisotropic materials, *Int. J. Solids Structures* **32** (11): 1607–1641.
4. Noat, P., et al. (1995) Anisotropic 3-D modelling of hot rolling and plane strain compression of Al Alloys, in *Simulation of Material Processing: Theory, Methods and Applications*, Shen and Dawson, eds.

SECTION 9.4

# Model of Inelastic Behavior Coupled to Damage

GEORGE Z. VOYIADJIS

*Department of Civil and Environmental Engineering, Louisiana State University, Baton Rouge, Louisiana*

Contents



## 9.4.1 INTRODUCTION

The coupling of damage and inelastic deformation in materials have been studied only recently by Ju [1], Johansson and Runesson [2], and Voyiadjis and Park [3]. Both Ju [1] and Johansson and Runesson [2] made use of the effective stress utilizing a scalar measure of isotropic damage. Voyiadjis and Park [3] made use of the effective configuration by invoking the kinematics of damage through the use of a second-order damage tensor. The damage mechanism of materials has been studied extensively by Lemaitre [4], Krajcinovic [5], Murakami [6], and Voyiadjis and Park [7].

## 9.4.2 THEORETICAL FORMULATION

Damage variables can be presented through the internal state variables of thermodynamics for irreversible processes in order to describe the effects


*Handbook of Materials Behavior Models*. ISBN 0-12-443341-3.

of damage. One of the most crucial aspects of continuum damage mechanics is the appropriate choice of the damage variable, since the accuracy and reliability of the developed damage model are mostly dependent on the suitable selection of the type and numbers of these variables. Damage is expressed as follows [8]:

$$\phi_{ij} = \sum_{k=1}^{3} \hat{\phi}_k n_i^k n_j^k \tag{1}$$

Since the elasto-viscoplastic response of the damaged material is considered here, both hardening effects due to viscoplasticity and damage together with previous above description of microcrack distribution can be introduced as hidden internal state variables in the thermodynamic state potential. The form of this potential in terms of observable and internal state variables can be given as follows:

$$\psi = \psi(\boldsymbol{\varepsilon}^e, T, \boldsymbol{\varphi}, p, \boldsymbol{\alpha}, \kappa, \boldsymbol{\Gamma}) \tag{2}$$

where the $p$ and $\alpha$ variables characterize the isotropic and kinematic hardening in plasticity and viscoplasticity, respectively, and the $\kappa$ and $\Gamma$ variables characterize, respectively, the isotropic and kinematic hardening in damage. In Eq. 2, $T$ characterizes the temperature, and $\varepsilon^e$ is the elastic component of the strain tensor. In order to account for loading rate dependency and to regularize the localization problems, a viscous anisotropic damage mechanism needs to be implemented. Such a model accounts for retardation of the microcrack growth at higher strain rates. The rate-dependent plastic strain rate [9–11] and the damage rate are given as follows:

$$\dot{\varepsilon}^{vp} = ||\dot{\varepsilon}^{vp}||\, \boldsymbol{n}^{vp} + ||\dot{\phi}||\, \boldsymbol{n}^{vpd} \tag{3a}$$

and

$$\dot{\phi} = ||\dot{\varepsilon}^{vp}||\, \boldsymbol{n}^{dvp} + ||\dot{\phi}||\, \boldsymbol{n}^{d} \tag{3b}$$

Superscripts do not indicate a tensorial character but only a particular state of the variable, such as elastic (e), plastic (p), viscoplastic (vp), damage (d), viscoplastic-damage (vpd), etc. Superscripts in this work do not imply tensorial indices but only describe the type of material inelasticity. In Eq. 3, $||\dot{\varepsilon}^{vp}||$ and $||\dot{\phi}||$ are the magnitudes of the plastic strain rate and damage rate which can be decomposed into a product of two functions [12] using the Zener parameters such that:

$$||\dot{\varepsilon}_{vp}|| = \vartheta^{vp}(T) Z^{vp} \geq 0 \tag{4a}$$

$$||\dot{\phi}|| = \vartheta^{d}(T) Z^{d} \geq 0 \tag{4b}$$

The unit tensors $\boldsymbol{n}^{vp}$, $\boldsymbol{n}^{vpd}$, $\boldsymbol{n}^{dvp}$ and $\boldsymbol{n}^{d}$ are used to identify the direction of flow of the plastic strain and damage, respectively. The dynamic potentials $F^{vp}$ and $G^{d}$ for viscoplasticity and damage given by Voyiadjis and Deliktas [2000] are used in this work. As pointed out earlier, the internal state variables are introduced in the material model to represent the true response of the material due to the variation of the microstructure when subjected to external forces. The anisotropic structure of the material is usually defined in two forms, either as material-inherited or deformation-induced. The anisotropic nature of the composite material is material-inherited anisotropy. However, at the local level its constituents are isotropic materials. Therefore, the use of a micromechanical model to analyze the composite material deals with deformation-induced anisotropy. This phenomenon is characterized in the theory by using internal variables for the hardening terms and by using the second-order tensorial form of the damage variable. The general form of the internal variables can be defined as follows [12]:

$$\dot{A}_k = \text{hardening} - \text{dynamic recovery} - \text{static recovery} \tag{5}$$

The hardening terms represent the strengthening mechanism, and the recovery terms represent the softening mechanism. The hardening and dynamic recovery terms evolve with the deformation due to either plasticity or damage or both. The static recovery term evolves with time. The evolution equations of the internal variables for the rate-dependent behavior are described as follows:

$$\dot{X} = \frac{3}{2} H^{vp} \left( \dot{\varepsilon}^{vp} - \frac{3}{2} \frac{||X||}{L^{vp}} ||\dot{\varepsilon}^{vp}|| \, \boldsymbol{d}^{vp} \right) - \vartheta^{vp} X \cdot \boldsymbol{b}^{vp} \tag{6a}$$

$$\dot{R} = Q^{vp} \left( 1 - \frac{R}{Q^{vp}} \right) ||\dot{\varepsilon}^{vp}|| - \vartheta^{vp} \gamma(r) \tag{6b}$$

$$\boldsymbol{b}^{vp} = \frac{X}{||X||} \tag{6c}$$

$$\boldsymbol{d}^{vp} = (1 - \rho^{vp}) \boldsymbol{b}^{vp} + \rho^{vp} \frac{\boldsymbol{n}^{vp} \cdot X \cdot \boldsymbol{n}^{vp}}{||X||} \tag{6d}$$

Similarly, the evolution equations for the hardening variables of damage can be written analogously to that of plasticity as follows:

$$\dot{\Gamma} = \frac{3}{2} H^{d} \left( \dot{\phi} - \frac{3}{2} \frac{||\Gamma||}{L^{d}} ||\dot{\phi}|| \, \boldsymbol{d}^{d} \right) - \vartheta^{d} \Gamma \cdot \boldsymbol{b}^{d} \tag{7a}$$

$$\dot{K} = Q^{d} \left( 1 - \frac{K}{Q^{d}} \right) ||\dot{\phi}|| - \vartheta^{d} \gamma(\kappa) \tag{7b}$$

$$b^d = \frac{\Gamma}{|\Gamma|} \tag{7c}$$

$$d^d = (1 - \rho^d) b^d + \rho^d \frac{n^d \cdot \Gamma \cdot n^d}{||\Gamma||} \tag{7d}$$

where $\rho^{vp}$ defines the nonproportionality condition. In these equations, $H^{vp}$, $L^{vp}, Q^{vp}, H^d, L^d$, and $Q^d$ are the model parameters.

In a general state of deformation and damage, the effective stress tensor $\bar{\sigma}$ is related to the Cauchy stress tensor $\sigma$ by the following linear transformation [13]:

$$\bar{\sigma} = M : \sigma \tag{8}$$

where $M$ is a fourth-order linear transformation operator called the damage effect tensor. Depending on the form used for $M$, it is very clear from Eq. 8 that the effective stress tensor $\bar{\sigma}$ is generally nonsymmetric. One of the symmetrization methods is given by [14].

The elastoplastic stiffness for the damaged material can be obtained by using the incremental relation of Hooke's law in the effective stress space as follows:

$$\dot{\bar{\sigma}} = \bar{E} : \dot{\bar{\varepsilon}}^e \tag{9}$$

The effective stress rate, $\dot{\bar{\sigma}}$, and the elastic component of the effective strain rate, $\dot{\bar{\varepsilon}}$, in Eq. 9 can be transformed into the damage configuration by using Eq. 8 such that

$$\dot{\sigma} = m^{-1} : \bar{E} : m^{-1} : \dot{\varepsilon}^e \tag{10}$$

The final constitutive equations are given as follows [11]:

$$\dot{\sigma} = E^d : (\dot{\varepsilon} - \chi^p : \dot{\sigma}), \quad \dot{\sigma} = (I + E^d : \chi^p)^{-1} : E^d : \dot{\varepsilon} = D : \dot{\varepsilon} \tag{11}$$

where $E^d$ represents the elastic damaged stiffness and is defined by

$$E^d = m^{-1} : \bar{E} : m^{-1} \tag{12}$$

$D$ in Eq. 11 represents the elastoplastic damaged stiffness [11].

## 9.4.3 DISCUSSION OF THE RESULTS FOR THE ELASTO-VISCOPLASTIC DAMAGE ANALYSIS

The special-purpose computer program DVP-CALSET, is used for the numerical simulation of uniaxial loading of laminated systems. The numerical results using the proposed model show good correlations with the experimental results on titanium metal matrix composites reinforced with silicon carbide fibers for the laminate layups of $(90)_{8s}$, which are tested under uniaxial tension. The proposed model is also validated by showing good agreement of the numerical results with the experimental observations on the characteristic behavior of the metal matrix composite at different strain rates and temperatures. The computational analysis of the viscoplastic damage model is performed for the laminate systems of $(90)_{8s}$ at elevated temperatures of 538°C and 649°C in Figure 9.4.1 and for different strain rates in Figure 9.4.2. The viscoplastic and damage model parameters are given in Tables 9.4.1 and 9.4.2

## 9.4.4 CONCLUSION

A coupled incremental damage and plasticity theory for rate-independent and rate-dependent composite materials is presented in this work. Damage is characterized kinematically here through a second-order damage tensor for each material constituent, and its physical interpretation is also presented. This is related to the microcrack porosity and crack density within the unit

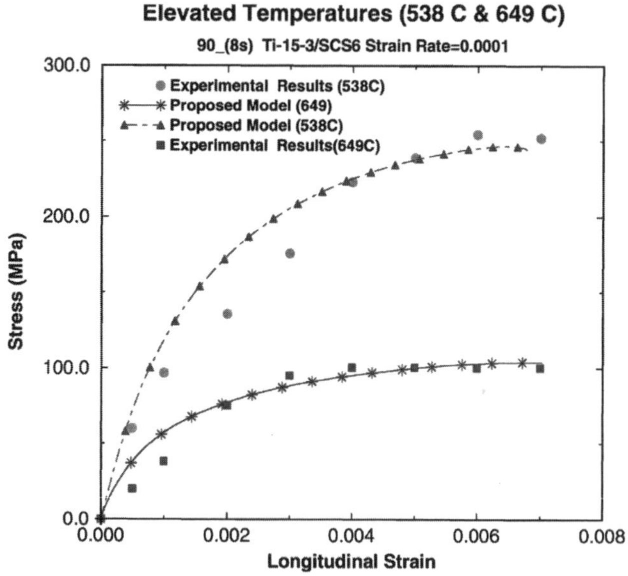


FIGURE 9.4.1 Comparison of viscoplastic damage model with experimental results [15] of the $(90)_s$, layup at different elevated temperatures of 538°C and 649°C.

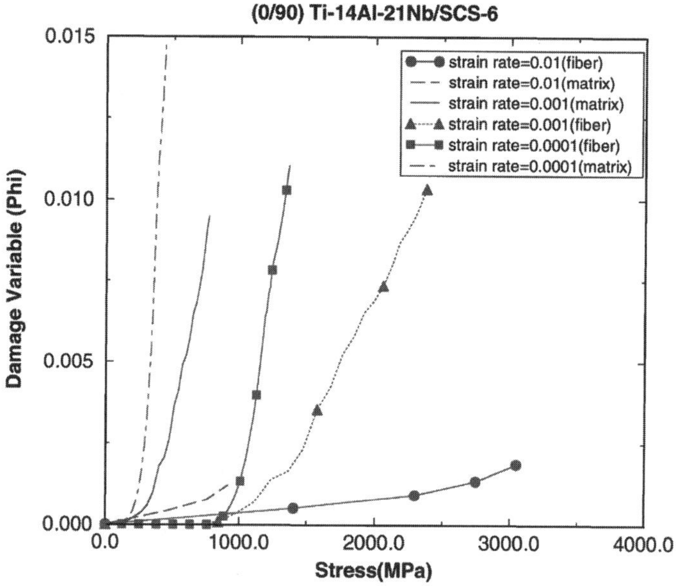


FIGURE 9.4.2 Strain rate effect on the damage variable $\phi$ of the $90_{(8s)}$ layup.

TABLE 9.4.1 Viscoelastic Model Parameters

| Properties | $T=21°C$ | $T=21°C$ | $T=21°C$ |
|---|---|---|---|
| $E^f$ (GPa) | 400 | 393 | 370 |
| $E^m$ (GPa) | 92.4 | 72.2 | 55 |
| $\sigma_y$ (MPa) | 689 | 45 | 15.5 |
| $H^{vp}$ (MPa) | 5000 | 50,000 | 50,000 |
| $L^{vp}$ (MPa) | 100 | 85 | 75 |
| $D^{vp}$ (MPa) | 840 | 450 | 85 |
| $n_1$ | 5.4 | 1.55 | 1.3 |

TABLE 9.4.2 Rate-Dependent Damage Parameters

| Properties | $T=538°C$ | $T=649°C$ |
|---|---|---|
| $\eta^m$ | 0.07 | 0.0047 |
| $\xi^m$ | 0.78 | 1.2 |
| $\nu^m$ | 0.0003 | 0.00003 |
| $\eta^f$ | 0.04 | 0.0045 |
| $\xi^f$ | 0.78 | 0.82 |
| $\nu^f$ | 0.00018 | 0.00005 |

cell and characterizes separately the damage in each material constituent. Damage is characterized here in terms of two fourth-order tensors, the damage in the matrix and the damage in the fiber. The damage in the fiber characterizes both the internal damage in the fiber due to the cracks and voids as well as the damage due to debonding. The present formulation seems to be robust and efficient for the type of problems presented in this work.

## REFERENCES


1. Ju, J. W. (1989). Energy based coupled elastoplastic damage theories: Constitutive modeling and computational aspect. *Int. J. Solids Struct.* **25**: 803–833.
2. Johansson, M., and Runesson, K. (1977). Viscoplastic with dynamic yield surface coupled to damage. *Comput. Mech.* **20**: 53–59.
3. Voyiadjis, G. Z., and Park, T. (1999). Kinematics description of damage for finite strain plasticity. *J. Eng. Sci.* **37**: 803–830.
4. Lemaitre, J. (1984). How to use damage mechanics. *Nucl. Eng. Design* **20**: 233–245.
5. Krajcinovic, D. (1983). Constitutive equations for damaging materials. *J. Appl. Mech.* **50**: 355–360.
6. Murakami, S. (1988). Mechanical modeling of material damage. *J. Appl. Mech.* **55**: 281–286.
7. Voyiadjis, G. Z., and Park, T. (1995). Local and interfacial damage analysis of metal–matrix composites. *Int. J. Appl. Mech.* **33**(11): 1595–1621.
8. Voyiadjis, G. Z., and Park, T. (1997). Anisotropic damage effect tensors for the symmetrization of the effective stress tensor. *J. Appl. Mech. ASME* **64**: 106–110.
9. Perzyna, P. (1963). The constitutive equations for rate sensitive plastic materials. *Appl. Math.* **20**: 321–332.
10. Perzyna, P. (1971). Thermodynamic theory of viscoplasticity. *Adv. Appl. Mech.* **11**: 313–354.
11. Voyiadjis, G. Z., and Deliktas, B. (2000). A coupled anisotropic damage model for the inelastic response of composite materials. *Comput. Methods Appl. Mech. Engr.* **183**: 159–199.
12. Freed, A. D., Chaboche, J. L., and Walker, K. P . (1991). A viscoplastic theory with thermodynamic considerations. *Acta Mechanica 90: 155–174.*
13. Murakami, S., and Ohno, N. (1981). A continuum theory of creep and creep damage, in *Creep in Structures*, pp. 422–444, Ponter, A. R. S., and Hayhurst, D. R., eds., Berlin: Springer.
14. Cordebois, J. P., and Sidoroff, F. (1979). Damage-induced elastic anisotropy, in *Mechanics of Behavior of Anisotropy Solids/N°295 Comportement Mechanique Des Solides Anisotropes*, pp. 19–22, Boehler, J. P., ed., MartinuesNijhoff.
15. Majumdar, B. S., and Newaz, G. M. (1992). Inelastic deformation of metal-matrix composite: Part I. Plasticity and damage mechanism. CR-189095, NASA.

SECTION 9.5

# Thermo-Elasto-Viscoplasticity and Damage

PIOTR PERZYNA

*Institute of Fundamental Technological Research, Polish Academy of Sciences, Świętokrzyska 21, 00–049 Warsaw, Poland*

Contents



## 9.5.1 BASIC ASSUMPTIONS AND DEFINITIONS

Let us assume that a continuum body is an open bounded set $\mathscr{B} \subset \mathbb{R}^3$, and let $\phi : \mathscr{B} \to \mathscr{S}$ be a $C^1$ configuration of $\mathscr{B}$ in $\mathscr{S}$. The tangent of $\phi$ is denoted $\mathbf{F} = T\phi$ and is called the deformation gradient of $\phi$.

Let $\{X^A\}$ and $\{x^a\}$ denote coordinate systems on $\mathscr{B}$ and $\mathscr{S}$, respectively. Then we refer to $\mathscr{B} \subset \mathbb{R}^3$ as the reference configuration of a continuum body with particles $X \in \mathscr{B}$ and to $\mathscr{S} = \phi(\mathscr{B})$ as the current configuration with points $\mathbf{x} \in \mathscr{S}$. The matrix $\mathbf{F}(\mathbf{X}, t) = \partial\phi(\mathbf{X}, t)/\partial\mathbf{X}$ with respect to the coordinate bases $\mathbf{E}_A(\mathbf{X})$ and $\mathbf{e}_a(\mathbf{x})$ is given by

$$F^a_A(\mathbf{X}, t) = \frac{\partial\phi^a}{\partial X^A}(\mathbf{X}, t) \tag{1}$$

where a mapping $\mathbf{x} = \phi(\mathbf{X}, t)$ represents a motion of a body $\mathscr{B}$.

*Handbook of Materials Behavior Models.* ISBN 0-12-443341-3.

We consider the local multiplicative decomposition

$$\mathbf{F} = \mathbf{F}^e \cdot \mathbf{F}^p \tag{2}$$

where $(\mathbf{F}^e)^{-1}$ is the deformation gradient that releases elastically the stress on the neighborhood $\phi[\mathcal{N}(\mathbf{X})]$ in the current configuration.

Let us define the total and elastic Finger deformation tensors

$$\mathbf{b} = \mathbf{F} \cdot \mathbf{F}^T, \quad \mathbf{b}^e = \mathbf{F}^e \cdot \mathbf{F}^{e^{\mathcal{T}}} \tag{3}$$

respectively, and the Eulerian strain tensors as follows:

$$\mathbf{e} = \frac{1}{2}(\mathbf{g} - \mathbf{b}^{-1}), \quad \mathbf{e}^e = \frac{1}{2}(\mathbf{g} - \mathbf{b}^{e^{-1}}), \tag{4}$$

where $\mathbf{g}$ denotes the metric tensor in the current configuration.

By the definition[1]

$$\mathbf{e}^p = \mathbf{e} - \mathbf{e}^e = \frac{1}{2}(\mathbf{b}^{e^{-1}} - \mathbf{b}^{-1}) \tag{5}$$

we introduce the plastic Eulerian strain tensor.

To define objective rates for vectors and tensors, we use the Lie derivative.[2] Let us define the Lie derivative of a spatial tensor field $\mathbf{t}$ with respect to the velocity field $\boldsymbol{\upsilon}$ as

$$\mathrm{L}_{\upsilon}\mathbf{t} = \phi_* \frac{\partial}{\partial t}(\phi^* \mathbf{t}) \tag{6}$$

where $\phi^*$ and $\phi_*$ denote the pull-back and push-forward operations, respectively.

The rate of deformation tensor is defined as follows:

$$\mathbf{d}^\flat = \mathrm{L}_{\upsilon}\mathbf{e}^\flat = \frac{1}{2}\mathrm{L}_{\upsilon}\mathbf{g} = \frac{1}{2}\left(g_{ac}\upsilon^c|_b + g_{cb}\upsilon^c|_a\right)\mathbf{e}^a \otimes \mathbf{e}^b \tag{7}$$

where the symbol $\flat$ denotes the index lowering operator and $\otimes$ the tensor product,

$$\upsilon^a|_b = \frac{\partial \upsilon^a}{\partial x^b} + \gamma^a_{bc}\upsilon^c \tag{8}$$

and $\gamma^a_{bc}$ denotes the Christoffel symbol for the general coordinate systems $\{x^a\}$. The components of the spin $\boldsymbol{\omega}$ are given by

$$\omega_{ab} = \frac{1}{2}\left(g_{ac}\upsilon^c|_b - g_{cb}\upsilon^c|_a\right) = \frac{1}{2}\left(\frac{\partial \upsilon_a}{\partial x^b} - \frac{\partial \upsilon_b}{\partial x^a}\right) \tag{9}$$

[1]For a precise definition of the finite elastoplastic deformation, see Perzyna [1].

[2]The algebraic and dynamic interpretations of the Lie derivative have been presented by Abraham *et al.* [2]; cf. also Marsden and Hughes [3].

Similarly,

$$\mathbf{d}^{e^b} = \mathrm{L}_v \mathbf{e}^{e^b}, \quad \mathbf{d}^{p^b} = \mathrm{L}_v \mathbf{e}^{p^b} \tag{10}$$

and

$$\mathbf{d} = \mathbf{d}^e + \mathbf{d}^p \tag{11}$$

Let $\boldsymbol{\tau}$ denote the Kirchhoff stress tensor related to the Cauchy stress tensor $\boldsymbol{\sigma}$ by

$$\boldsymbol{\tau} = J\boldsymbol{\sigma} = \frac{\rho_{Ref}}{\rho}\boldsymbol{\sigma} \tag{12}$$

where the Jacobian $J$ is the determinant of the linear transformation $\mathbf{F}(\mathbf{X}, t) = (\partial/\partial X)\phi(\mathbf{X}, t)$, and $\rho_{Ref}(\mathbf{X})$ and $\rho(\mathbf{x}, t)$ denote the mass density in the reference and current configuration, respectively.

The Lie derivative of the Kirchhoff stress tensor $\boldsymbol{\tau} \in \mathrm{T}^2(\mathscr{S})$ (elements of $\mathrm{T}^2[\mathscr{S}]$ are called tensors on $\mathscr{S}$, contravariant of order 2) gives

$$\begin{aligned}
\mathrm{L}_v\boldsymbol{\tau} &= \phi_* \frac{\partial}{\partial t}(\phi^* \boldsymbol{\tau}) \\
&= \left\{\mathbf{F} \cdot \frac{\partial}{\partial t}[\mathbf{F}^{-1} \cdot (\boldsymbol{\tau} \circ \phi) \cdot \mathbf{F}^{-1^T}] \cdot \mathbf{F}^T\right\} \circ \phi^{-1} \\
&= \dot{\boldsymbol{\tau}} - (\mathbf{d} + \boldsymbol{\omega}) \cdot \boldsymbol{\tau} - \boldsymbol{\tau} \cdot (\mathbf{d} + \boldsymbol{\omega})^T
\end{aligned} \tag{13}$$

where $\circ$ denotes the composition of mappings. In the coordinate system Eq. 13 reads

$$\begin{aligned}
(\mathrm{L}_v\boldsymbol{\tau})^{ab} &= F^a_A \frac{\partial}{\partial t}(F^{-1^A}_c \tau^{cd} F^{-1^B}_d) F^b_B \\
&= \frac{\partial \tau^{ab}}{\partial t} + \frac{\partial \tau^{ab}}{\partial x^c} v^c - \tau^{cb}\frac{\partial v^a}{\partial x^c} - \tau^{ac}\frac{\partial v^b}{\partial x^c}
\end{aligned} \tag{14}$$

Equation (14) defines the Oldroyd rate of the Kirchhoff stress tensor $\boldsymbol{\tau}$ (cf. Oldroyd [4]).

## 9.5.2 CONSTITUTIVE POSTULATES

Let us assume that (i) conservation of mass, (ii) balance of momentum, (iii) balance of moment of momentum, (iv) balance of energy, and (v) entropy production inequality hold.

We introduce the four fundamental postulates:

1. Existence of the free energy function. It is assumed that the free energy function is given by

$$\psi = \hat{\psi}(\mathbf{e}, \mathbf{F}, \vartheta; \boldsymbol{\mu}) \tag{15}$$

where $\mathbf{e}$ denotes the Eulerian strain tensor, $\mathbf{F}$ is the deformation gradient, $\vartheta$ is the temperature, and $\boldsymbol{\mu}$ denotes a set of the internal state variables.

To extend the domain of the description of the material properties and, in particular to take into consideration different dissipation effects, we have to introduce the internal state variables represented by the vector $\boldsymbol{\mu}$.

(2) Axiom of objectivity (spatial covariance). The constitutive structure should be invariant with respect to any diffeomorphism (any motion) $\boldsymbol{\xi}: \mathscr{S} \rightarrow \mathscr{S}$ (cf. Marsden and Hughes [3]). Assuming that $\boldsymbol{\xi}: \mathscr{S} \rightarrow \mathscr{S}$ is a regular, orientation-preserving map transforming $\mathbf{x}$ into $\mathbf{x}'$ and $T\boldsymbol{\xi}$ is an isometry from $T_{\mathbf{x}}\mathscr{S}$ to $T_{\mathbf{x}'}\mathscr{S}$, we obtain the axiom of material frame indifference (cf. Truesdell and Noll [5]).

(3) The axiom of the entropy production. For any regular motion of a body $\mathscr{B}$ the constitutive functions are assumed to satisfy the reduced dissipation inequality

$$\frac{1}{\rho_{Ref}}\boldsymbol{\tau} : \mathbf{d} - (\eta\dot{\vartheta} + \dot{\psi}) - \frac{1}{\rho\vartheta}\mathbf{q} \cdot \operatorname{grad} \vartheta \geq 0 \tag{16}$$

where $\rho_{Ref}$ and $\rho$ denote the mass density in the reference and actual configuration, respectively, $\boldsymbol{\tau}$ is the Kirchhoff stress tensor, $\mathbf{d}$ is the rate of deformation, $\eta$ is the specific (per unit mass) entropy, and $\mathbf{q}$ denotes the heat flow vector field. Marsden and Hughes [3] proved that the reduced dissipation inequality (Eq. 16) is equivalent to the entropy production inequality first introduced by Coleman and Noll [6] in the form of the Clausius–Duhem inequality. In fact, the Clausius–Duhem inequality gives a statement of the second law of thermodynamics within the framework of mechanics of continuous media; cf. Duszek and Perzyna [7].

(4) The evolution equation for the internal state variable vector $\boldsymbol{\mu}$ is assumed in the following form:

$$\mathrm{L}_{\upsilon}\boldsymbol{\mu} = \hat{\mathbf{m}}(\mathbf{e}, \mathbf{F}, \vartheta, \boldsymbol{\mu}) \tag{17}$$

where the evolution function $\hat{\mathbf{m}}$ has to be determined on the basis of a careful physical interpretation of a set of the internal state variables and analysis of available experimental observations.

The determination of the evolution function $\hat{\mathbf{m}}$ (in practice a finite set of the evolution functions) appears to be the main problem of the modern constitutive modeling.

## 9.5.3 FUNDAMENTAL ASSUMPTIONS

The main objective is to develop the rate-type constitutive structure for an elastic–viscoplastic material in which the effects of the plastic non-normality, plastic strain induced anisotropy (kinematic hardening), micro–damaged mechanism, and thermomechanical coupling are taken into consideration. To do this it is sufficient to assume a finite set of the internal state variables. For our practical purposes it is sufficient to assume that the internal state vector $\boldsymbol{\mu}$ has the form

$$\boldsymbol{\mu} = (\varepsilon^p, \xi, \boldsymbol{\alpha}) \tag{18}$$

where $\varepsilon^p$ is the equivalent viscoplastic deformation, i.e.,

$$\varepsilon^p = \int_0^t \left(\frac{2}{3}\mathbf{d}^p : \mathbf{d}^p\right)^{1/2} dt \tag{19}$$

The symbol $\xi$ is volume fraction porosity and takes account of microdamaged effects and $\boldsymbol{\alpha}$ denotes the residual stress (the back stress) and aims at the description of the kinematic hardening effects.

Let us introduce the plastic potential function $f = f(\tilde{J}_1, \tilde{J}_2, \vartheta, \boldsymbol{\mu})$, where $\tilde{J}_1, \tilde{J}_2$ denote the first two invariants of the stress tensor $\tilde{\boldsymbol{\tau}} = \boldsymbol{\tau} - \boldsymbol{\alpha}$.

Let us postulate the evolution equations as follows:

$$\mathbf{d}^p = \Lambda \mathbf{P}, \quad \dot{\xi} = \Xi, \quad \mathrm{L}_v \boldsymbol{\alpha} = \mathbf{A} \tag{20}$$

where for the elasto-viscoplastic model of a material we assume (cf. Perzyna [1, 8–10])

$$\Lambda = \frac{1}{T_m}\left\langle \Phi\left(\frac{f}{\kappa} - 1\right)\right\rangle \tag{21}$$

Here $T_m$ denotes the relaxation time for mechanical disturbances, the isotropic workhardening-softening function $\kappa$ is

$$\kappa = \hat{\kappa}(\varepsilon^p, \vartheta, \xi) \tag{22}$$

$\Phi$ is the empirical overstress function, the bracket $\langle\cdot\rangle$ defines the ramp function,

$$\mathbf{P} = \left.\frac{\partial f}{\partial \boldsymbol{\tau}}\right|_{\xi=const} \left(\left\|\frac{\partial f}{\partial \boldsymbol{\tau}}\right\|\right)^{-1} \tag{23}$$

and $\Xi$ and $\mathbf{A}$ denote the evolution functions which have to be determined.

## 9.5.4 INTRINSIC MICRODAMAGE PROCESS

An analysis of the experimental observations for cycle fatigue damage mechanisms at high temperature of metals performed by Sidey and Coffin [11] suggests that the intrinsic microdamage process very much depends on the strain rate effects as well as on the wave shape effects. In the tests in which duration of extension stress was larger than duration of compression stress (in single cycle), decreasing of the fatigue lifetime was observed and the fracture mode changed from a transgranular fracture for the fast–slow wave shape, to an intergranular single-crack fracture for equal ramp rates, to interior cavitation for the slow–fast test.

To take into consideration these observed time-dependent effects, it is advantageous to use the description of the intrinsic microdamage process presented by Perzyna [12, 13] and Duszek-Perzyna and Perzyna [14].

Let us assume that the intrinsic microdamage process consists of the nucleation and growth mechanism. [3]Physical considerations (cf. Curran *et al.* [16] and Perzyna [12, 13]) have shown that the nucleation of microvoids in dynamic loading processes which are characterized by a very short time duration is governed by the thermally activated mechanism. Based on this heuristic suggestion and taking into account the influence of the stress triaxiality on the nucleation mechanism, we postulate for rate-dependent plastic flow

$$\left(\dot{\xi}\right)_{nucl} = \frac{1}{T_m} h^*(\xi, \vartheta)\left[\exp\frac{m^*(\vartheta)|\tilde{I}_n - \tau_n(\xi, \vartheta, \in^p)|}{k\vartheta} - 1\right] \tag{24}$$

where $k$ denotes the Boltzmann constant, $h^*(\xi, \vartheta)$ represents a void nucleation material function which is introduced to take account of the effect of microvoid interaction, $m^*(\vartheta)$ is a temperature-dependent coefficient, $\tau_n(\xi, \vartheta, \varepsilon^p)$ is the porosity, temperature, and equivalent plastic strain dependent threshold stress for microvoid nucleation,

$$\tilde{I}_n = a_1 \tilde{J}_1 + a_2\sqrt{\tilde{J}_2'} + a_3\left(\tilde{J}_3'\right)^{1/3} \tag{25}$$

defines the stress-intensity invariant for nucleation, $a_i$ $(i = 1, 2, 3)$ are the material constants, $\tilde{J}_1$ denotes the first invariant of the stress tensor $\tilde{\boldsymbol{\tau}} = \boldsymbol{\tau} - \boldsymbol{\alpha}$, and $\tilde{J}_2'$ and $\tilde{J}_3'$ are the second and third invariants of the stress deviator $\tilde{\boldsymbol{\tau}}' = (\boldsymbol{\tau} - \boldsymbol{\alpha})'$.

For the growth mechanism we postulate (cf. Johnson [17]); Perzyna [12,13]; Perzyna and Drabik [18, 19])

$$\left(\dot{\xi}\right)_{grow} = \frac{1}{T_m}\frac{g^*(\xi, \vartheta)}{x_0}\left[\tilde{I}_g - \tau_{eq}(\xi, \vartheta, \in^p)\right] \tag{26}$$

where $T_m x_0$ denotes the dynamic viscosity of a material, $g^*(\xi, \vartheta)$ represents a void growth material function and takes account for void interaction,

[3]Recent experimental observation results (cf. Shockey *et al.* [15]) have shown that the coalescence mechanism can be treated as a nucleation and growth process on a smaller scale. This conjecture very much simplifies the description of the intrinsic microdamage process by only taking account of the nucleation and growth mechanisms.

$\tau_{eq}(\xi, \vartheta, \varepsilon^p)$ is the porosity, temperature, and equivalent plastic strain dependent void growth threshold stress,

$$\tilde{I}_g = b_1 \tilde{J}_1 + b_2 \sqrt{\tilde{J}_2'} + b_3 (\tilde{J}_3')^{\frac{1}{3}} \tag{27}$$

defines the stress-intensity invariant for growth, and $b_i$ $(i = 1, 2, 3)$ are the material constants.

Finally, the evolution equation for the porosity $\xi$ has the form

$$\dot{\xi} = \frac{h^*(\xi, \vartheta)}{T_m} \left[ \exp \frac{m^*(\vartheta)|\tilde{I}_n - \tau_n(\xi, \vartheta, \varepsilon^p)|}{k\vartheta} - 1 \right] + \frac{g^*(\xi, \vartheta)}{T_m \, x_0} \left[ \tilde{I}_g - \tau_{eq}(\xi, \vartheta, \varepsilon^p) \right] \tag{28}$$

This determines the evolution function $\Xi$.

## 9.5.5 KINEMATIC HARDENING

For a constitutive model describing the behavior of a material under cyclic loading processes, the evolution equation plays the crucial role for the back stress $\boldsymbol{\alpha}$, which is responsible for the description of the induced plastic strain anisotropy effects.

We shall follow some fundamental results obtained by Duszek and Perzyna [20]. Let us postulate

$$\mathrm{L}_v \boldsymbol{\alpha} = \mathbf{A}(\mathbf{d}^p, \tilde{\boldsymbol{\tau}}, \vartheta, \xi) \tag{29}$$

Making use of the tensorial representation of the function $\mathbf{A}$ and taking into account that there is no change of $\boldsymbol{\alpha}$ when $\tilde{\boldsymbol{\tau}} = 0$ and $\mathbf{d}^p = 0$, the evolution law (Eq. 29) can be written in the form (cf. Truesdell and Noll [5])

$$\begin{aligned} \mathrm{L}_v \boldsymbol{\alpha} = {} & \eta_1 \mathbf{d}^p + \eta_2 \tilde{\boldsymbol{\tau}} + \eta_3 \mathbf{d}^{p^2} + \eta_4 \tilde{\boldsymbol{\tau}}^2 + \eta_5 (\mathbf{d}^p \cdot \tilde{\boldsymbol{\tau}} + \tilde{\boldsymbol{\tau}} \cdot \mathbf{d}^p) \\ & + \eta_6 \left( \mathbf{d}^{p^2} \cdot \tilde{\boldsymbol{\tau}} + \tilde{\boldsymbol{\tau}} \cdot \mathbf{d}^{p^2} \right) + \eta_7 (\mathbf{d}^p \cdot \tilde{\boldsymbol{\tau}}^2 + \tilde{\boldsymbol{\tau}}^2 \cdot \mathbf{d}^p) + \eta_8 \left( \mathbf{d}^{p^2} \cdot \tilde{\boldsymbol{\tau}}^2 + \tilde{\boldsymbol{\tau}}^2 \cdot \mathbf{d}^{p^2} \right) \end{aligned} \tag{30}$$

where $\eta_1, \ldots, \eta_8$ are functions of the basic invariants of $\mathbf{d}^p$ and $\tilde{\boldsymbol{\tau}}$, the porosity parameter $\xi$, and temperature $\vartheta$.

A linear approximation of the general evolution law (Eq. 30) leads to the result

$$\mathrm{L}_v \boldsymbol{\alpha} = \eta_1 \mathbf{d}^p + \eta_2 \tilde{\boldsymbol{\tau}} \tag{31}$$

This kinetic law represents the linear combination of the Prager and Ziegler kinematic hardening rules (cf. Prager [21] and Ziegler [22]).

To determine the connection between the material functions $\eta_1$ and $\eta_2$, we take advantage of the geometrical relation (cf. Duszek and Perzyna [20])

$$(\mathrm{L}_v \boldsymbol{\alpha} - r \mathbf{d}^p) : \mathbf{Q} = 0 \tag{32}$$

where

$$\mathbf{Q} = \left[\frac{\partial f}{\partial \boldsymbol{\tau}} + \left(\frac{\partial f}{\partial \xi} - \frac{\partial \kappa}{\partial \xi}\right)\frac{\partial \xi}{\partial \boldsymbol{\tau}}\right]\left\|\frac{\partial f}{\partial \boldsymbol{\tau}} + \left(\frac{\partial f}{\partial \xi} - \frac{\partial \kappa}{\partial \xi}\right)\frac{\partial \xi}{\partial \boldsymbol{\tau}}\right\|^{-1} \tag{33}$$

and $r$ denotes the new material function.

Equation (32) leads to the result

$$\eta_2 = \frac{1}{T_m}\left\langle \Phi\left(\frac{f}{\kappa} - 1\right)\right\rangle [r(\xi, \vartheta) - \eta_1]\frac{\mathbf{P} : \mathbf{Q}}{\tilde{\boldsymbol{\tau}} : \mathbf{Q}} \tag{34}$$

Finally, the kinematic hardening evolution law takes the form

$$\mathrm{L}_v \boldsymbol{\alpha} = \frac{1}{T_m}\left\langle \Phi\left(\frac{f}{\kappa} - 1\right)\right\rangle \left[r_1(\xi, \vartheta)\mathbf{P} + r_2(\xi, \vartheta)\frac{\mathbf{P} : \mathbf{Q}}{\tilde{\boldsymbol{\tau}} : \mathbf{Q}}\tilde{\boldsymbol{\tau}}\right] \tag{35}$$

where

$$r_1(\xi, \vartheta) = \eta_1, \quad r_2(\xi, \vartheta) = r - \eta_1 \tag{36}$$

It is noteworthy to add that the developed procedure can be used as general approach for obtaining various particular kinematic hardening laws. As an example, let us assume that the evolution function $\mathbf{A}$ in Eq. 29 instead of $\mathbf{d}^p$ and $\tilde{\boldsymbol{\tau}}$ depends on $\mathbf{d}^p$ and $\boldsymbol{\alpha}$ only (cf. Agah–Tehrani *et al.* [23]). Then, instead of Eq. 35 we obtain

$$\mathrm{L}_v \boldsymbol{\alpha} = \frac{1}{T_m}\left\langle \Phi\left(\frac{f}{\kappa} - 1\right)\right\rangle [\zeta_1(\xi, \vartheta)\mathbf{P} - \zeta_2(\xi, \vartheta)\boldsymbol{\alpha}] \tag{37}$$

where

$$\zeta_1 = r_1, \quad \zeta_2 = -r_2(\xi, \vartheta)\frac{\mathbf{P} : \mathbf{Q}}{\boldsymbol{\alpha} : \mathbf{Q}} \tag{38}$$

When the infinitesimal deformations and rate-independent response of a material are assumed and the intrinsic microdamage effects are neglected then the kinematic hardening law (Eq. 37) reduces to that proposed by Armstrong and Frederick [24].

The kinematic hardening law (Eq. 37) leads to the nonlinear stress–strain relation with the characteristic saturation effect. The material function $\zeta_1(\xi, \vartheta)$ for $\xi = \xi_0$ and $\vartheta = \vartheta_0$ can be interpreted as an initial value of the kinematic hardening modulus, and the material function $\zeta_2(\xi, \vartheta)$ determines the character of the nonlinearity of kinematic hardening. The particular forms of the functions $\zeta_1$ and $\zeta_2$ have to take into account the degradation nature of the influence of the intrinsic microdamage process on the evolution of anisotropic hardening.

## 9.5.6 THERMODYNAMIC RESTRICTIONS AND RATE-TYPE CONSTITUTIVE RELATIONS

Suppose the axiom of the entropy production holds. Then the constitutive assumption (Eq. 15) and the evolution equations (Eq. 20) lead to the following results

$$\boldsymbol{\tau} = \rho_{Ref}\frac{\partial\hat{\psi}}{\partial\mathbf{e}}, \quad \eta = -\frac{\partial\hat{\psi}}{\partial\vartheta}, \quad -\frac{\partial\hat{\psi}}{\partial\boldsymbol{\mu}}\cdot \mathrm{L}_{\upsilon}\boldsymbol{\mu} - \frac{1}{\rho\vartheta}\mathbf{q}\cdot\mathrm{grad}\ \vartheta \geq 0. \tag{39}$$

The rate of internal dissipation is determined by

$$\vartheta\hat{\imath} = -\frac{\partial\hat{\psi}}{\partial\boldsymbol{\mu}}\cdot \mathrm{L}_{\upsilon}\boldsymbol{\mu} = -\left[\frac{\partial\hat{\psi}}{\partial\varepsilon^{p}}\sqrt{\frac{2}{3}} + \frac{\partial\hat{\psi}}{\partial\boldsymbol{\alpha}} : \left(r_1\mathbf{P} + r_2\frac{\mathbf{P}:\mathbf{Q}}{\tilde{\boldsymbol{\tau}}:\mathbf{Q}}\tilde{\boldsymbol{\tau}}\right)\right]\Lambda - \frac{\partial\hat{\psi}}{\partial\xi}\Xi. \tag{40}$$

Operating on the stress relation (first part of Eq. 39) with the Lie derivative and keeping the internal state vector constant, we obtain (cf. Duszek-Perzyna and Perzyna [14])

$$\mathrm{L}_{\upsilon}\boldsymbol{\tau} = \mathscr{L}^{e} : \mathbf{d} - \mathscr{L}^{th}\dot{\vartheta} - [(\mathscr{L}^{e} + \mathbf{g}\boldsymbol{\tau} + \boldsymbol{\tau}\mathbf{g}) : \mathbf{P}]\frac{1}{T_m}\left\langle\Phi\left(\frac{f}{\kappa} - 1\right)\right\rangle \tag{41}$$

where

$$\mathscr{L}^{e} = \rho_{Ref}\frac{\partial^2\hat{\psi}}{\partial\mathbf{e}^2}\quad \mathscr{L}^{th} = -\rho_{Ref}\frac{\partial^2\hat{\psi}}{\partial\mathbf{e}\partial\vartheta}. \tag{42}$$

Substituting $\dot{\psi}$ into the energy balance equation and taking into account the results (third part of Eq. 39 and Eq. 40) gives

$$\rho\vartheta\dot{\eta} = -\mathrm{div}\,\mathbf{q} + \rho\vartheta\hat{\imath} \tag{43}$$

Operating on the entropy relation (second part of Eq. 39) with the Lie derivative and substituting the result into Eq. 43, we obtain

$$\rho c_p\dot{\vartheta} = -\mathrm{div}\,\mathbf{q} + \vartheta\frac{\rho}{\rho_{Ref}}\frac{\partial\boldsymbol{\tau}}{\partial\vartheta} : \mathbf{d} + \rho\chi^{*}\boldsymbol{\tau} : \mathbf{d}^{p} + \rho\chi^{**}\dot{\xi} \tag{44}$$

where the specific heat

$$c_p = -\vartheta\frac{\partial^2\hat{\psi}}{\partial\vartheta^2} \tag{45}$$

and the irreversibility coefficients $\chi^*$ and $\chi^{**}$ are determined by

$$\chi^* = -\left[\left(\frac{\partial\hat{\psi}}{\partial\varepsilon^p} - \vartheta\frac{\partial^2\hat{\varepsilon}}{\partial\vartheta\partial\varepsilon^p}\right)\sqrt{\frac{2}{3}} + \left(\frac{\partial\hat{\psi}}{\partial\boldsymbol{\alpha}} - \vartheta\frac{\partial^2\hat{\psi}}{\partial\vartheta\partial\boldsymbol{\alpha}}\right) : \left(r_1\mathbf{P} + r_2\frac{\mathbf{P}:\mathbf{Q}}{\tilde{\boldsymbol{\tau}}:\mathbf{Q}}\tilde{\boldsymbol{\tau}}\right)\right]\frac{1}{\boldsymbol{\tau}:\mathbf{P}}$$

$$\chi^{**} = -\left(\frac{\partial\hat{\psi}}{\partial\xi} - \vartheta\frac{\partial^2\hat{\psi}}{\partial\vartheta\partial\xi}\right). \tag{46}$$

So, a set of the constitutive equations of the rate type has the form as follows

$$\mathrm{L}_v\boldsymbol{\tau} = \mathscr{L}^e : \mathbf{d} - \mathscr{L}^{th}\dot{\vartheta} - [(\mathscr{L}^e + \mathbf{g}\boldsymbol{\tau} + \boldsymbol{\tau}\mathbf{g}) : \mathbf{P}]\frac{1}{T_m}\left\langle\Phi\left(\frac{f}{\kappa} - 1\right)\right\rangle,$$

$$\rho c_p\dot{\vartheta} = -\operatorname{div}\mathbf{q} + \vartheta\frac{\rho}{\rho_{Ref}}\frac{\partial\boldsymbol{\tau}}{\partial\vartheta} : \mathbf{d} + \rho\chi^*\frac{1}{T_m}\left\langle\Phi\left(\frac{f}{\kappa} - 1\right)\right\rangle\boldsymbol{\tau} : \mathbf{P} + \rho\chi^{**}\dot{\xi},$$

$$\dot{\xi} = \frac{h^*(\xi,\vartheta)}{T_m}\left[\exp\frac{m^*(\vartheta)|\tilde{I}_n - \tau_n(\xi,\vartheta,\varepsilon^p)|}{k\vartheta} - 1\right] + \frac{g^*(\xi,\vartheta)}{T_m\sqrt{\kappa_0}}\left[\tilde{I}_g - \tau_{eq}(\xi,\vartheta,\varepsilon^p)\right],$$

$$\mathrm{L}_v\boldsymbol{\alpha} = \frac{1}{T_m}\left\langle\Phi\left(\frac{f}{\kappa} - 1\right)\right\rangle[\zeta_1(\xi,\vartheta)\mathbf{P} - \zeta_2(\xi,\vartheta)\boldsymbol{\alpha}]. \tag{47}$$

All the material functions and the material constants should be identified based on available experimental data.

## 9.5.7 IDENTIFICATION PROCEDURE

To do the proper identification procedure, we first make an assumption of the material functions (cf. Dornowski and Perzyna [25]).

The plastic potential function $f$ is assumed in the form (cf. Perzyna [26] and Shima and Oyane [27])

$$f = \left\{\tilde{J}_2' + [n_1(\vartheta) + n_2(\vartheta)\xi]\tilde{J}_1^2\right\}^{1/2} \tag{48}$$

where

$$n_1(\vartheta) = 0, \quad n_2(\vartheta) = \text{const} \tag{49}$$

The isotropic workhardening-softening function $\kappa$ is postulated as (cf. Perzyna [12] and Nemes and Eftis [28])

$$\kappa = \hat{\kappa}(\varepsilon^p,\vartheta,\xi) = \{\kappa_s(\vartheta) - [\kappa_s(\vartheta) - \kappa_0(\vartheta)]\exp[-\delta(\vartheta)\varepsilon^p]\}\left[1 - \left(\frac{\xi}{\xi_F}\right)^{\beta(\vartheta)}\right] \tag{50}$$

where

$$\kappa_s(\vartheta) = \kappa_s^* - \kappa_s^{**}\bar{\vartheta}, \quad \kappa_0(\vartheta) = \kappa_0^* - \kappa_0^{**}\bar{\vartheta},$$
$$\delta(\vartheta) = \delta^* - \delta^{**}\bar{\vartheta}, \quad \beta(\vartheta) = \beta^* - \beta^{**}\bar{\vartheta}, \quad \bar{\vartheta} = \frac{\vartheta - \vartheta_0}{\vartheta_0} \tag{51}$$

The overstress function $\Phi\left(\frac{f}{\kappa} - 1\right)$ is assumed in the form

$$\Phi\left(\frac{f}{\kappa} - 1\right) = \left(\frac{f}{\kappa} - 1\right)^m \tag{52}$$

The evolution equation for the kinematic hardening parameter $\boldsymbol{\alpha}$ is assumed in the form of Eq. 37 with

$$\zeta_1(\xi, \vartheta) = \zeta_1^* - \zeta_1^{**}\bar{\vartheta}, \; \zeta_2(\xi, \vartheta) = \zeta_2^* - \zeta_2^{**}\bar{\vartheta} \tag{53}$$

The evolution equation for the porosity $\xi$ is postulated as

$$\dot{\xi} = \dot{\xi}_{grow} = \frac{g^*(\xi, \vartheta)}{T_m\sqrt{\kappa_0(\vartheta)}}\left[\tilde{I}_g - \tau_{eq}(\xi, \vartheta, \varepsilon^p)\right] \tag{54}$$

where (cf. Dornowski [29])

$$g^*(\xi, \vartheta) = c_1(\vartheta)\frac{\sqrt{\kappa_0(\vartheta)}}{\kappa_0(\vartheta)}\frac{\xi}{1-\xi}$$

$$\tilde{I}_g = b_1\tilde{J}_1 + b_2\sqrt{\tilde{J}_2}$$

$$\tau_{eq}(\xi, \vartheta, \varepsilon^p) = c_2(\vartheta)(1-\xi)\ln\frac{1}{\xi}\{2\kappa_s(\vartheta) - [\kappa_s(\vartheta) - \kappa_0(\vartheta)]F(\xi_0, \xi, \vartheta)\}$$
$$c_1(\vartheta) = \text{const}, \quad c_2(\vartheta) = \text{const},$$

$$F(\xi_0, \xi, \vartheta) = \left(\frac{\xi_0}{1-\xi_0}\frac{1-\xi}{\xi}\right)^{\frac{2}{3}\delta} + \left(\frac{1-\xi}{1-\xi_0}\right)^{\frac{2}{3}\delta} \tag{55}$$

As in the infinitesimal theory of elasticity, we assume linear properties of the material, i.e.,

$$\mathcal{L}^e = 2\mu\mathbf{I} + \lambda(\mathbf{g}\otimes\mathbf{g}) \tag{56}$$

where $\mu$ and $\lambda$ denote the Lamé constants, and the thermal expansion matrix is postulated as

$$\mathcal{L}^{th} = (2\mu + 3\lambda)\theta\mathbf{g} \tag{57}$$

where $\theta$ is the thermal expansion constant.

To determine the material constants assumed, we take advantage of the experimental observations presented by Chakrabarti and Spretnak [30]. They

TABLE 9.5.1 Material Constants for AISI 4340 Steel

| | | | |
|---|---|---|---|
| $\kappa_s^* = 809\,\text{MPa}$ | $\kappa_s^{**} = 228\,\text{MPa}$ | $\kappa_0^* = 598\,\text{MPa}$ | $\kappa_0^{**} = 168\,\text{MPa}$ |
| $\delta^* = 14.00,$ | $\delta^{**} = 3.94$ | $\beta^* = 9.00$ | $\beta^{**} = 2.53$ |
| $\vartheta_0 = 293\,\text{K}$ | $\xi_F = 0.20$ | $\rho_{Ref} = 7850\,\text{kg/m}^3$ | $\mu = 76.92\,\text{GPa}$ |
| $\lambda = 115.38\,\text{GPa}$ | $\theta = 12 \times 10^{-6}\,\text{K}^{-1}$ | $T_m = 2.5\,\text{ms}$ | $m = 1$ |
| $\zeta_1^* = 15.00\,\text{GPa}$ | $\zeta_1^{**} = 4.22\,\text{GPa}$ | $\zeta_2^* = 69.60$ | $\zeta_2^{**} = 19.60$ |
| $c_1 = 0.202$ | $c_2 = 6.7 \times 10^{-2}$ | $b_1 = 1.00$ | $b_2 = 1.30$ |
| $\xi_0 = 6 \times 10^{-4}$ | $\bar{\chi} = 0.85$ | $\bar{\bar{\chi}} = 0$ | $c_p = 455\,\text{J/kg K}$ |

investigated the localized fracture mode for tensile steel sheet specimens simulating both plane stress and plane strain processes. The material used in their study was AISI 4340 steel. The principal variable in this flat specimen test was the width-to-thickness ratio. Variation in specimen geometry produces significant changes in stress state, directions of shear bands, and ductility. They found that fracture propagated consistently along the shear band localized region.

Let us now consider the adiabatic dynamic process for a thin steel plate under the condition of plane stress state. In fact, we idealize the initial boundary value problem investigated by Chakrabarti and Spretnak [30] by assuming the velocity-driven adiabatic process for a thin steel plate. The problem has been solved by using the finite difference method.

In numerical calculations it is assumed that

$$\dot{V}_0 = 1.5\,\text{m/s}, \quad t_0 = 50\,\mu\text{s}, \quad t_f = 800\,\mu\text{s}$$

The material of a plate is AISI 4340 steel.

Based on the best curve fitting of the experimental results obtained by Chakrabarti and Spretnak [30] for the stress-strain relation, the identification of the material constants has been done, (Table 9.5.1).

The application of the constitutive equations (Eq. 47) of a thermo-elasto-viscoplastic model of materials for the solution of various initial boundary value problems (evolution problems) has been recently presented; cf. Lodygowski and Perzyna [31] and Dornowski and Perzyna [25, 32].

## REFERENCES

1 Perzyna, P. (1995). Interactions of elastic-viscoplastic waves and localization phenomena in solids, in Nonlinear Waves in Solids, Proc. IUTAM Symposium, August 15–20, 1993, Victoria, Canada, Wegner, L. J., and Norwood, F. R., eds., ASME Book No AMR 137, pp. 114–121.

2 Abraham, R., Marsden, J. E., and Ratiu, T. (1988). *Manifolds, Tensor Analysis and Applications*, Berlin: Springer.
3 Marsden, J. E., and Hughes, T. J. R. (1983). *Mathematical Foundations of Elasticity* Englewood Cliffs: Prentice–Hall.
4 Oldroyd, J. (1950). On the formulation of rheological equations of state. *Proc. Roy. Soc. (London)* A **200**: 523–541.
5 Truesdell C., and Noll, W. (1965). The nonlinear field theories. *Handbuch der Physik*, Band III/3, pp. 1–579, Berlin: Springer.
6 Coleman, B. D., and Noll, W. (1963). The thermodynamics of elastic materials with heat conduction and viscosity. *Arch. Rational Mech. Anal.* **13**: 167–178.
7 Duszek, M. K., and Perzyna, P. (1991). The localization of plastic deformation in thermoplastic solids. *Int. J. Solids Structures* **27**: 1419–1443.
8 Perzyna, P. (1963). The constitutive equations for rate sensitive plastic materials. *Quart. Appl. Math.* **20**: 321–332.
9 Perzyna, P. (1966). Fundamental problems in viscoplasticity. *Advances in Applied Mechanics* **9**: 343–377.
10 Perzyna, P. (1971). Thermodynamic theory of viscoplasticity. *Advances in Applied Mechanics* **11**: 313–354.
11 Sidey, D., and Coffin, L. F. (1979). Low–cycle fatigue damage mechemism at high temperature, in *Fatigue Mechanism, Proc. ASTM STP 675 Symposium*, Kansas City, Mo., May 1978, Fong, J. T. ed., Baltimore, pp. 528–568.
12 Perzyna, P. (1986). Internal state variable description of dynamic fracture of ductile solids. *Int. J. Solids Structures* **22**: 797–818.
13 Perzyna, P. (1986). Constitutive modelling for brittle dynamic fracture in dissipative solids. *Arch. Mechanics* **38**: 725–738.
14 Duszek-Perzyna, M. K., and Perzyna, P. (1994). Analysis of the influence of different effects on criteria for adiabatic shear band localization in inelastic solids. in *Material Instabilities: Theory and Applications, ASME Congress*, Chicago, 9–11, November 1994, Batra, R. C. and Zbib, H. M. eds., AMD-Vol. 183/MD-Vol.50, pp. 59–85, New York: ASME.
15 Shockey, D. A., Seaman, L., and Curran, D. R. (1985). The microstatistical fracture mechanics approach to dynamic fracture problem. *Int. J. Fracture* **27**: 145–157.
16 Curran, D. R., Seaman, L., and Shockey, D. A. (1987). Dynamic failure of solids. *Physics Reports* **147**: 253–388.
17 Johnson, J. N. (1981). Dynamic fracture and spallation in ductile solids. *J. Appl. Phys.* **52**: 2812–2825.
18 Perzyna, P., and Drabik, A. (1989). Description of micro-damage process by porosity parameter for nonlinear viscoplasticity. *Arch. Mechanics* **41** 895–908.
19 Perzyna, P., and Drabik, A. (1999). Micro-damage mechanism in adiabatic processes. *Int. J. Plasticity* (submitted for publication).
20 Duszek, M. K., and Perzyna, P. (1991). On combined isotropic and kinematic hardening effects in plastic flow processes. *Int. J. Plasticity* **7**: 351–363.
21 Prager, W. (1955). The theory of plasticity: A survey of recent achievements (J. Clayton Lecture). *Proc. Inst. Mech. Eng.* **169**: 41–57.
22 Ziegler, H. (1959). A modification of Prager's hardening rule. *Quart. Appl. Math.* **17**: 55–65.
23 Agah-Tehrani, A., Lee, E. H., Malett, R. L., and Onat, E. T. (1987). The theory of elastic-plastic deformation at finite strain with induced anisotropy modelled isotropic-kinematic hardening. *J. Mech. Phys. Solids* **35**: 43–60.
24 Armstrong, P. J., and Frederick, C. O. (1966). A mathematical representation of the multiaxial Baushinger effect. *CEGB Report RD/B/N731*, Central Electricity Generating Board.

25 Dornowski, W., and Perzyna, P. (2000). Localization phenomena in thermo-viscoplastic flow processes under cyclic dynamic loadings. *CAMES* **7**: 117–160.
26 Perzyna, P. (1984). Constitutive modelling of dissipative solids for postcritical behaviour and fracture. *ASME J. Eng. Materials and Technology* **106**: 410–419.
27 Shima, S., and Oyane, M. (1976). Plasticity for porous solids. *Int. J. Mech. Sci.* **18**: 285–291.
28 Nemes, J. A., and Eftis, J. (1993). Constitutive modelling of the dynamic fracture of smooth tensile bars. *Int. J. Plasticity* **9**: 243–270.
29 Dornowski, W. (1999). Influence of finite deformation on the growth mechanism of microvoids contained in structural metals. *Arch. Mechanics* **51**: 71–86.
30 Chakrabarti, A. K., and Spretnak, J. W. (1975). Instability of plastic flow in the direction of pure shear. *Metallurgical Transactions* **6A**: 733–747.
31 Lodygowski, T., and Perzyna, P. (1997). Localized fracture of inelastic polycrystalline solids under dynamic loading processes. *Int. J. Damage Mechanics* **6**: 364–407.
32 Dornowski, W., and Perzyna, P. (1999). Constitutive modelling of inelastic solids for plastic flow processes under cyclic dynamic loadings. *Transaction of the ASME, J. Eng. Materials and Technology* **121**: 210–220.

SECTION 9.6

# High-Temperature Creep Deformation and Rupture Models

D. R. HAYHURST
*Department of Mechanical Engineering, UMIST, PO Box 88, Manchester M60 1QD, United Kingdom*

Contents



## 9.6.1 BACKGROUND

The mathematical modeling of primary, secondary, and tertiary creep is addressed in this paper, with emphasis being placed on both uniaxial and

*Handbook of Materials Behavior Models*. ISBN 0-12-443341-3.

multiaxial behaviour. The single state variable damage models due to Kachanov [1] and to Robotnov [2] are first presented and used to model aluminium alloys, copper, and austenitic stainless steels. Copper and the aluminium alloys have been selected since they represent materials with extreme types of multiaxial rupture behaviour, and austenitic stainless steel represents intermediate behaviour. The single state variable theory gives a good representation for these materials, since there is one dominant damage mechanism. Multistate damage variable theories [3] have been used to model those materials where either a more accurate representation is required than can be achieved by a single damage state theory, or the synergy between the mechanisms is so distinct that a multidamage state variable model is necessary. In this paper an aluminium alloy tested at 150°C is used to illustrate the former category and a Nimonic 80A superalloy and ferritic steels are used to represent the latter category. In the next sections, the single damage state variable theory is presented first, and is followed by the multidamage state variable models. The equations for each material are presented in turn, followed by the values of the constitutive parameters.

## 9.6.2 SINGLE DAMAGE STATE VARIABLE THEORIES

These theories were calibrated by Hayhurst [4, 5] against experimental uniaxial data, and, in addition, they were developed for multiaxial stress conditions by Hayhurst [4, 6]. Hayhurst [6] used isochronous rupture loci, as shown in Figure 9.6.1, to characterise multiaxial stress rupture behaviour. In this figure the plane stress conditions are normalised with respect to the uniaxial stress $\sigma_0$ required to give a specified lifetime. In the next section the behaviour of aluminium alloys is presented.

### 9.6.2.1 Aluminium Alloy at 150°C and 210°C

The behaviour of the alloy tested at 210°C has been investigated by Hayhurst [4, 5] for uniaxial stresses, and by Hayhurst [6] for multiaxial stresses. The alloy is specified as B.S.1472; it is a precipitation-hardened alloy tested after a 7-hour temperature soak period. The behaviour of a second alloy, tested at 150°C, has been investigated by Kowalewski, Hayhurst, and Dyson [7] for uniaxial stresses. The alloy is again specified as B.S.1472 and was tested following a 12-hour temperature soak.

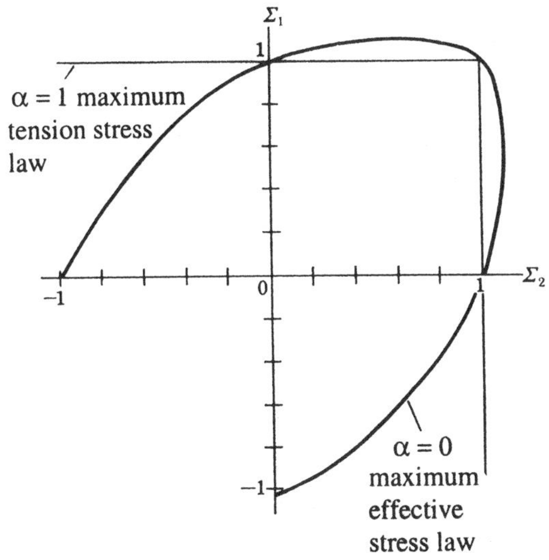


FIGURE 9.6.1 Schematic plot of plane stress isochronous rupture loci.

### 9.6.2.1.1 Uniaxial Behaviour

The uniaxial deformation, damage, and rupture behaviour for the materials tested at 150°C and 210°C is given by Eqs. 1 and 2, and the material constants, given in Table 9.6.1, may be traced to the references given previously.

$$\dot{\varepsilon} = G\sigma^n t^m/(1-\omega)^n \tag{1}$$

$$\dot{\omega} = M\sigma^\chi t^m/(1+\phi)(1-\omega)^\phi \tag{2}$$

The stress ranges, $\sigma$, over which the parameters hold are:

- 270 MPa > $\sigma$ > 220 MPa at 150°C; c.f. Kowalewski, Hayhurst, and Dyson [7]
- 70 MPa > $\sigma$ > 30 MPa at 210°C; c.f. Hayhurst, Dimmer, and Chernuka [8].

### 9.6.2.1.2 Multiaxial Behaviour

The multiaxial behaviour for materials tested at 150°C and 210°C is given by Eqs. 3 and 4.

$$\dot{\varepsilon}_{ij} = G\{\partial[\Psi^{n+1}(\sigma_{ke})/(n+1)]/\partial\sigma_{ij}\}t^m/(1-\omega)^n \tag{3}$$

$$\dot{\omega} = M\{\alpha\sigma_1 + (1-\alpha)\sigma_e\}^\chi t^m/(1+\phi)(1-\omega)^\phi \tag{4}$$

TABLE 9.6.1 Material Constants in Units of MPa, % Creep Strain, and Hours, for Aluminium Alloys at 150°C and 210°C

| Parameter | 150°C | 210°C |
|---|---|---|
| E/MPa | $71.10 \times 10^3$ | $60.03 \times 10^3$ |
| $n$ | 11.034 | 6.900 |
| $m$ | −0.3099 | −0.200 |
| $\chi$ | 8.220 | 6.480 |
| $\phi$ | 12.107 | 9.500 |
| $\alpha$ | 0 | 0 |
| $G$ | $3.511 \times 10^{-27}$ | $1.15 \times 10^{-15}$ |
| $M$ | $2.569 \times 10^{-22}$ | $1.79 \times 10^{-14}$ |
| $\omega_f$ | 0.99 | 0.99 |

where $\Psi(\sigma_{ke})$ is a homogeneous potential function of degree one in stress, $\sigma_1$ is the maximum principal tension stress, $\sigma_e(= 3s_{ij}s_{ij}/2)^{1/2})$ is the effective stress, $s_{ij}(= \sigma_{ij} - s_{ij}\sigma_{kk}/3)$ is the stress deviator, and the material constants are given in Table 9.6.1. The multiaxial stress rupture criterion for both aluminium alloys has been determined as close to $\alpha = 0$, i.e., a maximum effective stress rupture criterion. This has been supported for other alloys by Hayhurst, Brown, and Morrison [9], Hayhurst, Dimmer, and Morrison [10], and Hayhurst and Storakers [11].

### 9.6.2.2 Copper at 250°C

The material for which data are presented here is commercially pure copper which has been given a 12-hour temperature soak prior to testing at 250°C. The uniaxial behaviour is given by Eqs. 1 and 2, and the material data are provided in Table 9.6.2. Data are provided for two bars from different batches of manufacture which may be traced to Hayhurst, Dimmer, and Morrison [10] for bar 1, and to Hayhurst, Brown, and Morrison [9] for bar 2. The value of the multiaxial stress rupture parameter $\alpha = 0.70$ has been verified using notch bar tests by Hayhurst, Dimmer, and Morrison [10]. The stress range, $\sigma$, over which the parameters hold are:

- 45 MPa $> \sigma >$ 25 MPa at 250°C, bar 1; c.f. Hayhurst, Dimmer, and Churnuka [8]
- 65 MPa $> \sigma >$ 30 MPa at 250°C, bar 2; c.f. Hayhurst, Brown, and Morrison [9].

TABLE 9.6.2 Material Constants in Units of MPa, % Creep Strain, and Hours, for Two Different Copper Bars of the Same Composition Tested at 250°C

| Parameter | Copper bar 1 | Copper bar 2 |
|---|---|---|
| E/MPa | $66.24 \times 10^3$ | $66.24 \times 10^3$ |
| $n$ | 5.00 | 2.97 |
| $m$ | −0.43 | −0.79 |
| $\chi$ | 3.19 | 1.21 |
| $\phi$ | 6.00 | 3.83 |
| $\alpha$ | 0.70 | 0.70 |
| $G$ | $3.21 \times 10^{-10}$ | $1.28 \times 10^{-6}$ |
| $M$ | $1.89 \times 10^{-7}$ | $6.02 \times 10^{-4}$ |
| $\omega_f$ | 0.99 | 0.99 |

### 9.6.2.3 Austenitic Stainless Steel at 550°C

The material for which data are presented here is AISI 316 Stainless Steel which has been given a 12-hour temperature soak prior to testing at 550°C. The uniaxial behaviour is given by Eqs. 1 and 2, and the material data are provided in Table 9.6.3.

The value of the multiaxial stress rupture parameter $\alpha = 0.75$ has been verified using notch bar tests by Hayhurst, Dimmer, and Morrison [10]. It is shown in Figure 9.6.2 as an isochronous locus, where it is compared with the experimental results of Chubb and Bolton [12] for a similar material at 600°C. The stress range, $\sigma$, over which the parameters hold are:

- 350 MPa $> \sigma >$ 160 MPa at 250°C; c.f. Hayhurst, Dimmer, and Morrison [10].

## 9.6.3 MULTIDAMAGE STATE VARIABLE THEORIES

The need for synergistic multidamage state variable theories has been addressed by Dyson, Verma, and Szkopiak [13] and by Othman, Hayhurst, and Dyson [3]. The approach recognises the presence of competing softening mechanisms which interact through global deformation processes. Each physical mechanism is quantified by a single parameter, and its evolution is described by a single rate equation. This results in a set of coupled differential

TABLE 9.6.3 Material Constants in Units of MPa, % Creep Strain, and Hours, for an Austenitic Stainless Steel Tested at 550°C

| Parameter | 316 Stainless Steel |
|---|---|
| E/MPa | $169.617 \times 10^3$ |
| $v$ | 0.300 |
| $n$ | 1.737 |
| $m$ | −0.940 |
| $\chi$ | 0.478 |
| $\phi$ | 1.914 |
| $\alpha$ | 0.750 |
| $G$ | $1.383 \times 10^{-5}$ |
| $M$ | $2.774 \times 10^{-3}$ |
| $\omega_f$ | 0.99 |

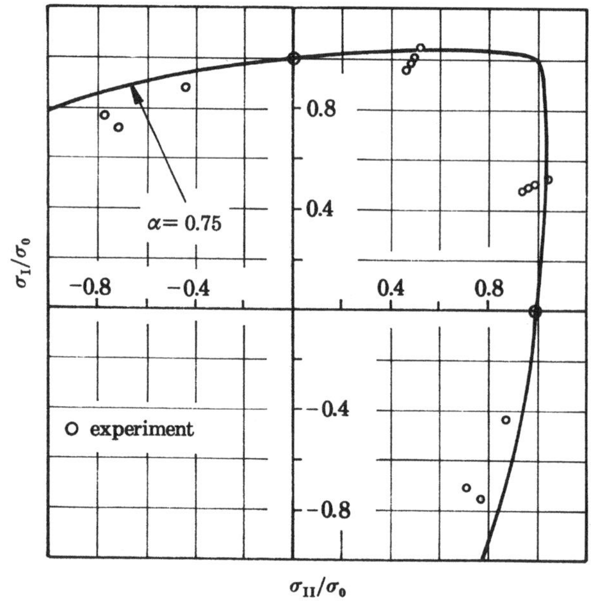


FIGURE 9.6.2 Comparison of isochronous loci for $\alpha = 0.75$ with experimental results obtained for 316 Stainless Steel at 600°C [12].

equations which, when solved together with strain rate equations, leads to the description of the creep curve. Presented in the next sections are the equations for the superalloy Nimonic 80A tested at 750°C, the aluminium alloy at 150°C, and ferritic steel tested at 640°C.

## 9.6.3.1 Nickel-Based Superalloy, Nimonic 80A Tested at 750°C

### 9.6.3.1.1 Uniaxial Behaviour

The uniaxial equations used by Othman, Hayhurst, and Dyson [3] to describe the damage evolution, creep strain, and failure of this material are as follows:

$$\dot{\varepsilon} = A(\sin hB\sigma)/\{(1-\omega_1)(1-\omega_2)^n\} \tag{5}$$

$$\dot{\omega}_1 = CA(1-\omega_1)(\sin hB\sigma)/(1-\omega_2)^n \tag{6}$$

$$\dot{\omega}_2 = DAN(\sin hB\sigma)/(1-\omega_1)(1-\omega_2)^n \tag{7}$$

$$n = B\sigma_e \coth(B\sigma_e) \tag{8}$$

$$\text{where } N = \begin{cases} 1 & \text{for} \quad \sigma_1 > 0 \\ 0 & \text{for} \quad \sigma_1 \le 0 \end{cases}$$

The two damage state variables represent physical mechanisms which operate simultaneously, i.e., softening which takes place due to grain boundary cavity nucleation and growth, $\omega_2$, and to the multiplication of mobile dislocations, $\omega_1$. The calibration of the equations is carried out as described by Othman, Hayhurst and Dyson [3], and the resulting material constants are given in Table 9.6.4.

### 9.6.3.1.2 Multiaxial Behaviour

The set of constitutive multiaxial equations for this material is given by the following:

$$\Psi = (A/B)(\cos h\, B\sigma_e) \tag{9}$$

$$\frac{d\varepsilon_{ij}}{dt} = \dot{\varepsilon}_{ij} = \frac{d\Psi}{ds_{ij}} \frac{1}{(1-\omega_1)(1-\omega_2)^n} \tag{10}$$

$$\dot{\varepsilon}_{ij} = \frac{3As_{ij}}{2\sigma_e} \frac{(\sin hB\sigma_e)}{(1-\omega_1)(1-\omega_2)^n} \tag{11}$$

$$\dot{\omega}_1 = CA(1-\omega_1)(\sin hB\sigma)/(1-\omega_2)^n \tag{12}$$

$$\dot{\omega}_2 = DAN(\sigma_1/\sigma)^{\nu}(\sin hB\sigma_e)/(1-\omega_1)(1-\omega_2)^n \tag{13}$$

$$\text{where } N = \begin{cases} 1 & \text{for} \quad \sigma_1 > 0 \\ 0 & \text{for} \quad \sigma_1 \le 0 \end{cases}$$

TABLE 9.6.4 Material Constants in Units of MPa, % Creep Strain, and Hours, for the Nickel-Based Superalloy Nimonic 90, Tested at 750°C

| Parameter | Nickel-based superalloy |
|---|---|
| E/MPa | $200 \times 10^3$ |
| $A$ | $2 \times 10^{-4}$ |
| $B$ | $16 \times 10^{-3}$ |
| $C$ | 300 |
| $D$ | 2 |
| $\nu$ | 2 |
| $\omega_f$ | 1/3 |

Note that neither uni- nor multiaxial equation sets include primary creep. The multiaxial rupture stress criterion $\nu$ has been calibrated using torsion test data by Dyson and Loveday [14]. The failure criterion in this model is $\omega_f = 1/3$, as defined by the creep constrained cavitation model, in contrast to $\omega_f = 0.99$ for the single damage state variable model. The equations and constitutive parameters are valid in the stress, $\sigma$, range:

- 600 MPa > $\sigma$ > 100 MPa at 750°C; c.f. Dyson and Loveday [14].

## 9.6.3.2 Aluminium Alloy Tested at 150°C

### 9.6.3.2.1 Uniaxial Behaviour

The uniaxial constitutive equations are given by:

$$\dot{\varepsilon} = A \sin h\left\{\frac{B\sigma(1-H)}{1-\Phi}\right\}/(1-\omega_2)^n \tag{14}$$

$$\dot{H} = (h\dot{\varepsilon}/\sigma)(1 - H/H^*) \tag{15}$$

$$\dot{\Phi} = (K_c/3)(1-\Phi)^4 \tag{16}$$

$$\dot{\omega}_2 = DAN \sin h\left\{\frac{B\sigma(1-H)}{1-\Phi}\right\}/(1-\omega_2)^n \tag{17}$$

$$n = \left\{\frac{B\sigma_e(1-H)}{(1-\Phi)}\right\}\coth\left\{\frac{B\sigma_e(1-H)}{(1-\Phi)}\right\} \tag{18}$$

$$\text{where } N = \begin{cases} 1 & \text{for} \quad \sigma_1 > 0 \\ 0 & \text{for} \quad \sigma_1 \le 0 \end{cases}$$

The creep constrained cavitation damage parameter $\omega_2$ is included in Eq. 17 together with two new state variables. The first, which is included in Eq. 15, describes primary creep using the variable $H$, which monotonically increases to its saturation value $H^*$ when primary creep has been concluded. The second variable $\Phi$ describes the physics of aging to lie within the range of 0 to 1, for mathematical convenience. The corresponding material constants are given in Table 9.6.5.

#### 9.6.3.2.2 Multiaxial Behaviour

The multiaxial generalisation takes place through a potential function based on the effective stress, and both the aging variable $\Phi$ and the dislocation density variable $H$ are treated as scalars. The evolution of the damage variable $\omega_2$ is dependent on the magnitude of the maximum principal tension stress, and the multiaxial stress state sensitivity is governed by the term $(\sigma_1/\sigma_e)^\nu$ in Eq. 24. This equation can be integrated to yield the isochronous loci given in Figure 9.6.3.

$$\Psi = (A/B)\cos h\left\{\frac{B\sigma_e(1-H)}{(1-\Phi)}\right\} \tag{19}$$

$$\frac{d\varepsilon_{ij}}{dt} = \dot{\varepsilon}_{ij} = \frac{d\Psi}{s_{ij}}\frac{1}{(1-\omega_2)^n} \tag{20}$$

$$\dot{\varepsilon}_{ij} = \frac{3A}{2\sigma_e}s_{ij}\sin h\left\{\frac{B\sigma_e(1-H)}{(1-\Phi)}\right\}/(1-\omega_2)^n \tag{21}$$

$$\dot{H} = \frac{hA}{\sigma_e(1-\omega_2)^n}(1-H/H^*)\sin h\left\{\frac{B\sigma_e(1-H)}{(1-\Phi)}\right\} \tag{22}$$

$$\dot{\Phi} = (K_c/3)(1-\Phi)^4 \tag{23}$$

$$\dot{\omega}_2 = DAN\left\{\frac{\sigma_1}{\sigma_e}\right\}^\nu \sin h\left\{\frac{B\sigma_e(1-H)}{(1-\Phi)}\right\} \tag{24}$$

$$\text{where } N = \begin{cases} 1 & \text{for} \quad \sigma_1/\sigma_e > 0 \\ 0 & \text{for} \quad \sigma_1/\sigma_e \leq 0 \end{cases}$$

It may be observed that the shapes of the loci due to the term $(\sigma_1/\sigma_e)^\nu$ in Eq. 24 are essentially the same as those given in Figures 9.6.1 and 9.6.2 for the function $[\alpha\sigma_1 + (1-\alpha)\sigma_e]^\chi$ given in Eq. 4.

It is worth noting that Kowalewski, Hayhurst, and Dyson [7] contrasted the quality of the creep curve representation provided by the single and multiple state variable theories. They observed that the latter gave a very

TABLE 9.6.5 Material Constants in Units of MPa, % Creep Strain, and Hours, for an Aluminium Alloy at 150°C

| Parameter | Aluminium alloy |
|---|---|
| E/MPa | $71.1 \times 10^3$ |
| $A$ | $2.960 \times 10^{-9}$ |
| $B$ | $7.167 \times 10^{-2}$ |
| $D$ | 6.630 |
| $h$ | $1.370 \times 10^5$ |
| $H^*$ | 0.2032 |
| $K_c$ | $19.310 \times 10^{-5}$ |
| $\omega_f$ | 0.3 |

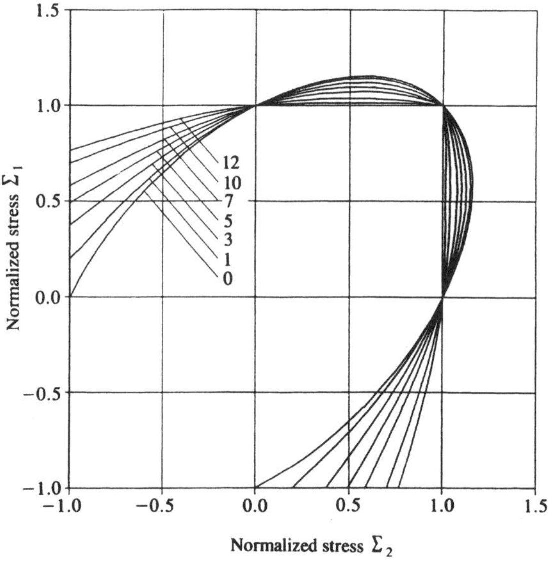


FIGURE 9.6.3 Isochronous rupture loci for biaxial plane stress determined by damage evolution due to creep constrained cavitation with aging using Eqs. 20–24 for $\sigma_0 = 262$ MPa. The loci are given for the range of values on the stress state sensitivity index $\nu$ marked on the figure $\nu = 12, 10, 7, \ldots, 0$.

accurate prediction of the tertiary part of the creep curve. The equations and constitutive parameters are valid in the stress range, $\sigma$, given by:

- 270 MPa $> \sigma >$ 220 MPa at 150°C; c.f. Kowalewski, Hayhurst, and Dyson [7].

### 9.6.3.3 Ferritic Steel and Associated Weld Materials at 640°C

Data are presented here for materials encountered in ferritic steel steam pipe butt welds. The parent pipe material considered is a 0.5Cr 0.5Mo 0.25V ferritic steel which has been characterised by Perrin and Hayhurst [15]; the weld material considered is derived from a 0.5Cr 0.5Mo 0.25V weld filler. The behaviour of the intermediary phases—heat affect zone (HAZ) and Type IV; see Figure 9.6.4, has been characterised using a technique used by Wang and Hayhurst [16]. The technique involves property ratios and a knowledge of the constitutive equations for the parent material. The constitutive equations are now considered.

#### 9.6.3.3.1 Uniaxial Behaviour

The equations developed by Perrin and Hayhurst [15] are as follows:

$$\dot{\varepsilon} = A \sin h \left\{ \frac{B\sigma(1-H)}{(1-\Phi)(1-\omega_2)} \right\} \tag{25}$$

$$\dot{H} = (h\dot{\varepsilon}/\sigma)(1 - H/H^*) \tag{26}$$

$$\dot{\Phi} = (K_c/3)(1-\Phi)^4 \tag{27}$$

$$\dot{\omega}_2 = DAN \sin h \left\{ \frac{B\sigma(1-H)}{(1-\Phi)(1-\omega_2)} \right\} \tag{28}$$

$$\text{where } N = \begin{cases} 1 & \text{for} \quad \sigma_1 > 0 \\ 0 & \text{for} \quad \sigma_1 \le 0 \end{cases}$$

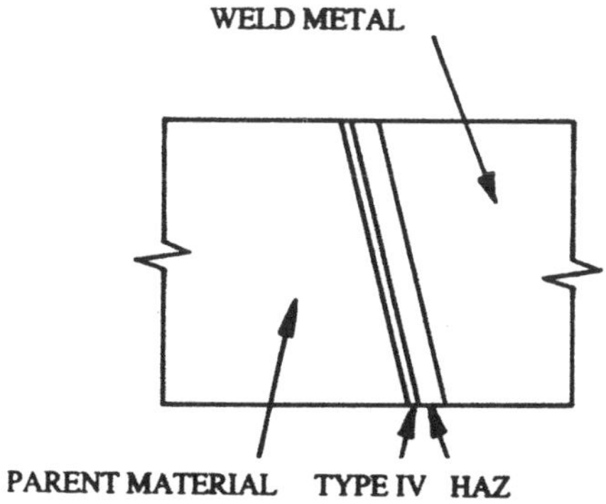


FIGURE 9.6.4 Schematic diagram showing material phases in across-weld.

Since all material phases of the weld are closely similar, the same equations and the associated physical processes are likely to be the same, and the uniaxial behaviour is described by Eqs. 25–28. Comparison of this equation set with Eqs. 14–17 for the aluminium alloy reveals that the only difference is in the term for $1/(1-\omega_2)$, which relates to creep deformation; they therefore have very similar modeling capabilities. The material constants are presented for these materials in Table 9.6.6.

It is worth noting that the HAZ properties are the same as those of the parent materials.

### 9.6.3.3.2 Multiaxial Behaviour

As in the cases of the superalloy and the aluminium alloy, the generalisation to multiaxial stress takes place using a potential function based on the effective stress and the multidamage state variables. The multiaxial stress rupture criterion is being expressed through the term $(\sigma_1/\sigma_e)^{\nu}$ in Eq. 33.

$$\Psi = (A/B)\cos h\left\{\frac{B\sigma_e(1-H)}{(1-\Phi)(1-\omega_2)}\right\} \tag{29}$$

$$\frac{d\varepsilon_{ij}}{dt} = \frac{d\Psi}{ds_{ij}} = \frac{3s_{ij}}{2\sigma_e}A\sin h\left\{\frac{B\sigma_e(1-H)}{(1-\Phi)(1-\omega_2)}\right\} \tag{30}$$

$$\dot{H} = \frac{hA}{\sigma_e}(1-H/H^*)\sin h\left\{\frac{B\sigma_e(1-H)}{(1-\Phi)(1-\omega_2)}\right\} \tag{31}$$

$$\dot{\Phi} = (K_c/3)(1-\Phi)^4 \tag{32}$$

$$\dot{\omega}_2 = DAN\left\{\frac{\sigma_1}{\sigma_e}\right\}^{\nu}\sin h\left\{\frac{B\sigma_e(1-H)}{(1-\Phi)(1-\omega_2)}\right\} \tag{33}$$

**TABLE 9.6.6 Material Constants in Units of MPa, % Creep Strain, and Hours, for Ferritic Steel Parent, Weld Metal, HAZ, and Type IV Materials at 640°C**

| Parameter | Parent | Weld Metal | HAZ | Type IV |
|---|---|---|---|---|
| E/MPa | $160 \times 10^3$ | $160 \times 10^3$ | $160 \times 10^3$ | $160 \times 10^3$ |
| $A$ | $1.6783 \times 10^{-8}$ | $3.3731 \times 10^{-8}$ | $1.6783 \times 10^{-8}$ | $9.4250 \times 10^{-7}$ |
| $B$ | 0.2434 | 0.2508 | 0.2434 | 0.1308 |
| $D$ | 1.2845 | 1.2845 | 1.28451 | 3.125 |
| $h$ | $2.4326 \times 10^5$ | $2.5128 \times 10^5$ | $2.4326 \times 10^5$ | $1.4750 \times 10^4$ |
| $H^*$ | 0.5929 | 0.6052 | 0.5929 | 0.6091 |
| $K_c$ | $4.7378 \times 10^{-4}$ | $9.4689 \times 10^{-4}$ | $4.7378 \times 10^{-4}$ | $2.5720 \times 10^{-3}$ |
| $\nu$ | 2.8 | 2.8 | 2.8 | 2.8 |
| $\omega_f$ | 1/3 | 1/3 | 1/3 | 1/2 |

$$\text{where } N = \begin{cases} 1 & \text{for} \quad \sigma_1/\sigma_e > 0 \\ 0 & \text{for} \quad \sigma_1/\sigma_e \leq 0 \end{cases}$$

The ranges of stress, $\sigma$, over which the equations are valid at 640°C are:

- 40 MPa $> \sigma >$ 20 MPa for parent
- 115 MPa $> \sigma >$ 20 MPa for weld
- 140 MPa $> \sigma >$ 20 MPa for HAZ
- 160 MPa $> \sigma >$ 55 MPa for Type IV

} 640°C; c.f.Perrin and Hayhurst [15]

## 9.6.4 IDENTIFICATION OF MATERIAL PARAMETERS

First, material data are required under a uniaxial state of stress, and care must be taken to achieve constancy of temperature control and to ensure axiality of loading. Second, data are required under at least one state of multiaxial stress, for example: biaxial tension cruciform test pieces [6, 17], Andrade shear discs [11], torsion tests [18], and axisymmetrically notched bars [10]. In this way, the multiaxial stress rupture criterion of the material can be determined and calibrated.

## 9.6.5 HOW TO USE THE MODEL

Any of the constitutive equations reported here may be used in nonlinear finite element codes for elasto-viscoplasticity. In order to solve the coupled multidamage state variable equation, it will usually be necessary to write specific user subroutines to enable the equations to be interfaced with the main finite element computer programme. In all cases it is recommended that the numerical routines be checked against published data to verify accuracy, stability, and convergence.

## REFERENCES

1. Kachanov, L. M. (1960). *The Theory of Creep* (English translation ed. A. J. Kennedy), chs IX, X, Boston Spa, Wetherby: British Library.
2. Rabotnov, Yu. M. (1969). *Creep Problems in Structural Members* (English translation ed. F. A. Leckie), ch. 6, Amsterdam: North Holland.

3. Othman, A. M., Hayhurst, D. R., and Dyson, B. F. (1993). Skeletal point stresses in circumferentially notched tension bars undergoing tertiary creep modelled with physically-based constitutive equations, *Proc. R. Soc. (London)* **441**: 343–358.
4. Hayhurst, D. R. (1970). Isothermal Creep Deformation and Rupture of Structures. Ph.D. thesis, Cambridge University.
5. Hayhurst, D. R. (1973). Stress redistribution and rupture due to creep in a uniformly stretched thin plate containing a circular hole. *J. Appl. Mech.* **40**: 244–250.
6. Hayhurst, D. R. (1972). Creep rupture under multi-axial states of stress. *J. Mech. Phys. Solids* **20**: 381–390.
7. Kowalewski, Z. L., Hayhurst, D. R., and Dyson, B. F. (1994). Mechanisms-based creep constitutive equations for an aluminium alloy. *J. Strain Analysis* **29**(4): 309–316.
8. Hayhurst, D. R., Dimmer, P. R., and Chernuka, M. W. (1975). Estimates of the creep rupture lifetime of structures using the finite element method. *J. Mech. Phys. Solids* **23**: 335–355.
9. Hayhurst, D. R., Brown, P. R., and Morrison, C. J. (1984). The role of continuum damage in creep crack growth. *Phil. Trans. R. Soc. (London)* **A311**: 130–158.
10. Hayhurst, D. R., Dimmer, P. R., and Morrison (1984). Development of continuum damage in the creep rupture of notched bars. *Phil. Trans. R. Soc. (London)* **A311**: 103–129.
11. Hayhurst, D. R., and Storakers, B. (1976). Creep rupture of the Andrade Shear Disc. *Proc. R. Soc. (London)* **A349**: 369–382.
12. Chubb, E. J., and Bolton, C. J. (1980). Stress state dependence of creep deformation and fracture in AISI type 316 stainless steel, in *Proc. Int. Conf. on Engng Aspects of Creep* 15–19 Sept. 1980, Sheffield 1, paper C201/80, p. 48, London: Institute of Mechanical Engineers.
13. Dyson, B. F., Verma, A. K., and Szkopiak, Z. C. (1981). The influence of stress state on creep resistance: Experiments and modelling. *Acta Metall.* **29**: 1573–1580.
14. Dyson, B. F., and Loveday, M. S. (1981). Creep facture in Nimonic 80A under tri-axial tensile stressing, in *Creep in Structures*, 1980 (IUTAM Symposium, Leicester, U.K.) (ed. A. R. S. Ponter and D. R. Hayhurst), 406–420. Berlin: Springer-Verlag.
15. Perrin, I. J., and Hayhurst, D. R. (1999). Continuum damage mechanics analyses of Type IV creep failure in ferritic steel crossweld specimens. *Int. J. Press. Vess and Piping* **76**: 599–617.
16. Wang, Z. P., and Hayhurst, D. R. (1994). The use of supercomputer modelling of high-temperature failure in pipe weldments to optimise weld and heat affected zone materials property selection. *Proc. R. Soc. (London)* **A446**: 127–148.
17. Hayhurst, D. R. (1973). A bi-axial-tension creep-rupture testing machine. *J. Strain Analy.* 8(2): 119–123.
18. Trampczynski, W. A., Hayhurst, D. R., and Leckie, F. A. (1981). Creep rupture of copper and aluminium under non-proportional loading. *J. Mech. Phys. Solids* **29**(5/6): 353–374.

SECTION **9.7**

# A Coupled Diffusion-Viscoplastic Formulation for Oxidasing Multiphase Materials

ESTEBAN P. BUSSO

*Department of Mechanical Engineering, Imperial College, University of London, London, SW7 2BX, United Kingdom*

Contents



## 9.7.1 VALIDITY

The coupled diffusion-constitutive approach is used to describe the effect of the phase transformations caused by local internal and external oxidation processes on the constitutive behaviour of multiphase metallic materials exposed to an oxidising environment.

## 9.7.2 BACKGROUND

The oxidation of multiphase metallic alloys is an important factor leading to local microcrack formation at or near bimaterial interfaces. A proper

*Handbook of Materials Behavior Models*. ISBN 0-12-443341-3.

description of the local deformation processes requires that the constitutive behaviour of the materials be coupled with the dominant diffusion and oxidation mechanisms. Without a proper description of such phenomena, the crucial evolving stress conditions, as the material undergoes the transition from an unoxidised to a fully oxidised state, cannot be predicted [1]. Examples of material systems which undergo selective oxidation are those present in thermal barrier coatings, widely used in both aerospace and land-based gas turbines, and single crystal silicon wafers used in the manufacture of LSI devices by the microelectronics industry.

The coupled oxidation-constitutive material framework recently proposed by Busso [2] (see also [3] and [4]) accounts for the diffusion of oxidant species through a multiphase material and the subsequent oxidation of one of its phases (e.g., Al-rich or Si-based), and incorporates the effect that the local volumetric expansion of the newly formed oxide has in the generation of inelastic volumetric strains and residual stresses. It relies on the level of oxygen concentration at each material point to identify whether oxide precipitates have formed, in which case the material at that location is treated, through a self-consistent approach, as an homogeneous equivalent material consisting of multimetallic phases and oxide products.

## 9.7.3 DESCRIPTION OF THE MODEL

A hypoelastic formulation for the oxidising metal which accounts for nonisothermal effects can be expressed in terms of the Jaumann derivative of the overall aggregate stress, **T**, as

$$\overset{\triangledown}{\mathbf{T}} = \mathscr{L}[\mathbf{D} - \mathbf{D}^{in}] - 3\kappa\alpha\dot{\theta}\,\mathbf{1} \tag{1}$$

where $\overset{\triangledown}{\mathbf{T}} = \dot{\mathbf{T}} - \mathbf{WT} + \mathbf{TW}$, and **D** is the stretching tensor, $\dot{\theta}$ is the time rate of change of the absolute temperature, $\alpha$ is the thermal expansion coefficient, and **W** is the total material spin. The inelastic stretching tensor associated with the deformation, $\mathbf{D}^{in}$, is expressed in terms of the average inelastic stretching rate due to slip, $\mathbf{D}^{cr}$, and a transformation term, $\mathbf{D}^{tr}$, which represents the nonrecoverable deformation rate induced by the oxidation of one of the metallic phases. Then,

$$\mathbf{D}^{in} = \mathbf{D}^{cr} + \mathbf{D}^{tr} \tag{2}$$

The average inelastic stretching rate tensor $\mathbf{D}^{cr}$ is defined as the volume fraction weighted sum of the inelastic strain rate tensor of each individual phase, $\dot{\mathbf{e}}_i^{cr}, \mathbf{D}^{cr} = \sum_i f_i\, \dot{\mathbf{e}}_i^{cr}$, where $f_i$ is the current volume fraction of phase $i$.

Moreover, $\mathbf{D}^{tr}$ is assumed to be controlled by the rate of change of an internal state variable associated with the oxidation kinetics, $f$. Here, $f$ is related to the oxide volume fraction and varies from 0, before the oxidation begins, to 1, when the internal oxidation is complete. Thus

$$\mathbf{D}^{tr} = f_2^{ini}\,\dot{f}\left\{P\frac{\mathbf{T}'}{\tilde{S}} + e_V^T\,\mathbf{1}\right\} \tag{3}$$

where $f_2^{ini}$ is the initial volume fraction of the oxidation-prone phase, $\mathbf{1}$ is the second-order unit tensor, $P$ is a coefficient which depends on the shape of the oxide particles, and $\tilde{S}$ and $\mathbf{T}'$ are the norm and deviatoric component of the aggregate stress tensor $\mathbf{T}$. The dilatational term in Eq. 3 represents the volume change due to oxidation, and the deviatoric one accounts for any deviations from isotropy associated with the oxide formation.

The formulation considers that oxidation begins at each material point once the local oxygen concentration, $C_O \geq C_{Ocr}$, where $C_{Ocr}$ is a temperature-dependent critical concentration.

The evolution of the internal variable $f$ is based on nucleation kinetics,

$$\dot{f} = (1-f)\,\dot{N}_p\,V_p \quad \text{for} \quad C_O \geq C_{Ocr}, \quad \text{else}\,\dot{f} = 0 \tag{4}$$

where $\dot{N}_p$ is the rate of increase of oxide precipitates per unit volume, and $V_p$ is the average volume of each oxide particle.

The local behaviour of the coexisting metallic phases and oxide products is described through a homogenization procedure based on self-consistent relations which depend on an interphase accommodation tensorial variable, $\mathbf{A}_i$, for each phase $i$. The evolution of the stress rate tensor of each individual oxidising and non-oxidising phase, $\sigma_i$, is given by

$$\dot{\sigma}_i = \dot{\mathbf{T}} + 2\mu(1-\beta)\{\mathbf{D}^{cr} - \dot{\mathbf{e}}_i^{cr} - \dot{\mathbf{A}}_i\} \tag{5}$$

Here $\beta$ is Eshelby's elastic accommodation factor and $\mu$ is the shear modulus. In this equation, $\dot{\mathbf{A}}_i$ is the time rate of change of $\mathbf{A}_i$ which is, in turn, expressed in terms of a corresponding phase accommodation tensorial variable $\mathbf{S}_i$,

$$\dot{\mathbf{A}}_i = \hat{H}(f,\theta)\left\{\sum\nolimits_{k=1}^{m} f_k\,\mathbf{S}_k|\dot{\bar{e}}_k^{cr}| - \mathbf{S}_i\,|\dot{\bar{e}}_i^{cr}|\right\} \tag{6}$$

where $\hat{H}(f,\theta)$ is a dimensionless homogenization function, and the index $m$ denotes the total number of coexisting phases. The phase accommodation variable evolves according to the following relation:

$$\dot{\mathbf{S}}_i = \dot{\mathbf{e}}_i^{cr} - \hat{H}(f,\theta)\mathbf{S}_i|\dot{\bar{e}}_i^{cr}| \tag{7}$$

## 9.7.4 IDENTIFICATION OF MATERIAL PARAMETERS

The true mean dilatational strain caused by the internal oxidation of one of the metallic phases, $e_V^T$ in Eq. 3, can be determined from the corresponding chemical reaction. Let $V_0$ be the material volume before the chemical reaction takes place, and $V$ that of the oxidized material. Then,

$$e_V^T = \frac{1}{3} Ln \frac{V}{V_0} \tag{8}$$

The critical local oxygen concentration level, $C_{Ocr}$ (Eq. 4), can be numerically calibrated from oxidation data, such as oxide thickness vs. oxidation time, at different temperatures. The time constant defined by $\tau = 1/(\dot{N}_p V_p)$ can be determined from the knowledge of the kinetics of the oxidation reaction, and its value is expected to be temperature-dependent.

The thermoelastic properties of the homogenised material are obtained from the thermoelastic behaviour of each individual phase using existing homogenisation (e.g., self-consistent) relations. The inelastic strain rate tensor for each individual phase is defined by a power law relation of the form

$$\dot{\mathbf{e}}_i^{cr} = \frac{3}{2} A_i \exp\left\{ -\frac{Q_i}{R\,\theta} \right\} \tilde{\sigma}_i^{n_i} \frac{\sigma_i'}{\tilde{\sigma}_i} \tag{9}$$

where $\sigma_i{}'$ is the deviatoric stress tensor for the phase $i$, and $\tilde{\sigma}_i$ is the corresponding equivalent stress, $\tilde{\sigma}_i = \left(\frac{3}{2}\,\sigma_i' : \sigma_i'\right)^{1/2}$. The material constants $A_i$, $Q_i$, and $n_i$ in Eq. 9 are determined from bulk material creep data for each individual phase.

The dimensionless homogenization function, $\hat{H}(f, \theta)$, needs to be calibrated from experimental data and detailed unit cell finite element calculations of the oxide-metal system at each stage of the oxidation process, that is, for a range of values of $f$ and temperatures. Finally, the oxide shape parameter $P$ can be deduced from the transformation strain ratios extracted from oxidation experiments, e.g., thin film curvature vs. oxide thickness measurements.

## 9.7.5 HOW TO USE THE MODEL

Use of the coupled formulation in complex boundary value problems representative of service generally requires its numerical implementation into the finite element method. This can be done using either fully explicit or implicit integration schemes [2, 3, 4].

## 9.7.6 LIST OF MATERIAL PARAMETERS

The formulation was calibrated in Reference [3] for an oxidising plasma sprayed thermal barrier coating (TBC) system consisting of a typical NiCoCrAlY metallic coating and a 8% Y-stabilised zirconia top layer. It was then used to investigate the effect of the thermally grown oxide on the stresses at or near the metal – thermally grown oxide interface, which are known to be responsible for microcrack nucleation.

The microstructure of a typical NiCoCrAlY alloy in the 950–1000°C temperature range consists of a 52.6 vol.% of $\beta$-NiAl and a 47.4 vol.% $\gamma$-Ni solid solution phase. Thus $f_1^{ini} = 0.474$, and $f_2^{ini} = 0.526$. The average thermoelastic properties of the metallic coating and the alumina oxide at the oxidising temperature of 950°C are:

$$E_{ox} = 340\ \text{GPa} \quad \nu_{ox} = 0.18 \quad \alpha_{ox} = 9.15\times 10^{-6}\ 1/°\text{C}$$
$$E_{coat} = 133\ \text{GPa} \quad \nu_{coat} = 0.30 \quad \alpha_{coat} = 17.0\times 10^{-6}\ 1/°\text{C}$$

and the creep constants for the Ni solid solution phase (1), and the polycrystalline NiAl phase (2) are:

$$n_1 = 4.6 \quad Q_1 = 227\ \text{KJ/mol} \quad A_1 = 5.283\ 1/(\text{s MPa}^{n_1})$$
$$n_2 = 5.5 \quad Q_2 = 245\ \text{KJ/mol} \quad A_2 = 3.73\times 10^{-3}\ 1/(\text{s MPa}^{n_2})$$

From the primary oxidation reaction,

$$\tfrac{2}{3}\,\text{NiAl} + \tfrac{1}{2}\,\text{O}_2 \rightarrow \tfrac{1}{3}\,\text{Al}_2\text{O}_3 + \tfrac{2}{3}\,\text{Ni} \qquad (10)$$

one finds that $e_V^T = 0.122$. Also, typically, $\tau = 0.5$ to 1 hour at 950°C, and $P = 0.24$ for the oxidation of zirconium (value not available for NiCoCrAlY).

The average critical concentration relative to the value specified at the coating surface found to accurately describe the oxidation data at 950°C was $C_{Ocr} = 1.45 \times 10^{-3}$. The diffusivity of oxygen in both the unoxidised and oxidised coating was taken as $D_O = 2.64 \times 10^{-20}$ m$^2$/s. Finally, typical values for the homogenisation function $H$ at 950°C calibrated from unit cell finite element calculations are given in Table 9.7.1.

Figure 9.7.1 shows a comparison between the metallic coating creep data and the corresponding behaviour predicted by the self-consistent model. Also included are the individual phases' creep data. Figure 9.7.2, on the other

TABLE 9.7.1 Values of the Dimensionless $f$ Homogenisation Function $H$ at 950°C for a Material with a Maximum Oxide Volume Fraction of 100%

| $f$ | 0.00 | 0.20 | 0.35 | 0.57 | 0.61 |
|---|---|---|---|---|---|
| $H$ | 4961 | 2834 | 1707 | 434 | 0 |

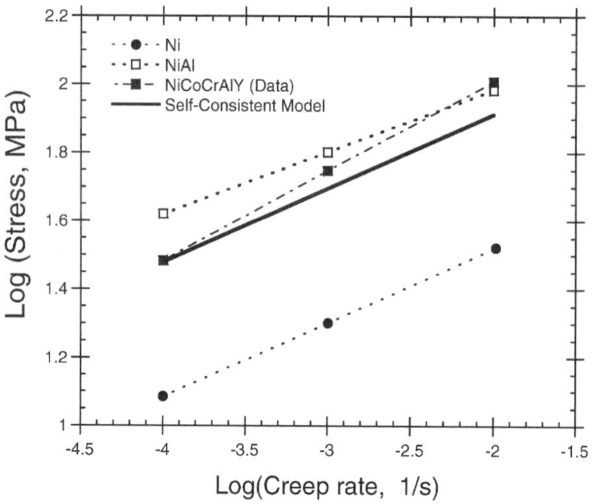


FIGURE 9.7.1. Comparison between the two-phase NiCoCrAlY alloy creep data and the corresponding behaviour predicted by the self-consistent model. Also included are the individual phases' creep data.

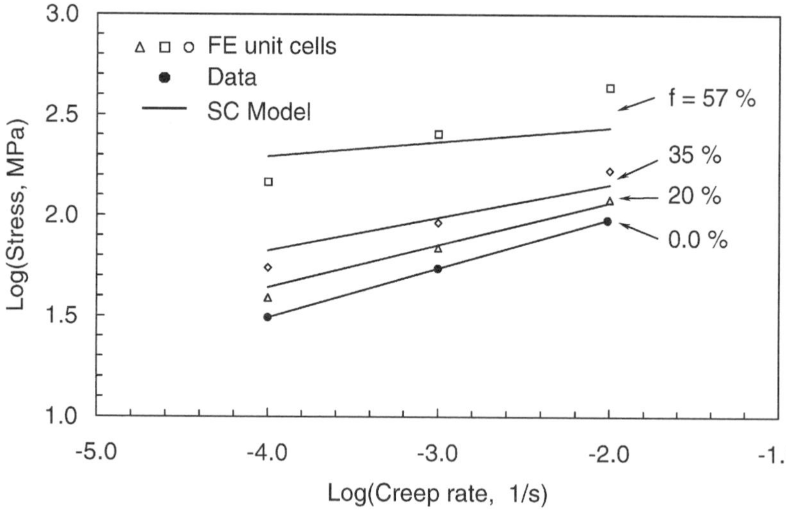


FIGURE 9.7.2. Comparison between accurate reference creep solutions for the two-phase NiCoCrAlY alloy containing different oxide volume fractions ($f$) with the corresponding predictions obtained from the coupled diffusion-viscoplastic model.

hand, presents a comparison between accurate reference creep solutions for the two-phase NiCoCrAlY alloy containing different oxide volume fractions with the corresponding predictions obtained from the coupled diffusion-viscoplastic model.

## REFERENCES


1. Busso, E. P. (1997). Coupled phenomena in bimaterial interface regions at high temperatures, in *Mecanismes et Mecanique des Materiaux Multifunctionnels*, pp. 49–52, Perreux, D., and L'Excellent, C., eds., MECAMAT, Paris, France, Proc. Colloque National MECAMAT 97, Aussois, France.
2. Busso, E. P. (1999). Oxidation induced stresses in ceramic–metal interfaces. *Journal de Physique IV*, **9**: 287–296.
3. Busso, E. P., Lin, J, Sakurai, S., and Nakayama, M. (2000). A Mechanistic Study of Oxidation-Induced Degradation in a Plasma-Sprayed Thermal Barrier Coating System. Part I: Model Formulation. *Acta Materialia*, **49**(9): 1515–1528.
4. Busso, E. P., Lin, J., and Sakurai, S. (2000). A Mechanistic Study of Oxidation-Induced Degradation in a Plasma-Sprayed Thermal Barrier Coating System. Part II: Life Prediction Model. *Acta Materialia*, **49**(9): 1529–1536.

SECTION **9.8**

# Hydrogen Attack

ERIK VAN DER GIESSEN and SABINE M. SCHLÖGL
*University of Groningen, Applied Physics, Micromechanics of Materials, Nyenborgh 4, 9747 AG Groningen, The Netherlands*

Contents



## 9.8.1 VALIDITY

The model described in this article is intended to describe the secondary and tertiary stages of creep at elevated temperatures under hydrogen-rich conditions, as typically encountered in petrochemical installations. The associated damage is grain boundary cavitation leading to intergranular fracture. This phenomenon, known as hydrogen attack (HA), occurs predominantly in low-carbon, low-alloy ferritic steels (typically Cr-Mo steels).

## 9.8.2 BACKGROUND

At temperatures exceeding approximately 0.3 times the melting temperature, steels creep at a steady strain rate which is often conveniently described by the


*Handbook of Materials Behavior Models*. ISBN 0-12-443341-3.

Norton power law

$$\dot{\varepsilon}_e^C = B\sigma_e^n, \tag{1}$$

with the creep exponent $n$ and the temperature-dependent parameter $B(T)$. For uniaxial tension, $\sigma_e$ and $\dot{\varepsilon}_e^C$ are the applied stress and the creep strain rate, respectively. Under multiaxial stress, they denote the Mises equivalent quantities:

$$\sigma_e = \sqrt{\frac{3}{2}\sigma'_{ij}\sigma'_{ij}}, \qquad \sigma'_{ij} = \sigma_{ij} - \sigma_m\delta_{ij}, \qquad \sigma_m = \frac{1}{3}\sigma_{kk} \tag{2}$$

$$\dot{\varepsilon}_e^C = \sqrt{\frac{2}{3}D_{ij}D_{ij}} \tag{3}$$

with $\sigma_{ij}$ denoting Cauchy stress and $D_{ij}$ the stretching or strain rate.

The termination of the secondary creep regime in low-carbon ferritic steels in hydrogen-rich environments is caused by the initiation and growth of grain boundary cavities, filled with a methane-hydrogen gas mixture [6]. The so-called hydrogen attack initiates by hydrogen molecules in the gas atmosphere dissociating and the hydrogen atoms diffusing into the steel. Some get trapped at discontinuities occurring mainly between grain boundary carbides and matrix, where the hydrogen reacts with the carbon in the steel to generate methane. The methane molecules are too large to diffuse away and generate an internal pressure inside the cavity which drives subsequent growth. The deformation mechanisms involved during cavity growth are grain boundary diffusion and dislocation creep. Their relative contributions are determined by the cavity density, the internal gas pressure

$$p_m = p_{CH_4} + p_{H_2} \tag{4}$$

and the stress state in the material, $\sigma_m$, $\sigma_e$ and the stress $\sigma_n$ normal to the grain boundary. For cavities with a spherical-caps shape (see Fig. 9.8.1), radius $a$, and mean spacing $2b$, the rate of change of the cavity volume,

$$V^{cav} = \frac{4}{3}\pi a^3 h(\psi), \qquad h(\psi) = \left[(1+\cos\psi)^{-1} - \frac{1}{2}\cos\psi\right] \Big/ \sin\psi \tag{5}$$

can be expressed as [7]

$$\dot{V}^{cav} = \max\left[\dot{V}^L_{diff} + \dot{V}^L_{cr},\ \dot{V}^H_{diff} + \dot{V}^H_{cr}\right] \tag{6}$$

with the maximum criterion in this expression originating from two modes in which the cavity can grow, depending on its size and the stress state. In Eq. 6,

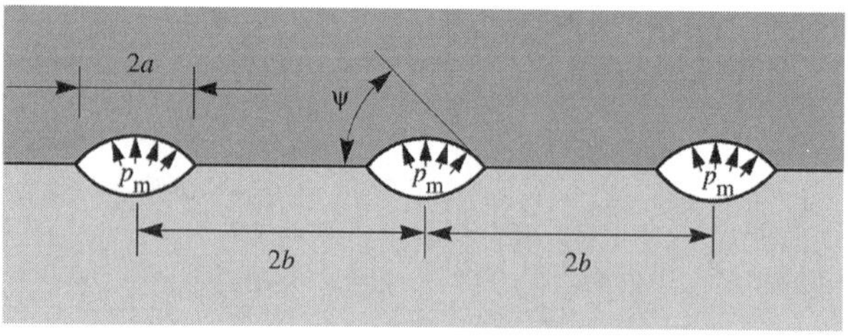


FIGURE 9.8.1 Spherical-cap-shaped cavities on a grain boundary.

the diffusive (*diff*) and creep (*cr*) contributions are given by

$$\dot{V}^L_{diff} = 4\pi\mathscr{D}\frac{p_m + \sigma_n}{\ln(1/f) - \frac{1}{2}(3-f)(1-f)},$$
$$f = \max\left[\left(\frac{a}{b}\right)^2, \left(\frac{a}{a+1.5L}\right)^2\right], \qquad L = (\mathscr{D}\sigma_e\dot{\varepsilon}^C_e)^{1/3} \tag{7}$$

$$\dot{V}^L_{cr} = 2\pi\dot{\varepsilon}^C_e a^3 h(\psi)\left[\alpha_n\left(\frac{p_m+\sigma_m}{\sigma_e}\right) + \beta_n\right]^n$$
$$\alpha_n = 3/(2n), \qquad \beta_n = (n-1)(n+0.4319)/n^2 \tag{8}$$

$$\dot{V}^H_{diff} = 4\pi\mathscr{D}\frac{p_m + \sigma_n}{\ln(1/f) - \frac{1}{2}(3-f)(1-f)}, \qquad f = \left(\frac{a}{b}\right)^2 \tag{9}$$

$$\dot{V}^H_{cr} = 2\pi\dot{\varepsilon}^C_e a^3 h(\psi)\left\{\frac{1}{1-(0.87a/b)^{3/n}}\right\}^n$$
$$\times\left[\alpha_n\left(\frac{p_m+\sigma_m}{\sigma_e}\right) + \frac{1}{n}\frac{\text{sign}(\sigma_n-\sigma_m)}{\text{sign}(\sigma_m+p_m)}\right]^n \tag{10}$$

provided that $p_m + \sigma_m > \sigma_e$ (which is typically the case in HA).

The enhanced straining in the tertiary regime is primarily due to cavity growth. The contribution of growth to the overall strain is controlled by the average rate of separation between the grains adjacent to the grain boundary,

$$\dot{\delta} = \dot{V}^{cav}/\pi b^2 = \dot{\delta}(p_m, \sigma_m, \sigma_e, \sigma_n; a, b) \tag{11}$$

The cavities grow until coalescence, when microcracks are formed along the grain boundaries, and this finally leads to intergranular failure. A working definition for cavity coalescence is when $a/b = 0.7$.

## 9.8.3 DESCRIPTION OF THE MODEL

Hydrogen attack implies a coupling between the transport of hydrogen and carbon, the chemistry of methane formation and evolution of the microstructure, as well as the mechanics of cavity growth and creep. Full details are not yet understood, but the common assumption is that transport and chemistry occur on time scales that are much shorter than the expected lifetime of the components that are typically subject to hydrogen attack. The nucleation time of cavities is also neglected. An upper bound of the attack is then obtained by assuming that the methane gas inside the cavities is in thermodynamic equilibrium at all instants. This allows a simpler model which consists of a thermodynamic part (describing the methane pressure as a function of temperature and the composition of the material) and a damage mechanics part (describing the damage evolution under a given methane pressure and applied stress state).

### 9.8.3.1 Thermodynamic Model

The methane pressure stems from the reaction of carbides with the hydrogen and depends on the various phases in the steel. In case of an alloy carbide $M_xC_y$ containing Cr, Mo, V, and Fe, the chemical reaction is of the type

$$(\mathrm{Cr}_{y_{\mathrm{Cr}}}\mathrm{Mo}_{y_{\mathrm{Mo}}}\mathrm{V}_{y_{\mathrm{V}}}\mathrm{Fe}_{y_{\mathrm{Fe}}})_x\mathrm{C}_y + 2y\mathrm{H}_2 \leftrightarrows y\mathrm{CH}_4 + xy_{\mathrm{Cr}}\mathrm{Cr} + xy_{\mathrm{Mo}}\mathrm{Mo} + xy_{\mathrm{V}}\mathrm{V} + xy_{\mathrm{Fe}}\mathrm{Fe} \quad (12)$$

where $y_{\mathrm{Cr}}$, $y_{\mathrm{Mo}}$, $y_{\mathrm{V}}$, and $y_{\mathrm{Fe}}$ are the concentration parameters of Cr, Mo, V, and Fe, respectively, in the carbide $M_xC_y$ $(y_{\mathrm{Cr}} + y_{\mathrm{Mo}} + y_{\mathrm{V}} + y_{\mathrm{Fe}} = 1)$. An upper limit to the methane pressure is obtained by assuming this reaction to be in equilibrium, so that

$$y\mu_{\mathrm{CH}_4} + xy_{\mathrm{Cr}}\mu_{\mathrm{Cr}} + xy_{\mathrm{Mo}}\mu_{\mathrm{Mo}} + xy_{\mathrm{V}}\mu_{\mathrm{V}} + xy_{\mathrm{Fe}}\mu_{\mathrm{Fe}} - 2y\mu_{\mathrm{H}_2} - \mu_{\mathrm{M}_x\mathrm{C}_y=0} \quad (13)$$

where $\mu_x$ denotes the chemical potential of the component $x$.

Knowing the chemical potentials of the metallic components at a given temperature (see later) as well as the pressure-dependent chemical potential of $H_2$,

$$\mu_{\mathrm{H}_2} = \mu^0_{\mathrm{H}_2} + RT \ln p_{\mathrm{H}_2} \quad (14)$$

($\mu^0_{\mathrm{H}_2}$ is the chemical potential of one mol $H_2$ in the standard state where $p_{\mathrm{H}_2} = p^0 = 1$ atm), the methane chemical potential $\mu_{\mathrm{CH}_4}$ is solved from Eq. 13. The methane pressure can then be solved from the expression

$$\mu_{\mathrm{CH}_4} = \mu^0_{\mathrm{CH}_4} + RT \ln f_{\mathrm{CH}_4}, \qquad f_{\mathrm{CH}_4} = p_{\mathrm{CH}_4} \exp\{C(T)p_{\mathrm{CH}_4}\} \quad (15)$$

where $C(T)$ can be found in [2, 5].

The chemical potentials of Cr, Mo, V, and Fe in Eq. 13 are those of these elements dissolved in the ferritic (bcc) matrix. The thermodynamic treatment of such a system is complex, but the solution is available in the literature [5], which gives $\mu_{Cr}$, $\mu_{Mo}$, $\mu_V$, and $\mu_{Fe}$ as a function of the composition and temperature. The chemical potential $\mu_{M_xC_y}$ of the reacting carbide also depends on its crystal structure, composition, and temperature. Neglecting changes in the carbide composition during hydrogen attack, this chemical potential is also now known [5] for all most likely carbides in Cr-Mo steels.

### 9.8.3.2 Damage Mechanics Model

At the macroscopic continuum level, the constitutive equation in terms of Cartesian coordinates reads

$$\overset{\triangledown}{\sigma}_{ij} = \mathscr{L}_{ijkl}(D_{kl} - D_{kl}^{CC} - \alpha \dot{T} \delta_{kl}) \tag{16}$$

where $\overset{\triangledown}{\sigma}_{ij}$ denotes the Jaumann derivative of Cauchy stress and $D_{ij}^{CC}$ the strain-rate contribution due to creep and grain boundary cavitation. $\mathscr{L}_{ijkl}$ is the tensor of elastic moduli,

$$\mathscr{L}_{ijkl} = \frac{E}{1+v}\left[\frac{1}{2}\left(\delta_{ik}\delta_{jl} + \delta_{il}\delta_{jk}\right) + \frac{v}{1-2v}\,\delta_{ij}\delta_{kl}\right], \tag{17}$$

in terms of Young's modulus $E$ and Poisson's ratio $v$, and $\alpha$ is the cubic thermal expansion coefficient. If there were no cavitation, $D_{ij}^{CC}$ would simply be equal to the power-law creep rate

$$D_{ij}^{C} = \dot{\varepsilon}_e^{C}\,\frac{3}{2}\frac{\sigma'_{ij}}{\sigma_e} \tag{18}$$

with $\dot{\varepsilon}_e^C$ determined by the macroscopic effective stress $\sigma_e$ through Eq. 1. If there is cavitation, $D_{ij}^{CC}$ also accounts for the macroscopic rate of deformation caused by cavitation through the average separation rate $\dot{\delta}$ through Eq. 11. The way in which this is done depends on the mode of cavitation, as mentioned in the previous section.

When creep deformations of the grains are significant, the effect of grain boundary cavitation can be described in terms of a penny-shaped crack model for which $D_{ij}^{CC}$ reads

$$\begin{aligned} D_{ij}^{CC_p} = \dot{\varepsilon}_e^{C}\Bigg[&\frac{3}{2}\frac{\sigma'_{ij}}{\sigma_e}\left(1 + \rho\sum_{K=1}^{3}\left(\frac{n-1}{n+1}\right)\left(\frac{n_k^K\sigma_{kl}n_l^K - \sigma_n^K}{\sigma_e}\right)^2\right) \\ &+\rho\sum_{K=1}^{3} n_i^K n_j^K\left(\frac{2}{n+1}\,\frac{n_k^K\sigma_{kl}n_l^K - \sigma_n^K}{\sigma_e}\right)\Bigg]. \end{aligned} \tag{19}$$

This expression accounts for the contribution of three orthogonal families of cavitating facets with unit normal vector $n_i^K$ ($K = 1, \ldots, 3$). The parameter $\rho$ is related to the density of cavitating facets and is given by $\rho = 0.168 \times (n+1)/\sqrt{1+3/n}$ when all are potentially cavitated. The term $n_k^K \sigma_{kl} n_l^K$ is the component of the macroscopic stress normal to grain facets $K$. The $\sigma_n^K$ in Eq. 19 is the normal stress transmitted by the facets $K$. This stress is determined from the condition that the cavitation rate $\dot{\delta}(\sigma_n^K)$ caused by cavitation under a facet normal stress of $\sigma_n^K$ according to Eq. 11 be equal to the opening rate $\dot{\delta}_p^K$ of a facet-size, penny-shaped "crack" (radius $R$) in a creeping solid bridged by the same stress,

$$\dot{\delta}_p^K = \frac{4}{\pi}\left(1+\frac{3}{n}\right)^{-1/2} \frac{n_k^K \sigma_{kl} n_l^K - \sigma_n^K}{\sigma_e} \dot{\varepsilon}_e^C 2R \tag{20}$$

In the other extreme mode of deformation, where creep of the grains is negligible, the macroscopic strain rate is

$$D_{ij}^{CCrg} = \dot{\varepsilon}_e^C \frac{3}{2}\frac{\sigma'_{ij}}{\sigma_e} + \kappa \frac{\dot{\delta}_{rg}}{R}\delta_{ij} \tag{21}$$

and consists of a pure (yet small) creep part and a dilatational part due to isotropic cavitation on all facets. The cavitation rate $\dot{\delta}_{rg}$ is equal to the separation rate from Eq. 11 with the facet normal stress $\sigma_n$ being taken equal to the macroscopic hydrostatic stress $\sigma_m$. The coefficient $\kappa$ depends on the grain shape and is typically equal to 0.3.

The two extreme modes are combined simply by selecting one (Eq. 19) or the other (Eq. 21). A possible criterion for selecting the mode is the maximum value of the cavitation rates predicted in each of the two modes. If $\dot{\delta}_{rg}$ delivers the largest value, $D^{CC}$ to be substituted into Eq. 16 is taken to be equal to that in Eq. 21; otherwise the expression in Eq. 19 is used. Other criteria are conceivable but have not been used so far.

## 9.8.4 PARAMETER IDENTIFICATION

The model requires various types of parameters related to the following

- Composition of the steel, including its carbides. The composition of the carbides can be determined by EDS chemical analysis and their crystal structure by diffraction analysis.
- Microstructure, including the grain size and the average half-spacing $2b$ between carbides on grain facets. The parameter $R$ is obtained from the average grain volume $V_{gr}$ as $R \approx (0.042\, V_{gr})^{1/3}$.

- Material properties, including creep and diffusion parameters. At a given temperature, the creep parameters $B$ and $n$ in Eq. 1 are obtained by fitting to the minimum creep rate in a standard creep curve. According to Frost and Ashby [1], the temperature dependence of $B$ is conveniently incorporated by writing

$$B(T) = B(T_0)\,\frac{T_0}{T}\exp\left[-\frac{Q_V}{k}\left(\frac{1}{T}-\frac{1}{T_0}\right)\right] \tag{22}$$

in terms of the value of $B$ at another temperature $T_0$ and the activation energy $Q_V$ (J/atom) ($k$ is Boltzmann's constant). Values of the latter for various steels are listed in the literature [1]. The diffusion parameter $\mathscr{D}$ in Eqs. 7 and 9 is defined as

$$\mathscr{D} = \frac{D_B\delta_B\Omega}{kT}\exp\left[-\frac{Q_B}{kT}\right] \tag{23}$$

in terms of the grain boundary diffusion coefficient $D_B$ and the corresponding activation energy $Q_B$ (J/atom). Furthermore, $\delta_B$ is the boundary thickness and $\Omega$ the atomic volume. The values of $D_B\delta_B$, $Q_B$, and $\Omega$ can be found in the literature [3].

## 9.8.5 HOW TO USE THE MODEL

The damage model (Eq. 16) has the standard form of a viscoplastic or creeplike constitutive equation. It can therefore be incorporated into existing codes through a modification of the creeplike term $D_{ij}^{CC}$. This modification will need to include a solution procedure for the nonlinear matching condition (Eq. 20); a straightforward secant method has been found to be effective [8]. Because of the high nonlinearity of creep, numerical stability requires attention just as in any creep analysis. A relatively simple rate tangent operator derived in [8] has proved to be efficient in the integration of Eq. 16 [4].

The calculation of the methane pressure is straightforward in principle but requires knowledge of the temperature (either constant or an outcome of a transient thermal analysis) and the composition at each point in the body. The necessary equations as summarized in Section 9.8.3.1 can be found in full detail in [5]. The Fortran code of the expressions is available from the authors upon request.

A simplified version of the model can be used when the component or material is under autoclave conditions, i.e., at a constant temperature without external loading. In that case, the voids are subjected only to the gas pressure $p_m$ and the material does not creep at a macroscopic scale. Then, the void growth equations (Eqs. 6–10) can be simply used to integrate the cavity size *a* in time until cavity coalescence takes place.

## 9.8.6 LIST OF PARAMETERS

Most of the parameters in the model are strongly dependent on the material and therefore cannot be listed. Following are physical constants and parameters that are applicable for a wide range of materials:

- $k = 1.38 \times 10^{-23}\,\mathrm{J/K}$
- $\Omega = 1.18 \times 10^{-29}\,\mathrm{m}^3$
- $\kappa \approx 0.3$
- $\psi \approx 70-75^\circ$

## REFERENCES

1. Frost, H. J., and Ashby, M. F. (1982). *Deformation-Mechanism Maps*, Oxford: Pergamon Press.
2. Odette, G. R., and Vagarali, S. S. (1982). An equation-of-state for methane for modeling hydrogen attack in ferritic steels. *Metall. Trans. A*, **13A**: 299–303.
3. Parthasarathy, T. A. (1985). Mechanisms of hydrogen attack of carbon and 2.25Cr-1Mo steels. *Acta Metall.*, **33**: 1673–1681.
4. Schlögl, S. M., and Van der Giessen, E. (1999). Hydrogen attack in a welded reactor. *J. de Phys. IV* 9: 137–146.
5. Schlögl, S. M., Van Leeuwen, Y., and Van der Giessen, E. (2000). On methane generation and decarburization in low alloy Cr-Mo steels during hydrogen attack. *Metall. Mater. Trans.* **A31**: 125–137.
6. Shewmon, P. (1987). Synergism between creep ductility and grain boundary bubbles. *Acta Metall.* **35**: 1317–1324.
7. Van der Burg, M. W. D., Van der Giessen, E., Needleman, A., and Tvergaard, V. (1995). Void growth due to creep and grain boundary diffusion at high triaxialities. *J. Mech. Phys. Solids* **43**: 123–165.
8. Van der Burg, M. W. D., Van der Giessen, E., and Tvergaard, V. (1998). A continuum damage analysis of hydrogen attack in in a 2.25Cr-1Mo vessel. *Mat. Sci. and Eng.* **A241**: 1–13.

SECTION 9.9

# Hydrogen Transport and Interaction with Material Deformation: Implications for Fracture

PETROS SOFRONIS

*Department of Theoretical and Applied Mechanics, University of Illinois at Urbana-Champaign, 104 South Wright Street, Urbana, Illinois*

Contents





*Handbook of Materials Behavior Models.* ISBN 0-12-443341-3.

## 9.9.1 NON-HYDRIDE-FORMING SYSTEMS

### 9.9.1.1 VALIDITY

The model applies to a variety of systems, such as fcc, bcc, hcp, pure metals (e.g., Ni, Fe, Mo, W), solid solutions (e.g., steels), precipitation-strengthened systems (e.g., aluminum alloys), and intermetallics (e.g., $Ni_3Al$ and FeAl). It can be used to calculate the local hydrogen concentration qualified by hydrostatic stress and trapping induced by plastic straining. Then by using information on whether the fracture is strain- or stress-controlled, one can predict the location of the first microcracking event in a specimen.

### 9.9.1.2 BACKGROUND

The interaction between solute hydrogen atoms and an applied stress field results from the hydrogen-induced volume and local moduli changes that accompany the introduction of the solute hydrogen in the lattice [11, 23]. In regions of tensile hydrostatic stress and softened elastic moduli, interstitial hydrogen has a lower chemical potential. As a consequence, diffusion through normal interstitial lattice sites (NILS) is generated toward these regions, tending to eliminate the gradients of the chemical potential. Regions with compressive hydrostatic stress or hardened elastic moduli are depleted. Transported hydrogen through NILS diffusion can interact with and accumulate at various microstructural heterogeneities, such as dislocations, grain boundaries, inclusions, voids, surfaces, and impurity atoms described as traps, and from there initiate fracture [7]. Trapping characterizes the fact that interstitial solute atoms often find interstices associated with lattice imperfections to be energetically preferable to NILS. It is emphasized that trap characteristics are evolutionary in nature. Deformation affects dislocation density and structure, changes the void population and size, and influences the behavior of inclusions and grain boundaries in their activity as traps.

### 9.9.1.3 DESCRIPTION OF THE MODEL

The governing equation of transient hydrogen diffusion that accounts for trapping and hydrostatic stress drift is [9, 22]

$$\frac{D}{D_{eff}}\frac{dC_L}{dt} = DC_{L,ii} - \left(\frac{DV_H C_L}{3RT}\sigma_{kk,i}\right)_{,i} - \alpha\theta_T \frac{\partial N_T}{\partial \varepsilon^p}\frac{d\varepsilon^p}{dt} \qquad (1)$$

where $(\ )_{,i} = \partial(\ )/\partial x_i$, $d/dt$ is the time derivative, $C_L = \theta_L \beta N_L$ is the hydrogen concentration in NILS measured in atoms per unit volume, $\theta_L$ is the occupancy of NILS, $\beta$ is the number of NILS per solvent atom, $N_L = N_A/V_M$, $N_A = 6.0232 \times 10^{23}$ atoms per mole is Avogadro's number, $V_M$ is the molar volume of the host lattice, $C_T = \theta_T \alpha N_T$ is hydrogen concentration in number of atoms per unit volume in trapping sites, $\theta_T$ is the occupancy of the trapping sites, $\alpha$ is the number of trapping sites per trap, $N_T$, which is a function of the local effective plastic strain, i.e., $N_T = N_T(\varepsilon^p)$, denotes the trap density measured in number of traps per unit volume, $D$ is the hydrogen diffusion constant through NILS, $V_H$ is the partial molar volume of hydrogen in solid solution, $R$ is the gas constant equal to 8.31 J mole$^{-1}$ K$^{-1}$, $T$ is the absolute temperature, $D_{eff}$ is an effective diffusion constant given by

$$D_{eff} = D/(1 + \partial C_T/\partial C_L), \qquad (2)$$

$\sigma_{ij}$ is the Cauchy stress, and the standard summation convention over the range is implied for a repeated index. The relationship between $C_L$ and $C_T$ in Eq. 2 is established through Oriani's equilibrium theory, which requires

$$\frac{\theta_T}{1-\theta_T} = \frac{\theta_L}{1-\theta_L} K, \qquad (3)$$

where $K = \exp(W_B/RT)$ is the equilibrium constant, and $W_B$ is the trap binding energy. As a result,

$$\frac{\partial C_T}{\partial C_T} = \frac{K\dfrac{\alpha N_T}{\beta N_L}}{\left[1 + \dfrac{K-1}{\beta N_L}\right]^2} \qquad (4)$$

## 9.9.1.4 Identification of Parameters

Information on binding energy values and the experimental techniques used to determine the binding energy of hydrogen to vacancies, solutes, solute–defect complexes, dislocations, internal interfaces, and isolated metal clusters can be found in the work of Meyers *et al.* [16]. By way of example, the trap density and binding energy for pure iron can be determined by correlating experimental results on measured permeation transients and time lags [8] with full numerical solutions to the diffusion equation (Eq. 1). The analysis should be repeated at several levels of macroscopic uniaxial strain to obtain the functional relationship $N_T = N_T(\varepsilon^p)$. When trapping is associated with dislocations, a model estimation for $N_T$ can be devised by correlating $N_T$ with the dislocation density and the lattice structure. Thus, in a bcc lattice

$N_T = \sqrt{2}\rho/a$, where $\rho(\varepsilon^p)$ is the dislocation density, $a$ is the lattice parameter, and the assumption of one trap per atomic plane threaded by a dislocation was made.

The flow stress in unixial tension is required as a function of hydrogen concentration.

### 9.9.1.5 How to Use the Model

Numerical implementation in a finite element formalism is given in the work by Sofronis and McMeeking [22], Lufrano *et al.* [14], and Krom *et al.* [9]. Solution to Eq. 1 requires knowledge of the local values of he hydrostatic stress and effective plastic strain. This is obtained by coupling the solution process of Eq. 1 to that of the elastoplastic deformation of the material. Solution to the latter is carried out by considering the hydrogen dilatational effect [17] through a contribution to the deformation rate tensor (symmetric part of the velocity gradient)

$$D_{ij}^H = \frac{d}{dt}\left\{\ln\left[1 + \frac{(c - c_0)\Delta v}{3\Omega}\right]\right\}\delta_{ij} \quad (5)$$

where $c$ is the total hydrogen concentration (in NILS and trapping sites) measured in hydrogen atoms per solvent atom, $c_0$ is the corresponding initial hydrogen concentration in the absence of any straining, $\Delta v$ is the volume change per atom of hydrogen introduced into solution that is directly related to the partial molar volume of hydrogen $V_H = \Delta v N_A$ in solution, and $\Omega$ is the mean atomic volume of the host metal atom. In the case of small strain plasticity, Eq. 5 reduces to $\dot{\varepsilon}_{ij}^H = (\dot{c}\Delta v/3\Omega)\delta_{ij}$.

So far, research in non-hydride-forming systems [2, 3] has established two viable mechanisms of embrittlement, namely: (i) hydrogen-enhanced localized plasticity and (ii) hydrogen-induced decohesion. However, a fracture criterion which would allow a relevant model to have predictive capabilities has not yet been devised. Hence, the following discussion should be viewed only as suggestive and by no means as reflecting all of the fundamental physics underlying the embrittlement effect.

Once the solution to the coupled initial boundary value problem is obtained for a given initial concentration of hydrogen, the following cases may be assessed [24]: (i) in the case of brittle intergranular fracture, the location of the first microcracking event should be expected to occur at locations where both stresses and hydrogen accumulation are high; (ii) in the case of ductile fracture by void nucleation at inclusion, the event will likely take place again in regions where stresses and hydrogen concentrations are high; (iii) in the case of fracture occurring predominantly by plastic flow localization, microcracking can be expected to initiate at locations where the effective plastic strain and the local hydrogen concentrations are high.

### 9.9.1.6 Parameters

#### 9.9.1.6.1 Iron and steels [7, 22]

$W_b = 60$ kJ/mole; $N_T = N_T(\varepsilon^p)$ is calculated from experimental data by Kumnick and Johnson [10] and a relevant plot can be found in the work of Sofronis and McMeeking [22]; parameter $\alpha$ can vary from 1 to several powers of 10 [6]; $\beta = 1$ (maximum NILS concentration of 1 hydrogen atom per solvent lattice atom); $V_M = 7.116\,\mathrm{cm^3/mole}$; $V_H = 2.0\,\mathrm{cm^3/mole}$; and for $T \geq 325\,K$ the diffusion constant is given by $D = 2.1 \times 10^{-7}(\mathrm{m^2/s})\exp(-6.88\,\mathrm{kJmole^{-1}}/RT)$ [26]. For a stress-free solid, the initial NILS concentration $C_0$ in equilibrium with hydrogen gas at pressure $p$ and temperatures well above 300 K is given by Sievert's law as $C_0 = 1.989 \times 10^{26}\sqrt{p}(\mathrm{atoms/m^3})\exp(-28.6\mathrm{kJmole^{-1}}/RT)$.

#### 9.9.1.6.2 Niobium [12, 13]

$W_b = 29.2$ kJ/mole; a dislocation-based model [24] for the trap density is formulated (cf. Section 9.9.1.4) by assuming that the dislocation density measured in length per cubic meter varies linearly with effective plastic strain [5], such that

$$\rho = \begin{cases} \rho_0 + \gamma\varepsilon^p & \text{for} \quad \varepsilon^p < 0.5 \\ 10^{16} & \text{for} \quad \varepsilon^p \geq 0.5 \end{cases}, \tag{6}$$

where $\rho_0 = 10^{10}$ line length per cubic meter denotes the dislocation density for the annealed material, and $\gamma = 2.0 \times 10^{16}$ line length per cubic meter is a proportionality constant; parameter $\alpha$ can vary from 1 to several powers of 10; $\beta = 1$ (maximum NILS concentration of 1 hydrogen atom per solvent lattice atom); $V_M = 10.852\,\mathrm{cm^3/mole}$; $V_H = 1.88\,\mathrm{cm^3/mole}$; for $T \geq 273$ K the diffusion constant is given by $D = 5.0 \times 10^{-8}(\mathrm{m^2/s})\exp(-10.215\mathrm{kJmole^{-1}}/RT)$; and since niobium is a high-solubility system, NILS initial concentrations for the stress-free lattice can vary from 0 to 1 H/Nb.

## 9.9.2 HYDRIDE-FORMING SYSTEMS

### 9.9.2.1 Validity

The model applies to systems that fail predominantly by hydride formation and cleavage. These are systems in which hydrides are either stable or can be stabilized by the application of a stress field. Examples are the IVb and Vb metals and their alloys (Ti, Zr, V, Nb, Ta), as well as a number of other metals such as Mg and Al. This hydride mechanism is supported by microscopic observations [21, 25] and thermodynamic calculations [4].

### 9.9.2.2 Background

In hydride-forming systems, embrittlement occurs by hydride formation in severe stress raisers such as crack tips and is followed by cleavage of the brittle hydride [3, 12, 13]. The phenomenon is intermittent, with the crack propagating through the hydride and stopping when it reaches the matrix. Subsequently, new hydride forms either autocatalytically [21] or because of external loading, and then cleavage reinitiates [27]. Stress-induced hydride formation is a consequence of the volume dilatation, of the order of ~15%, which accompanies hydride precipitation [18]. It has been experimentally shown that hydrides form in regions of hydrostatic tensile stress even at temperatures which are above the solvus temperature in the absence of stress. The hydride formation is a result of the enhanced hydrogen concentration in the area of tensile stress [11] and the decreased chemical potential of the hydride relative to the solid solution in the same stress field [4, 18].

### 9.9.2.3 Description of the Model

The terminal solid solubility (also termed the solvus concentration), measured in hydrogen atoms per solute, of hydrogen in solution in a material under external stress $\sigma_{ij}$ is given by

$$c_s^{\sigma} = B \exp\left[\left(\Delta G_{\alpha-\beta}^{\text{chem}} + \Delta G_{\alpha-\beta}^{\text{sur}} + \Delta G_{\alpha-\beta}^{\text{mech}}\right)/RT\right], \tag{7}$$

where $B$ is an experimentally determined constant, $\Delta G_{\alpha-\beta}^{\text{chem}}$ is the chemical Gibbs energy change in the hydride formation, $\Delta G_{\alpha-\beta}^{\text{sur}}$ is the free energy needed for the creation of the interface, and $\Delta G_{\alpha-\beta}^{\text{mech}}$ is the total mechanical free energy of hydride formation. The mechanical energy $\Delta G_{\alpha-\beta}^{\text{mech}}$ involves the elastic work done on the system (hydride + matrix) and is stored as elastic energy, the plastic work of accommodation, and the work done by the applied stress against the matrix displacement upon hydride formation [1, 13].

The dominant mode for diffusion is through the solid solution, while trapping is ignored. Diffusion through the hydride phase is extremely slow. Let $C_L$ denote the hydrogen concentration in the lattice expressed in hydrogen atoms per unit volume of solid solution phase when the hydride volume fraction $f$ is less than 1. At a given time $t$ with the composition of the composite (solid solution + hydride phase) defined pointwise by its hydride volume fraction $f$ such that $0 \leq f \leq 1$, the hydrogen diffusion equation [12, 22] is written as follows:

$$Q\frac{\partial C_L}{\partial t} = \left(D_c C_{L,i} - \frac{D_c C_L V_H}{3RT}\partial_{kk,i}\right)_{,i}, \tag{8}$$

where

$$Q = \begin{cases} 1-f & \text{if} \quad 0 \leq f < 1 \\ 1 & \text{if} \quad f = 1 \end{cases}, \tag{9}$$

$D_c$ is an effective diffusion constant for the hydrogen diffusing through the composite material given by

$$D_c = \begin{cases} (1-f)D_s & \text{if} \quad 0 \leq f < 1 \\ D_h & \text{if} \quad f = 1 \end{cases}, \tag{10}$$

and $D_s$ and $D_h$ represent the diffusion constants of hydrogen in solution with metal and in the hydride, respectively.

### 9.9.2.4 Identification of Parameters

Standard solubility techniques are used to determine the solvus concentration as a function of temperature in the absence of externally applied stress. Those may be either direct by measuring the hydrogen concentration in solid solution [20] or indirect, e.g., through measurements of electrical resistivity or dilatometry or internal friction [1, 18]. The mechanical energy of hydride formation $\Delta G^{\mathrm{mech}}_{\alpha-\beta}$ is calculated by solving a separate elastoplastic boundary value problem (no hydrogen diffusion involved) in which the formation of a hydride precipitate is effected by a transformation strain (volume dilatation) in a plastically deforming matrix under externally applied loads that span the range of possible values from purely hydrostatic to purely deviatoric. The calculated values of the mechanical energy are then tabulated in order to be used as a subroutine module for the numerical solution of Eq. 8 [13]. Correlating the experimentally measured solvus concentration as a function of temperature with the calculated value of the accommodation energy $\Delta G^{\mathrm{mech}}_{\alpha-\beta}$ in the absence of external applied stress [13], one determines the value of the sum of the chemical and free energy, $\Delta G^{\mathrm{chem}}_{\alpha-\beta} + \Delta G^{\mathrm{sur}}_{\alpha-\beta}$, for hydride formation at a specific temperature. It should be noted that this new approach to the energetics of hydride formation is an improvement over previous models, which, based on purely elastic considerations, separated the mechanical energy into an accommodation component and an interaction component.

### 9.9.2.5 How to Use the Model

A finite element solution of Eq. 8 provides the local hydrogen concentration and the hydride volume fraction in a specimen under load as a function of time. At each time step, the local hydrogen concentration in the solid solution phase and the associated change in the hydride volume fraction are calculated by the lever rule in a continuum sense pointwise. The details of the numerical implementation are given in the work by Lufrano *et al.* [12]. Part of the solution procedure is the pointwise calculation of the terminal solid solubility of hydrogen as a function of stress. This is done through Eq. 7 once $\Delta G^{\mathrm{chem}}_{\alpha-\beta} + \Delta G^{\mathrm{sur}}_{\alpha-\beta}$ is known and the module for the calculation of $\Delta G^{\mathrm{mech}}_{\alpha-\beta}$ is established.

Solution to the transient diffusion problem needs to be coupled to a code for the solution of the elastoplastic boundary value problem that accounts for the hydrogen and hydride-induced volume dilatation. As in the case of non-hydride-forming systems, the part of the deformation rate tensor denoting the hydrogen effect is given by Eq. 5 with the term $(c - c_0)\Delta v/\Omega$ replaced by

$$e^{\mathrm{T}} = \begin{cases} (c - c_0)\theta_h & \text{if} \quad f = 0 \\ (1 - f)(c - c_0)\theta_h + f(\theta_{\mathrm{hyd}} - c_0\theta_h) & \text{if} \quad f \neq 0 \end{cases}. \tag{11}$$

In Eq. 11, $c$ and $c_0$ are the local and initial hydrogen concentrations in hydrogen atoms per metal atom, respectively, $\theta_h$ is the lattice local dilatation when a hydrogen atom dissolves in solution with the metal [17], and $\theta_{\mathrm{hyd}}$ is the volume dilatation of a material element that is 100% hydride.

An averaging approach is used to estimate the hydride size directly ahead of a crack tip during the numerical simulation. Suppose that at a given time $t$ the region directly ahead of the crack tip along the axis to symmetry has a hydride volume fraction distribution $f(r)$, where $r$ is the distance from the tip and $f = 0$ for $r > r_0$. In view of the large values of hydrogen diffusivity, an individual hydride particle in this region could be as large as $a = \bar{f} r_0$ in length, where $\bar{f}$ is the average hydride volume fraction in the region $0 \leq r \leq r_0$. The fracture toughness in the presence of hydrogen $K_{IC}$ is defined [13] as the level of the applied load measured in terms of the applied stress-intensity factor at the moment when the hydride particle size $a$ at some location $r$ will reach a critical size at which the fracture of the hydride is energetically favored in a Griffith sense, namely

$$a = a_{\mathrm{crit}}(r) = \frac{2\gamma_s E}{\pi(1 - \nu^2)\sigma^2(r)}. \tag{12}$$

The parameter $\gamma_s$ is the surface energy of the hydride phase, $E$ is Young's modulus, $\nu$ is Poisson's ratio, and $\sigma$ is the local stress ahead of the crack in the direction normal to the axis of symmetry. Of course, such a prediction for the fracture toughness is a conservative one, since the presence of a void ahead of a blunting crack tip due to cracking of a hydride particle does not necessarily lead to fracture.

## 9.9.2.6 Parameters

### 9.9.2.6.1 Niobium [1, 12, 13]

$\theta_h = 0.174$; $\theta_{\text{hyd}} = 0.12$; the activation energy for diffusion through the hydride phase can be assumed equal to three times the activation energy for diffusion through the solid solution phase; $\gamma_s = 5.04$ Joules/m$^2$; and terminal solid solubility is given by

$$c_s^\sigma = 3.74 \exp\left(\frac{-12.6\,\text{kJ/mole}}{RT}\right) \exp\left(\frac{\Delta G_{\alpha-\beta}^{\text{mech}}}{RT}\right) \tag{13}$$

### 9.9.2.6.2 Zirconium [15, 19, 20]

$\theta_h = 0.12$ based on $V_H = 1.67\,\text{cm}^3$/mole; $\theta_{\text{hyd}} = 0171$; $V_M = 13.85\,\text{cm}^3$/mole; $\gamma_s$ can be calculated by considering that the fracture toughness of the pure hydride phase is $\sim 5.0\,\text{MPa}\sqrt{\text{m}}$; $D = 2.17 \times 10^{-7}(\text{m}^2/\text{s})\exp(-35.1\text{kJmole}^{-1}/RT)$; the activation energy for diffusion through the hydride phase can be assumed equal to three times the activation energy for diffusion through the solution solution phase; and terminal solid solubility is given by

$$c_s^\sigma = 10.32 \exp\left(\frac{-34.0\,\text{kJ/mole}}{RT}\right) \exp\left(\frac{\Delta G_{\alpha-\beta}^{\text{mech}}}{RT}\right). \tag{14}$$

# ACKNOWLEDGEMENTS

This work was supported by the Department of Energy under grant DEFGO2-96ER45439. The author would like to thank Prof. H. K. Birnbaum for many helpful discussions on the subject.

# REFERENCES

1. Birnbaum, H. K., Grossbeck, M. L., and Amano, M. (1976). Hydride precipitation in Nb and some properties of NbH. *J. Less Comm. Met.* **49**: 357–370.

2. Birnbaum, H. K., and Sofronis, P. (1994). Hydrogen-enhanced localized plasticity: A mechanism for hydrogen related fracture. *Mater. Sci. Eng.* **A176**: 191–202.
3. Birnbaum, H. K., Robertson, I. M., Sofronis, P., and Teter, D. (1997). Mechanisms of hydrogen related fracture: A review, in *Corrosion Deformation Interactions CDΓ96* (Second International Conference, Nice, France, 1996), pp. 172–195, Magnin, T., ed., The Institute of Materials, Great Britain.
4. Flannagan, T. B., Mason, N. B., and Birnbaum, H. K. (1981). The effect of stress on hydride precipitation. *Scr. Met.* **15**: 109–112.
5. Gilman, J. J. (1969). *Micromechanics of Flow in Solids*, New York: McGraw-Hill Book Company, pp. 185–199.
6. Hirth, J. P., and Carnahan, B. (1978). Hydrogen adsorption at dislocations and cracks in Fe. *Acta Metall.* **26**: 1795–1803.
7. Hirth, J. P. (1980). Effects of hydrogen on the properties of iron and steel. *Met. Trans.* **11A**: 861–890.
8. Johnson, H. H., and Lin, R. W. (1981). Hydrogen and deuterium trapping in iron, in *Hydrogen Effects in Metals*, pp. 3–23, Bernstein, I. M., and Thompson A. W., eds., Metallurgical Society of AIME.
9. Krom, A. H. M., Koers, R. W. J., and Bakker, A. (1999). Hydrogen transport near a blunting crack. *J. Mech. Phys. Solids* **47**: 971–992.
10. Kumnick, A. J., and Johnson, H. H. (1980). Deep trapping states for hydrogen in deformed iron. *Acta Metall.* **28**: 33–39.
11. Li, J. C. M., Oriani, R. A., and Darken, L. S. (1966). The thermodynamics of stressed solids. *Z. Physik Chem. Neue Folge* **49**: 271–291.
12. Lufrano, J., Sofronis P., and Birnbaum, H. K. (1996). Modeling of hydrogen transport and elastically accommodated hydride formation near a crack tip. *J. Mech. Phys. Solids* **44**: 179–205.
13. Lufrano, J., Sofronis, P., and Birnbaum, H. K. (1998). Elastoplastically accommodated hydride formation and embrittlement. *J. Mech. Phys. Solids* **46**: 1497–1520.
14. Lufrano, J., Symons, D., and Sofronis, P. (1998). Hydrogen transport and large strain elastoplasticity near a notch in alloy X-750. *Eng. Fracture Mech.* **59**: 827–845.
15. Lufrano, J., and Sofronis, P. (2000). Micromechanics of hydride formation and cracking in zirconium alloys. *Computer Modeling in Engineering Science* **1**: 119–131.
16. Meyers, S. M., *et al.* (1992). Hydrogen interaction with defects in crystalline solids. *Rev. Mod. Phys.* **64**: 559–617.
17. Peisl, H. (1978). Lattice strains due to hydrogen in metals, in *Hydrogen in Metals I, Topics in Applied Physics*, pp. 53–74, vol. 28, Alefeld, G., and Volkl, J., eds., New York: Springer-Verlag.
18. Puls, M. P. (1984). Elastic and plastic accommodation effects on metal-hydride solubility. *Acta Metall.* **32**: 1259–1269.
19. Puls, M. P. (1990). Effects of crack tip stress states and hydride-matrix interaction stresses on delayed hydride cracking. *Metall. Trans.* **21A**: 2905–2917.
20. Shi, S.-Q., Shek, G. K., and Puls, M. P (1995). Hydrogen concentration limit and critical temperatures for delayed hydride cracking in zirconium alloys. *J. Nucl. Mater.* **218**: 189–201.
21. Shih, D. S., Robertson, I. M., and Birnbaum, H. K. (1988). Hydrogen embrittlement of $\alpha$ titanium: in situ TEM studies. *Acta Metall.* **36**: 111–124.
22. Sofronis, P., and McMeeking, R. M. (1989). Numerical analysis of hydrogen transport near a blunting crack tip. *J. Mech. Phys. Solids* **37**: 317–350.
23. Sofronis, P., and Birnbaum, H. K. (1995). Mechanics of the hydrogen-dislocation-impurity interactions—I. Increasing shear modulus. *J. Mech. Phys. Solids* **43**: 49–90.
24. Taha, A., and Sofronis, P. (2001). A micromechanics approach to the study of hydrogen transport and embrittlement. *Eng. Fracture Mech.* **68**: 803–837.

25. Takano, S., and Suzuki, T. (1974). An electron-optical study of $\beta$-hydride and hydrogen embrittlement of vanadium. *Acta Metall.* **22**: 265–274.
26. Völkl, J., and Alefeld, G. (1978). Diffusion of hydrogen in metals, in *Hydrogen in Metals I, Topics in Applied Physics*, pp. 53–74, vol. 28, Alefeld, G., and Volkl, J., eds., New York: Springer-Verlag.
27. Westlake, D. G. (1969). A generalized model for hydrogen embrittlement. *Trans. ASM* **62**: 1000–1006.

SECTION **9.10**

# Unified Disturbed State Constitutive Models

CHANDRA S. DESAI
*Department of Civil Engineering and Engineering Mechanics, The University of Arizona, Tucson, Arizona*

Contents



## 9.10.1 VALIDITY

The disturbed state concept (DSC) is a unified approach for constitutive modeling of materials and interfaces and joints under thermomechanical and environmental loading. It allows one to consider, in a hierarchical framework, various behavioral features of materials, such as elastic, plastic, and creep strains, microcracking leading to degradation or damage, and stiffening or healing. Degradation and stiffening are incorporated by using the idea of disturbance (D). The DSC and its specialized versions have been found to provide a satisfactory characterization of a wide range of materials, such as geologic, concrete, asphalt concrete, ceramics, metals, allows (solders), and silicon.

Details of the theoretical development of the models, their use and validation for various materials and interfaces and joints, and their

*Handbook of Materials Behavior Models*. ISBN 0-12-443341-3.

implementation in a computer (finite element) procedure and validation for a wide range of field and laboratory-simulated practical boundary value problems are given in various publications; only typical studies are listed under References.

## 9.10.2 FORMULATION

The basic incremental DSC equations are derived as

$$d\underset{\sim}{\sigma}^a = (1 - D)\underset{\sim}{C}^i d\underset{\sim}{\varepsilon}^i + D\underset{\sim}{C}^c d\underset{\sim}{\varepsilon} + dD\left(\underset{\sim}{\sigma}^c - \underset{\sim}{\sigma}^i\right) \quad (1)$$

where $a$, $i$, and $c$ denote observed or actual, relative intact (RI) and fully adjusted (FA) responses, respectively, $\underset{\sim}{\sigma}$ and $\underset{\sim}{\varepsilon}$ are the stress and strain vectors, $\underset{\sim}{C}$ is the constitutive matrix, $D$ is the disturbance, $dD$ is the increment or rate of disturbance, and $d$ denotes increment. Although $D$ is often treated as scalar, its tensorial form can be introduced in the DSC equations. In the DSC, it is considered that at any stage during deformation, a material element is composed of a mixture of two or more reference materials. For the dry material, one of the reference materials is in the RI state, whose behavior is expressed by using such continuum theories as elasticity, plasticity, or elasto-viscoplasticity. The RI material continuously transforms to the material in the FA state because of the internal self-adjustment of the material's microstructure. It is the consequence of relative particle motions and/or microcracking; in the limit, the FA material is assumed to approach an invariant state. Behavior of the FA can be characterized by using various assumptions, e.g., (1) it can carry hydrostatic or isotropic stress and no shear stress and act like a *constrained liquid*, and (2) it can continue to carry a limiting shear stress under a given hydrostatic stress and continue to deform in shear without any change in volume, as in the critical state concept [15] and act like a *constrained liquid-solid*. If the material in the FA state is treated as a "void" as in the continuum damage concept [16], it can carry no stress at all. Since the material in the FA state is surrounded by the RI material, the foregoing two idealizations are considered to be more realistic compared to the "void" assumption in the damage concept. Furthermore, the DSC model allows for the coupling and interaction between the material parts in the RI and FA states. Such coupling is not allowed in the damage concept. As a result, the DSC model implicitly allows for the neighborhood or nonlocal effects, and external enrichments such as gradient and Cosserat theories and microcrack interaction are not required. The disturbance, $D$, acts as the coupling and interpolation mechanism and leads to the observed behavior in terms of the behavior of the RI and FA material parts.

**Specializations**: If $D = 0$, Eq. 1 leads to the continuum models in which the RI constitutive matrix, $\underset{\sim}{C}^i$, can be based on linear (nonlinear) elasticity, plasticity, or elasto-viscoplasticity. If $D \neq 0$, Eq. 1 provides for microcracking and degradation or stiffening (healing). If the terms related to the FA state (c) are ignored, Eq. 1 leads to the classical damage model.

### 9.10.3 IDENTIFICATION OF MATERIAL PARAMETERS

The DSC model allows for the flexibility to choose specialized versions, such as the elastic, elastoplastic, viscoplastic, and disturbance (degradation). Hence, the user needs to specify parameters related to the version chosen for a given material and application need. The basic parameters are listed in following text, along with the explanation for various quantities:

- **Linear elastic**: Young's modulus, $E$, and Poisson's ratio, $v$; or shear modulus, $G$, and bulk modulus, $K$;
- **Plasticity**
  *Classical*: von Mises: yield stress, $\sigma_y$, or cohesion, $c$. Drucker-Prager: cohesion, $c$, angle of friction, $\phi$;
  *Continuous yielding*: Hierarchical Single Surface (HISS) (see "Function"): Ultimate (failure): slope of ultimate envelope, $\gamma$; yield surface shape, $\beta$; bonding stress, $R$; Transition (compaction to dilation): $n$; Hardening or yielding: parameters $a_1$ and $\eta_1$;
- **Creep**: fluidity, $\Gamma$; flow function, $N$;
- **Disturbance**: $A$, $Z$, $D_u$;
- **Thermal effect**:
  Coefficient of thermal expansion, $\alpha_T$;
  Parameter dependence: $p_r$ and $\lambda$;
- **Yield function**: $F = \bar{J}_{2D} - (-\alpha \bar{J}_1^n + \gamma \bar{J}_1^2)(1 - \beta S_r)^{-0.5} = 0$;
- **Growth function**: $\alpha = a_1 / \xi^{n_1}$;
- **Creep (viscoplastic [17])**: $d\underset{\sim}{\varepsilon}^{vp} = \Gamma \langle \phi \rangle (\partial F / \partial \underset{\sim}{\sigma})$ and $\phi = (F/F_o)^N$; multicomponent DSC or overlay models allow for viscoelastic and viscoplastic creep;
- **Disturbance**: $D = D_u(1 - e^{-A\xi_D^Z})$;
- **Temperature dependence**: $p(T) = p(T_r)(T/T_r)^c$;

where $J_1$ and $J_{2D}$ are the first invariant of the stress tensor, $\sigma_{ij}$, and the second invariant of the deviatoric stress tensor, $S_{ij}$, the overbar denotes nondimensional value with respect to the atmospheric pressure, $p_a$, $\bar{J}_1 = J_1 + 3R$, $R$ is the bonding stress, $S_r = (\sqrt{27}/2)J_{3D} \cdot J_{2D}^{-3/2}$, $J_{3D}$ is the third invariant of $S_{ij}$, $\xi$ and $\xi_D$ are the trajectories of total and deviatoric plastic strains, respectively, $p_a$ is any parameter (see preceding), $T_r$ is the reference temperature (e.g., 300 K), and $T$ is any temperature. The yield surface, $F$, plots as continuous in various stress spaces, and the associated plasticity models are referred to as HISS (hierarchical single surface).

### 9.10.3.1 Tests for Finding Parameters

Details of the procedures for finding the preceding parameters from laboratory tests are given in Desai [1]. The parameters have physical meanings in that they are related to specific states during deformations. The parameters are found from uniaxial tension and compression, shear, biaxial, triaxial ($\sigma_1, \sigma_2 = \sigma_3$), multiaxial ($\sigma_1 \neq \sigma_2 \neq \sigma_3$), creep and relaxation, and/or cyclic thermomechanical tests. The disturbance parameters are found on the basis of the degradation response under static or cyclic tests in terms of stress, volumetric (void ratio), effective stress (pore water pressure) for saturated materials, and nondestructive properties such as ultrasonic P- and S-wave velocities or attenuation.

The parameters are expressed in terms of temperature and strain rates.

## 9.10.4 IMPLEMENTATION

The DSC model is implemented in linear and nonlinear finite element procedures for the solution of problems under static, dynamic, and repetitive loading, and involving dry and porous saturated materials. The finite element procedures have been used to predict the behavior of a number of practical problems in civil and mechanical engineering, electronic packaging, and semiconductor systems.

The incremental constitutive equations, Eq. 1, are expressed as $d\underset{\sim}{\sigma}^a = \underset{\sim}{C}^{DSC}\, d\underset{\sim}{\varepsilon}^i$, where $\underset{\sim}{C}^{DSC}$ is the constitutive matrix given by

$$\underset{\sim}{C}^{DSC} = (1-D)\,\underset{\sim}{C}^i + D(1+\alpha)\,\underset{\sim}{C}^c + \underset{\sim}{R}^T\,\underset{\sim}{\sigma}^r \tag{2}$$

where $\alpha$ is the relative motion parameter and $\underset{\sim}{\sigma}^r$ is the relative stress vector based on $\underset{\sim}{\sigma}^c - \underset{\sim}{\sigma}^i$. The incremental finite element equations with $\underset{\sim}{C}^{DSC}$ at step

$n$ are derived as

$$\int_V (1 - D_n) \underset{\sim}{B}^T d \underset{\sim}{\sigma}^{i}_{n+1} dV + \int_{\sim} D_n \underset{\sim}{B}^T d\sigma^{c}_{n+1} dV \\ + \int_{\sim} \underset{\sim}{B}^T \left( \underset{\sim}{\sigma}^{c}_{n} - \underset{\sim}{\sigma}^{i} \right) dD_n - dV = d \underset{\sim}{Q}_{n+1} \tag{3}$$

which in terms of displacement increment $d\underset{\sim}{q}^i$ are expressed as

$$\int_{\sim} \underset{\sim}{B}^T \underset{\sim}{C}^{DSC}_{n} \underset{\sim}{B} dV \, d\underset{\sim}{q}^{i}_{n+1} = d \underset{\sim}{Q}_{n+1} \tag{4}$$

where $\underset{\sim}{B}$ is the transformation matrix and $d\underset{\sim}{Q}$ is the vector of applied loads. Various techniques for the incremental-iterative solutions for Eq. 4 are given by Desai [1].

The finite element procedures have been used to successfully predict the observed behavior of a wide range of simulated and field problems in civil and mechanical engineering and electronic packaging.

### 9.10.5 PARAMETERS

Parameters for typical engineering materials such as a clay, sands, rocks, concrete, rock salt, ceramic composite, and solder alloy (40/60, Pb/Sn) are given in Appendix 1. Parameters for many other materials and interfaces and joints are given elsewhere, e.g., Desai [1].

## APPENDIX 1 DSC/HISS PARAMETERS FOR TYPICAL MATERIALS

In view of space limitations, details of the physical properties of the materials and associated tests are not presented; they can be obtained from the references cited.

**Parameters for Saturated Marine Clay-DSC Model**

| Parameter | Value | Parameter | Value |
|---|---|---|---|
| E, MPa (psi) | 10 (1500) | $\bar{m}$ | 0.0694 |
| $\nu$ | 0.35 | $\lambda$ | 0.169 |
| $\gamma$ | 0.047 | $e_{oc}$ | 0.903 |
| $\beta$ | 0.00 | A | 1.73 |
| $n$ | 2.80 | Z | 0.309 |

## Parameters for Saturated Marine Clay-DSC Model (*Continued*)

| Parameter | Value | Parameter | Value |
|---|---|---|---|
| $h_1$ | 0.0001 | $D_u$ | 0.75 |
| $h_2$* | 0.78 | | |

* $\alpha = h_1/\xi_v^{h_2}$, $\alpha$ is the hardening in growth function, $e_{oc}$ is the initial void ratio, and $\xi_v$ is the trajectory of volumetric strains.

## Parameters for Sands: HISS Models

| | Dry sand | | | | | |
|---|---|---|---|---|---|---|
| Parameter | Ottawa | | Leighton Buzzard | Munich | Saturated sand |
| E, MPa | 262 | | 79 | 63 | 140 |
| (psi) | (38,000) | | (11,500) | (9200) | (20,420) |
| $v$ | 0.37 | | 0.29 | 0.21 | 0.15 |
| $\gamma$ | 0.124 | | 0.102 | 0.105 | 0.636 |
| $\beta$ | 0.494 | | 0.362 | 0.747 | 0.60 |
| $n$ | 3.0 | | 2.5 | 3.20 | 3.0 |
| | | $h_1$** | 0.135 | 0.1258 | |
| $a_1$* | $2.5 \times 10^{-3}$ | $h_2$ | 450 | 1355 | $a_1 = 0.16 \times 10^{-4}$ |
| $\eta_1$ | 0.370 | $h_3$ | 0.0047 | 0.001 | |
| | | $h_4$ | 1.02 | 1.11 | $\eta_1 = 1.17$ |
| R | 0.00 | | 0.00 | 0.00 | 0.00 |
| $\kappa$ | 0.265 | | 0.290 | 0.35 | — |

* $\alpha = a_1/\xi^{\eta_1}$

** $\alpha = \bar{h}_1 \exp[-\bar{h}_2\xi(1 - \xi_D/(\bar{h}_3 + \bar{h}_4\xi_D))]$

Note: $\alpha$ is the hardening or growth function; $\kappa$ is the nonassociative parameter.

## Parameters for Rocks and Concrete

| Parameter | Rocks | | |
|---|---|---|---|
| | Soapstone | Sandstone | Concrete |
| E, MPa | 9150 | 25500 | 7000 |
| (psi) | $(1328 \times 10^3)$ | $(3700 \times 10^3)$ | $(10^6)$ |
| $v$ | 0.0792 | 0.11 | 0.14 |
| $\gamma$ | 0.0470 | 0.0774 | 0.113 |
| $\beta_0$* | 0.750 | 0.767 | 0.8437 |
| $\beta_1$* | 0.0465 | 0.0020 | 0.027 |
| $n$ | 7.0 | 7.20 | 7.00 |
| $a_1$ | $0.177 \times 10^{-2}$ | $0.467 \times 10^{-2}$ | $9 \times 10^{-3}$ |

## Parameters for Rocks and Concrete (*Continued*)

| Parameter | Rocks | | |
|---|---|---|---|
| | Soapstone | Sandstone | Concrete |
| $\eta_1$ | 0.747 | 0.345 | 0.44 |
| R, MPa | 1.067 | 2.90 | 2.72 |
| (psi) | (155) | (420) | (395) |
| $\kappa$ | | | |
| $A$ | | | 668 |
| $Z$ | | | 1.50 |
| $D_u$ | | | 0.875 |

* $\beta = \beta_0 e^{-\beta_1 J_1}$

** $F_s = [\exp(\beta_1/\beta_0 J_1) - \beta S_r]^{-0.5}$

## Parameters for Rock Salt at Different Temperatures

| Temperature parameter | 296 K | 336 K | 350 K | 473 K | 573 K | 673 K |
|---|---|---|---|---|---|---|
| E (GPa) | 34.13 | 32.19 | 31.59 | 27.49 | 25.15 | 23.35 |
| $\nu$ | 0.279 | 0.287 | 0.290 | 0.310 | 0.324 | 0.336 |
| $\alpha_T$ (1/K) ($\times$ $10^{-5}$) | 3.8 | 4.2 | 4.3 | 5.5 | 6.4 | 7.2 |
| $\gamma$ | 0.0516 | 0.0384 | 0.0349 | 0.0173 | 0.0111 | 0.0076 |
| $\beta$ | 0.690 | 0.620 | 0.590 | 0.450 | 0.380 | 0.33 |
| $n$ (average) | — | 3.92 | — | — | — | — |
| $a_1$ ($\times 10^{-9}$) | 1.80 | 0.97 | 0.95 | | | |
| $\eta_1$ (average) | — | — | 0.474 | — | — | — |
| $\Gamma$($\times 10^{-3}$/day) | 4.95 | 6.11 | 6.54 | 10.77 | 14.8 | 19.35 |
| $N$ (average) | — | 3.0 | — | — | — | — |

Temperature dependence:

$$E(T) = 33.92\left(\frac{T}{300}\right)^{-0.462}; \nu(T) = 0.28\left(\frac{T}{300}\right)^{0.224}$$

$$\alpha_T(T) = 3.85 \times 10^{-5}\left(\frac{T}{300}\right)^{0.778};$$

$$\gamma(T) = 0.05\left(\frac{T}{300}\right)^{-2.326}; \beta(T) = 0.68\left(\frac{T}{300}\right)^{-0.91}$$

$$\alpha(T) = \left(\frac{a_1}{\xi^{\eta_1}}\right)_{300}\left(\frac{T}{300}\right)^{-0.334}; \Gamma(T) = 5.0\left(\frac{T}{300}\right)^{1.70}$$

## Elasticity and Plasticity Parameters for Solder (Pb40/Sn60) at $\dot{\varepsilon}$=0.002/sec: $\delta_0$-model

| Temperature (K) | 208 | 273 | 348 | 373 |
|---|---|---|---|---|
| E (GPa) | 26.1 | 24.1 | 22.45 | 22.00 |
| $\nu$ | 0.380 | 0.395 | 0.408 | 0.412 |
| $\alpha_T$ (1/K) $\times\ 10^{-6}$ | 2.75 | 2.93 | 3.11 | 3.16 |
| $\gamma$ | 0.00083 | 0.00082 | 0.00082 | 0.00081 |
| $\beta$ | 0.0 | 0.0 | 0.0 | 0.0 |
| $n$ | 2.1 | 2.1 | 2.1 | 2.1 |
| $a_1$ ($\times 10^{-6}$) | 8.3 | 2.93 | 1.25 | 0.195 |
| $\eta_1$ | 0.431 | 0.553 | 0.626 | 0.849 |
| $\eta_1$ (average) | | 0.615 | | |
| $\sigma_y$, yield stress (MPa) | 37.241 | 31.724 | 20.690 | 15.172 |
| Bonding stress, $R$ (MPa) | 395.80 | 288.20 | 175.20 | 122.10 |

## Creep Parameters for Pb40/Sn60 Solder at Different Temperatures

| Temperature | 298 K | 313 K | 333 K | 373 K | 393 K |
|---|---|---|---|---|---|
| Fluidity parameter ln($\Gamma$) | 0.578 | 2.058 | 3.475 | 4.61 | 6.96 |
| Parameter | | | | | |
| N | 2.665 | 2.645 | 2.667 | 2.448 | 2.74 |
| Average | 2.67 | 2.67 | 2.67 | 2.67 | 2.67 |

## Disturbance Parameters for Pb40/Sn60 Solder at Different Temperatures ($D_u$ = 1.00)

| Temperature | 223°K | 308°K | 398°K | 423°K |
|---|---|---|---|---|
| Z | 0.7329/0.8697 | 0.5214/0.6031 | 0.6973/0.5914 | 0.6612/0.7224 |
| Average | 0.676 | 0.676 | 0.676 | 0.676 |
| A | 0.056/0.072 | 0.188/0.1298 | 0.0496/0.146 | 0.197/0.169 |

## Temperature Dependence of Parameters for Pb40/Sn60 Solder

| Parameter | $p_{300}$ | $c$ |
|---|---|---|
| E | 23.45 (GPa) | −0.292 |
| $\nu$ | 0.40 | 0.14 |
| $\alpha_T$ | $3 \times 10^{-6}$ (1/K) | 0.24 |
| $\gamma$ | 0.00082 | −0.034 |
| $\alpha$ | $0.05 \times 10^{-4}$ | −5.5 |
| R | 240.67 (MPa) | −1.91 |

## Temperature Dependence of Parameters for Pb40/Sn60 Solder

| Parameter | $p_{300}$ | $c$ |
|---|---|---|
| $\Gamma$ | 1.80/sec | 6.185 |
| $A$ | 0.102 | 1.55 |

Note: Other parameters are not affected significantly by temperature; hence, their average values are used.

# REFERENCES


1. Desai, C. S. (1999). *Mechanics of Materials and Interfaces: The Disturbed State Concept*, Boca Raton, Florida: CRC Press (in press).
2. Desai, C. S. (1995). Constitutive modelling using the disturbed state as microstructure self-adjustment concept, Chapter 8 in *Continuum Models for Materials with Microstructure*, Mühlhaus, H. B., ed., John Wiley.
3. Desai, C. S., Basaran, C., and Zhang, W. (1997). Numerical algorithms and mesh dependence in the disturbed state concept. *Int. J. Num. Meth. Eng.* **40**: 3059–3083.
4. Desai, C. S., Chia, J., Kundu, T., and Prince, J. L. (1997). Thermomechanical response of materials and interfaces in electronic packaging: Parts I and II. *J. Elect. Packaging, ASME* **119**: 294–309.
5. Desai, C. S., Dishongh, T. J., and Deneke, P. (1998). Disturbed state constitutive model for thermomechanical behavior of dislocated silicon with impurities. *J. Appl. Physics* **84**: 11.
6. Desai, C. S., and Ma, Y. (1992). Modelling of joints and interfaces using the disturbed state concept. *Int. J. Num. Analyt. Meth. Geomech.* **16**: 623–653.
7. Desai, C. S., and Salami, M. R. (1987). A constitutive model and associated testing for soft rock. *Int. J. Rock Mech. Min. Sc.* **24**(5): 299–307.
8. Desai, C. S., Samtani, N. C., and Vulliet, L. (1995). Constitutive modeling and analysis of creeping slopes. *J. Geotech. Eng., ASCE*, **121**(1): 43–56.
9. Desai, C. S., Somasundaram, S., and Frantziskonis, G. (1986). A hierarchical approach for constitutive modelling of geologic materials. *Int. J. Num. Analyt. Meth. Geomech.* **10**(3): 225–252.
10. Desai, C. S., and Toth, J. (1996). Disturbed state constitutive modeling based on stress-strain and nondestructive behavior. *Int. J. Solids Struct.* **33**(11): 1619–1650.
11. Desai, C. S., and Varadarajan, S. (1987). A constitutive model for quasistatic behavior of rock salt. *J. Geophys. Res.* **92**(B11): 11445–11456.
12. Desai, C. S., and Whitenack, R. (2000). Review of models and the disturbed state concept for thermomechanical analysis in electronic packaging. *J. Electronic Packaging, ASME* (in press).
13. Desai, C. S., Zaman, M. M., Lightner, J. G., and Siriwardane, H. J. (1984). Thin-layer element for interfaces and joints. *Int. J. Num. Analyt. Meth. Geomech.* **8**(1): 19–43.
14. Katti, D. R., and Desai, C. S. (1994). Modeling and testing of cohesive soil using the disturbed state concept. *J. Eng. Mech., ASCE*, **121**: 648–658.
15. Roscoe, K. H., Schofield, A. N., and Wroth, C. P. (1958). On yielding of soils. *Geotechnique* **8**: 22–53.
16. Kachanov, L. M. (1986). *Introduction to Continuum Damage Mechanics*, Dordrecht: Martinus Nijhoff Publishers.
17. Perzyna, P. (1966). Fundamental problems in viscoplasticity. *Adv. Appl. Mech.* **9**: 247–277.

SECTION **9.11**

# Coupling of Stress-Strain, Thermal, and Metallurgical Behaviors

TATSUO INOUE
*Department of Energy Conversion Science, Graduate School of Energy Science, Kyoto University, Yoshida-Honmachi, Sakyo-ku, Kyoto, Japan*

Contents



## 9.11.1 INTRODUCTION

Coupling among metallic structures, including the molten state, temperature, and stress and/or strain occurring in processes accompanied by phase transformation, sometimes is one of the predominant effects of such industrial processes as quenching, welding, casting, and so on. Figure 9.11.1 shows the schematic representation of the effect of metallo-thermomechanical coupling with the induced phenomena [1–5]. When the temperature distribution in a material varies, thermal stress (①) is caused in the body, and the induced phase transformation (②) affects the structural distribution, which is known as melting or solidification in solid–liquid transition and pearlite or martensite transformation in the solid

*Handbook of Materials Behavior Models.* ISBN 0-12-443341-3.

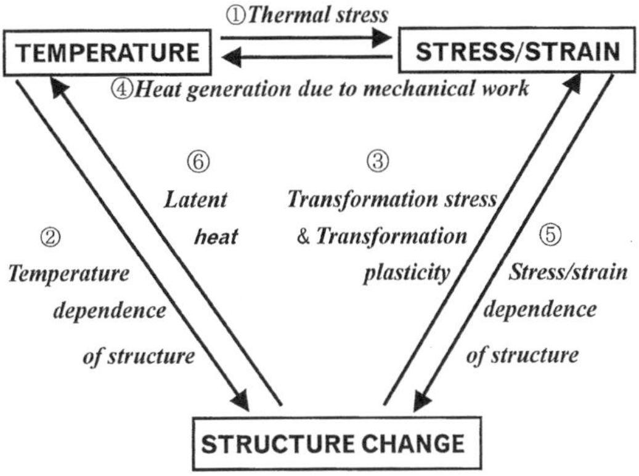


FIGURE 9.11.1 Metallo-thermo-mechanics: coupling between temperature, stress/strain, and metallic structure.

phase. Local dilatation due to structural changes in the body bring out the transformation stress (③) and interrupt the stress or strain field in the body.

In contrast to these phenomena, which are well known in ordinal analysis, arrows in the opposite direction indicate coupling in the following manner. Part of the mechanical work done by the existing stress in the material is converted into heat (④) which may be predominant in the case of inelastic deformation, thus disturbing the temperature distribution. The acceleration of phase transformation by stress or strain, which is called stress- or strain-induced transformation (⑤) has been treated by metallurgists as one of leading parameters of transformation kinetics. The arrow numbered by (⑥) corresponds to the latent heat due to phase transformation, which is essential in determining the temperature. The purpose of this section is to present the governing equations relevant to simulating such processes involving phase transformation when considering the effect of the coupling mentioned. Formulation of the fundamental equations for stress-strain relationships, heat conduction, and transformation kinetics based on continuum thermodynamics will be done in the first part, and a list of some examples of the numerical simulation of temperature, stress-strain, and metallic structures in the processes of quenching, welding, and casting will be presented.

## 9.11.2 CONTINUUM THERMODYNAMICS AND FUNDAMENTAL FRAMEWORK

Consider a material undergoing structural change due to phase transformation as a mixture of $N$ kinds of constituents [6]. Denoting the volume fraction of the $I$th constituent as $\xi_I$ (see Fig. 9.11.2), the physical and mechanical properties $x$ of the material are assumed to be a linear combination of the properties $x_I$ of the constituent as

$$x = \sum_{I=1}^{N} x_I \xi_I \tag{1}$$

with

$$\sum_{I=1}^{N} \xi_I = 1 \tag{2}$$

where $\sum_{I=1}^{N}$ is the summation for suffix $I$ from 1 to $N$. All material parameters appearing in following text are defined in the manner of Eq. 1.

The Gibbs free-energy density function $G$ is defined as

$$G = U - T\eta - \frac{1}{\rho}\sigma_{ij}\varepsilon_{ij}^{e} \tag{3}$$

where $U$, $T$, $\eta$, and $\rho$ are the internal energy density, temperature, entropy density, and mass density, respectively. Elastic strain rate $\dot{\varepsilon}_{ij}^{e}$ in Eq. 3 is defined as the subtraction of inelastic strain rate $\dot{\varepsilon}_{ij}^{\mathrm{i}}$ from total strain rate $\dot{\varepsilon}_{ij}$, that is,

$$\dot{\varepsilon}_{ij}^{e} = \dot{\varepsilon}_{ij} - \dot{\varepsilon}_{ij}^{\mathrm{i}} \tag{4}$$

The thermodynamic state of a material is assumed to be determined by stress $\sigma_{ij}$, temperature $T$, temperature gradient $g_i(= \operatorname{grad} T)$, and a set of internal variables of inelastic strain $\varepsilon_{ij}^{\mathrm{i}}$, back stress $\alpha_{ij}$, and hardening parameter $\kappa$ related to inelastic deformation, together with the volume fraction of the constituents $\xi_I$. Then, the general form of the constitutive equation can be

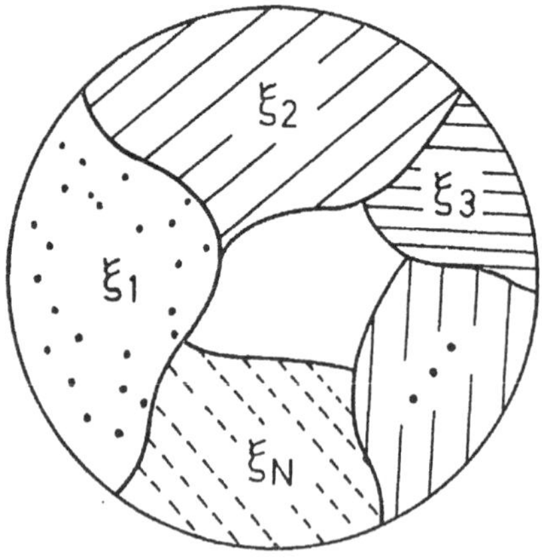


FIGURE 9.11.2 Concept of mixture.

expressed as

$$G = G\left(\sigma_{ij}, T, g_i, \varepsilon^{i}_{ij}, \alpha_{ij}, \kappa, \xi_I\right) \tag{5}$$

$$\eta = \eta\left(\sigma_{ij}, T, g_i, \varepsilon^{i}_{ij}, \alpha_{ij}, \kappa, \xi_I\right) \tag{6}$$

$$q_i = q_i\left(\sigma_{ij}, T, g_i, \varepsilon^{i}_{ij}, \alpha_{ij}, \kappa, \xi_I\right) \tag{7}$$

$$\varepsilon^{e}_{ij} = \varepsilon^{e}_{ij}\left(\sigma_{ij}, T, g_i, \varepsilon^{i}_{ij}, \alpha_{ij}, \kappa, \xi_I\right) \tag{8}$$

Here, $q_i$ is the heat flux. The evolution equations for the internal variables are defined in the same form as Eqs. 5–8, i.e.,

$$\dot{\varepsilon}^{i}_{ij} = \dot{\varepsilon}^{i}_{ij}\left(\sigma_{ij}, T, g_i, \varepsilon^{i}_{ij}, \alpha_{ij}, \kappa, \xi_I\right) \tag{9}$$

$$\dot{\alpha}_{ij} = \dot{\alpha}_{ij}\left(\sigma_{ij}, T, g_i, \varepsilon^{i}_{ij}, \alpha_{ij}, \kappa, \xi_I\right) \tag{10}$$

$$\dot{\kappa} = \dot{\kappa}\left(\sigma_{ij}, T, g_i, \varepsilon^{i}_{ij}, \alpha_{ij}, \kappa, \xi_I\right) \tag{11}$$

$$\dot{\xi}_I = \dot{\xi}_I\left(\sigma_{ij}, T, g_i, \varepsilon^{i}_{ij}, \alpha_{ij}, \kappa, \xi_I\right) \tag{12}$$

When the strong form of the Clausius–Duhem inequality

$$-\rho(\dot{G} + \eta\dot{T}) - \dot{\sigma}_{ij}\varepsilon^{e}_{ij} + \sigma_{ij}\dot{\varepsilon}^{i}_{ij} \geq 0 \tag{13}$$

with

$$g_i q_i \leq 0 \tag{14}$$

is applied, the constitutive relationships in Eqs. 5–8 are reduced to

$$G = G\left(\sigma_{ij}, T, \varepsilon^{i}_{ij}, \alpha_{ij}, \kappa, \xi_I\right) \tag{15}$$

$$\varepsilon^{e}_{ij} = \rho\frac{\partial}{\partial\sigma_{ij}}G\left(\sigma_{ij}, T, \varepsilon^{i}_{ij}, \alpha_{ij}, \kappa, \xi_I\right) \tag{16}$$

$$\eta = -\rho\frac{\partial}{\partial T}G\left(\sigma_{ij}, T, \varepsilon^{i}_{ij}, \alpha_{ij}, \kappa, \xi_I\right) \tag{17}$$

$$q_i = -k\left(\sigma_{ij}, T, \varepsilon^{i}_{ij}, \alpha_{ij}, \kappa, \xi_I\right)g_i \tag{18}$$

Here, Fourier's law has been used in Eq. 18 with thermal conductivity $k$.

## 9.11.3 STRESS-STRAIN CONSTITUTIVE EQUATION

To obtain an explicit expression for the elastic strain in Eq. 16, the Gibbs free-energy $G$ is assumed to be determined by that of constituent $G_I$ in the form of Eq. 1 as

$$G\left(\sigma_{ij}, T, \varepsilon^{i}_{ij}, \alpha_{ij}, \kappa, \xi_I\right) = \sum_{I=1}^{N} \xi_I G_I\left(\sigma_{ij}, T, \varepsilon^{i}_{ij}, \alpha_{ij}, \kappa\right) \tag{19}$$

When $G_I$ is divided into the elastic and inelastic parts as

$$G_I\left(\sigma_{ij}, T, \varepsilon^{i}_{ij}, \alpha_{ij}, \kappa, \xi_I\right) = G^{e}_I(\sigma_{ij}, T) + G^{i}_I\left(T, \varepsilon^{i}_{ij}, \alpha_{ij}, \kappa\right) \tag{20}$$

we can derive the elastic strain from Eq. 16 by expanding elastic part $G^e_I$ around the natural state, $\sigma_{ij} = 0$ and $T = T_0$, in terms of the representation theorem for an isotropic function;

$$\begin{aligned} \varepsilon^{e}_{ij} = & \left(\sum_{I=1}^{N} \frac{1+\nu_I}{E_I}\xi_I\right)\sigma_{ij} - \left(\sum_{I=1}^{N} \frac{\nu_I}{E_I}\xi_I\right)\delta_{ij}\sigma_{kk} \\ & + \delta_{ij}\int_{T_0}^{T} \sum_{I=1}^{N} \alpha_I \xi_I dT + \delta_{ij}\sum_{I=1}^{N} \beta_I(\xi_I - \xi_{I0}) \end{aligned} \tag{21}$$

Here, $E_I$, $\nu_I$, $\alpha_I$, and $\beta_I$ correspond to Young's modulus, Poisson's ratio, thermal expansion coefficient, and dilatation of the $I$th constituent, respectively.

### 9.11.3.1 Plastic Strain Rate

Assume that the evolution of back stress $\alpha_{ij}$ of the yield surface and hardening parameter $\kappa$ can be determined by

$$\dot{\alpha}_{ij} = C(T, \kappa, \xi_I)\dot{\varepsilon}^{i}_{ij} \tag{22}$$

$$\dot{\kappa} = \dot{\bar{\varepsilon}}^{i} = \left(\tfrac{2}{3}\dot{\varepsilon}^{i}_{ij}\dot{\varepsilon}^{i}_{ij}\right)^{1/2} \tag{23}$$

where $\dot{\bar{\varepsilon}}^{i}$ represents the equivalent inelastic strain rate. To take into account the effect of changing structural fraction $\xi_I$, we take the form of the yield function as

$$F = F\left(\sigma_{ij}, T, \varepsilon^{i}_{ij}, \alpha_{ij}, \kappa,\ \xi_I\right) = \left[\tfrac{3}{2}\left(s_{ij} - \alpha_{ij}\right)^2\right]^{1/2} - K(T, \kappa, \xi_I) \tag{24}$$

where $s_{ij}(= \sigma_{ij} - \frac{1}{3}\delta_{ij}\sigma_{kk})$ represents the deviatoric stress component. Employing the normality rule and the consistency relationship, the final form of the time-independent inelastic strain rate, or plastic strain rate, reads:

$$\dot{\varepsilon}^{i}_{ij} = \Lambda \frac{\partial F}{\partial \sigma_{ij}} = \hat{G}\left( \frac{\partial F}{\partial \sigma_{kl}}\dot{\sigma}_{kl} + \frac{\partial F}{\partial T}\dot{T} + \sum_{I=1}^{N} \frac{\partial F}{\partial \xi_I}\dot{\xi}_I \right) \frac{\partial F}{\partial \sigma_{ij}} \tag{25}$$

with

$$\frac{1}{\hat{G}} = -\left\{ \frac{\partial F}{\partial \sigma_{mn}} \frac{\partial F}{\partial \varepsilon^{i}_{mn}} + \frac{\partial F}{\partial \sigma_{mn}} \frac{\partial F}{\partial \alpha_{mn}} + \left( \frac{2}{3} \frac{\partial F}{\partial \sigma_{mn}} \frac{\partial F}{\partial \sigma_{mn}} \right)^{1/2} \frac{\partial F}{\partial \kappa} \right\} \tag{26}$$

Equation 25 means that plastic strain is induced not only by stress, but also by the temperature and phase change.

### 9.11.3.2 Viscoplastic Strain Rate

The elastic-plastic constitutive relationship is suitable for describing the material behavior at relatively low temperature. However, time dependency or viscosity might be predominate at a higher temperature level, particularly when the material behaves like a viscous liquid beyond its melting point. In order to analyze such processes as welding and casting, in which melting and solidification of the metal are essential phenomena, adequate formulation of the viscoplastic constitutive model is needed. Malvern [7] and Perzyna [8] proposed a viscoplastic constitutive equation for time-dependent inelastic strain rate $\dot{\varepsilon}^{i}_{ij}$ in the form

$$\dot{\varepsilon}^{i}_{ij} = \frac{1}{3\mu} \langle \psi(F) \rangle \frac{\partial F}{\partial \sigma_{ij}} \tag{27}$$

with the static yield function

$$F = \frac{f\left(\sigma_{ij}, T, \varepsilon^{i}_{ij}\right)}{K} - 1 \tag{28}$$

where $\mu$ and $K$ denote the stress, and the coefficient of viscosity and the static flow, and

$$\langle \psi(F) \rangle = \begin{cases} 0, & \text{if } \psi(F) \le 0 \\ \psi(F), & \text{if } \psi(F) > 0 \end{cases} \tag{29}$$

Equation 27 indicates that the inelastic strain rate is induced in an outer direction normal to static yield surface $F$, and that the magnitude of the strain rate depends on the ratio of excess stress $(f - K)$ to flow stress $K$. If we adopt the flow rule (Eq. 27) to the liquid state, the flow stress tends to vanish ($K \rightarrow 0$) and the yield surface $F$ expands infinitely ($F \rightarrow \infty$), which implies that the strain rate is infinite at low stress. To compensate for such an inconsistency occurring in a liquid, a modification to Eq. 28 is made such that

$$F = f\left(\sigma_{ij}, T, \varepsilon^{i}_{ij}, \alpha_{ij}\right) - K(T, \kappa, \xi_I) \tag{30}$$

When we take the simple forms of functions $\psi$ and $F$,

$$\psi(F) = F \tag{31}$$

$$F = \left[\tfrac{3}{2}\left(s_{ij} - \alpha_{ij}\right)\left(s_{ij} - \alpha_{ij}\right)\right]^{1/2} - K(T, \kappa, \xi_I) \tag{32}$$

Equation 27 can be reduced to

$$\dot{\varepsilon}^{i}_{ij} = \frac{1}{2\mu}\left\langle 1 - \frac{K(T, \kappa, \xi_I)}{\left[3(s_{kl} - \alpha_{kl})(s_{kl} - \alpha_{kl})\right]^{1/2}}\right\rangle \left(s_{ij} - \alpha_{ij}\right) \tag{33}$$

This constitutive relationship may be relevant to a liquid–solid transition region with high viscosity, as well as to a normal time-independent plastic body [9]. For instance, when flow stress $K$ equals zero in Eq. 33, the total strain rate $\dot{\varepsilon}_{ij}$ $(= \dot{\varepsilon}^{e}_{ij} + \dot{\varepsilon}^{i}_{ij})$ is given by

$$\dot{\varepsilon}_{ij} = \frac{1+\nu}{E}\dot{\sigma}_{ij} - \frac{\nu}{E}\dot{\sigma}_{kk}\delta_{ij} + \frac{1}{2\mu}s_{ij} \tag{34}$$

when the effect of temperature and phase change is neglected for simplicity. This equation is equivalent to the Maxwell constitutive model for a viscoelastic body.

When the elastic component of shear deformation is small enough compared with the viscoplastic component, as is usual for a viscous fluid, the Newtonian fluid model

$$\sigma_{ij} = 2\mu\dot{\varepsilon}_{ij} - \tfrac{2}{3}\mu\dot{\varepsilon}_{kk}\delta_{ij} - p\delta_{ij}, \quad p = -\tfrac{1}{3}\sigma_{kk} \tag{35}$$

is obtainable from Eq. 34 by neglecting the elastic shear strain rate. Furthermore, when the elastic volume dilatation $\varepsilon^{e}_{kk}$ is removed from Eq. 35, we have

$$\sigma_{ij} = 2\mu\dot{\varepsilon}_{ij} - p\delta_{ij} \tag{36}$$

which represents the model for an incompressible Newtonian fluid. In the limiting case for an inviscid material ($\mu = 0$), $\psi(F)$ in Eq. 27 tends to infinity and

$$\frac{1}{3\mu}\psi(F) \equiv \Lambda \tag{37}$$

should hold to give the form

$$\dot{\varepsilon}^{i}_{ij} = \Lambda\frac{\partial F}{\partial \sigma_{ij}} \tag{38}$$

Parameter $\Lambda$ can be easily determined by applying the consistency relationship, and thus we get back again to the previous discussion for time-independent plastic strain. From the considerations just mentioned, the constitutive relationship developed in Eq. 33 seems to be useful for a wide range of metals, from inelastic solids to viscous fluids.

## 9.11.4 HEAT CONDUCTION EQUATION

When we adopt Eqs. 15–18 to the energy conservation law

$$\rho\dot{U} - \sigma_{ij}\dot{\varepsilon}_{ij} + \frac{\partial q_i}{\partial x_I} = 0 \tag{39}$$

the equation of heat conduction

$$\rho c\dot{T} - k\frac{\partial^2 T}{\partial x_i \partial x_i} + \rho\sum_{I=1}^{N} l_I\xi_I + T\frac{\partial \varepsilon^e_{ij}}{\partial T}\dot{\sigma}_{ij} + \left(\rho\frac{\partial H}{\partial \varepsilon^i_{ij}}\dot{\varepsilon}^i_{ij} + \rho\frac{\partial H}{\partial \alpha_{ij}} + \rho\frac{\partial H}{\partial \kappa}\dot{\kappa} - \sigma_{ij}\dot{\varepsilon}^i_{ij}\right) = \rho r \tag{40}$$

holds with enthalpy density $H(= G + T\eta)$ and latent heat $l_I$ due to the increase of the $I$th phase

$$l_I = \frac{\partial H}{\partial \xi_I} \tag{41}$$

The fifth term on the left-hand side of Eq. 40 denotes the heat generation by inelastic dissipation, which is significant when compared with the elastic work represented by the fourth term, and the third term arises from the latent heat through phase changes. Hence, it can be seen that Eq. 40 corresponds to the ordinal equation of heat conduction, provided that these terms are neglected.

## 9.11.5 KINETICS OF PHASE TRANSFORMATION

During phase transformation, a given volume of material is assumed to be composed of several kinds of constituent $\xi_I$ as expressed in Eq. 1. We choose four kinds of volume fraction: liquid $\xi_L$, austenite $\xi_A$, pearlite $\xi_P$, and martensite $\xi_M$, and other structures induced by precipitation by recovery effect, say, during the annealing process. When austenite is cooled in equilibrium, bainite, ferrite, and carbide are produced in addition to pearlite, but for brevity all these structures resulting from a diffusion type of transformation are called pearlite. The nucleation and growth of pearlite in an austenitic structure are phenomenologically governed by the mechanism for a diffusion process, and Johnson and Mehl [10] proposed a formula for volume fraction $\xi_P$ as

$$\xi_P = 1 - \exp(-V_e) \tag{42}$$

where $V_e$ means the extended volume of the pearlitic structure given by

$$V_e = \int_0^t \tfrac{4}{3}\pi R(t-\tau)^3 n d\tau \tag{43}$$

Here, $R(t)$ is the radius of the pearlite particle at time $t$, and $n$ denotes the number of nucleating particles per unit time. Bearing in mind that the value of $R$ is generally a function of stress as well as temperature, Eq. 43 may be reduced to

$$V_e = \int_0^t f(T, \sigma_{ij})(t-\tau)^3 d\tau \tag{44}$$

Function $f(T, 0)$ can be determined by fitting the temperature-time-transformation (TTT) diagram or continuous-cooling transformation (CCT) diagram without stress, and $f(T, \sigma_{ij})$ may be given by the start-time or finish-time data for pearlite transformation with an applied stress, an example of which is shown in Figure 9.11.3 [12].

The empirical relationship for the austenite-martensite transformation is also obtainable by modifying the kinetic theory of Magee [11]. Assume that the growth of a martensite structure is a linear function of the increase of the difference $\Delta G$ in free energy between austenite and martensite as

$$d\xi_M = -\bar{v}(1-\xi_M)\phi d(\Delta G) \tag{45}$$

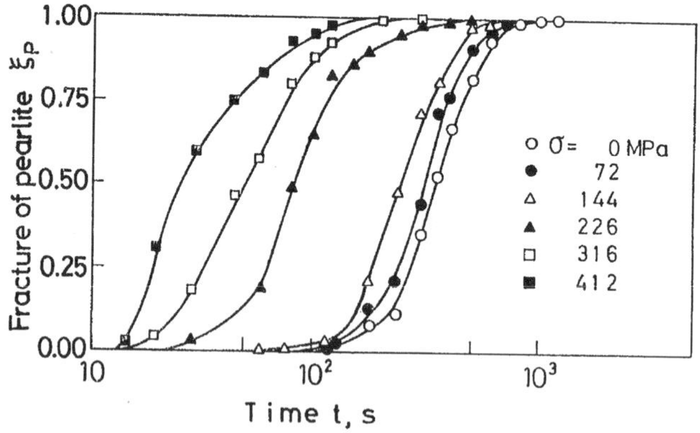


FIGURE 9.11.3 Dependence of stress on pearlite reaction (AISI 4340 steel, 343°C).

Regarding the free-energy $G$ as a function of temperature and stress, we can obtain the form of $\xi_M$ by integrating Eq. 45 as

$$\xi_M = 1 - \exp\left[\phi_1(T - M_s) + \phi_2(\sigma_{ij})\right] \tag{46}$$

The function $\phi_2(\sigma_{ij})$ is identified by the such data as shown in Figure 9.11.4 [14].

For the case of solidification, we employ the well-known lever rule (see Fig. 9.11.5) as an example, and the volume fraction of austenite is

$$\xi_A = \frac{(T_L - T)/m_L}{(T - T_A)/m_A + (T_L - T)/m_L} \tag{47}$$

where $T_L$ and $T_A$ denote the liquidus and solidus temperatures, respectively, and $m_L$ and $m_A$ are gradients of the liquidus and solidus temperatures with respect to the carbon content in the phase diagram.

## 9.11.6 SOME SIMULATED RESULTS OF ENGINEERING PROCESSES

Simulation of some typical engineering processes involving phase transformation has been made by the authors. Following is the list of topics

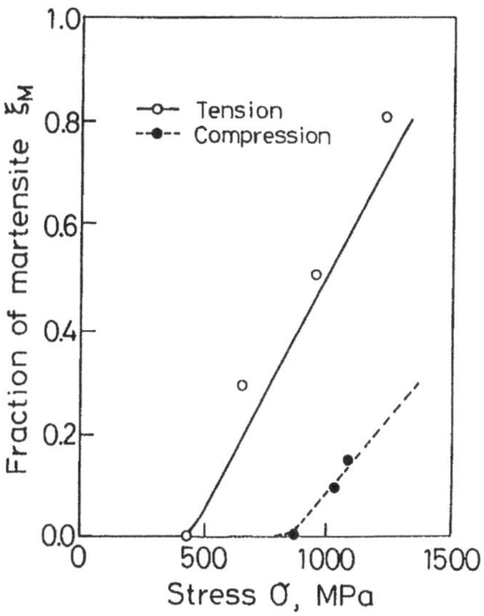


FIGURE 9.11.4 Dependence of stress on martensitic reaction (29%C steel, 247°C).

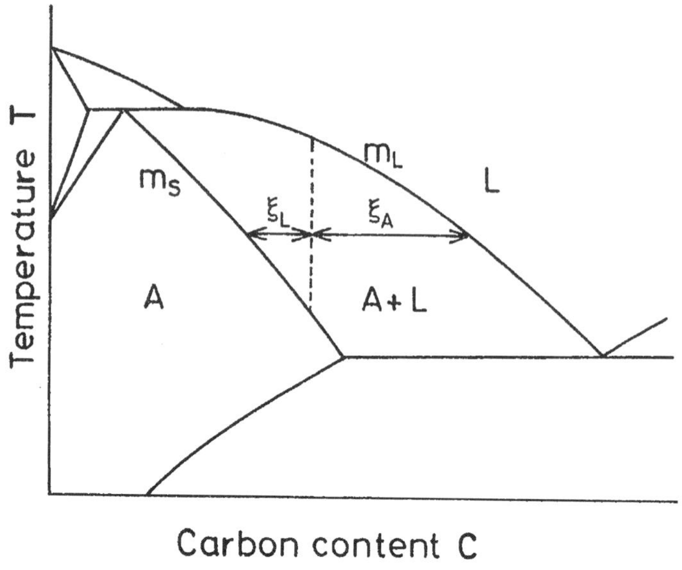


FIGURE 9.11.5 Schematic phase diagram.

with references. The results of heat treatment are those simulated by the finite element CAE system "HEARTS" (HEAt tReaTment Simulation program), which is available for the analysis of the heat treatment process in this stage [14–17].

Quenching

- Quenching of infinite cylinder [14]
- Carburized quenching of ring and gear wheel [14, 18].
- Induction hardening of ring and gear wheel [19, 20]
- Quenching of Japanese sword [21, 22, 23]

Welding

- Butt-welding of plates [24]
- Bead-on-plate [25]

Casting

- Continuous casting of a slab [26]
- Thin slab casting [27]
- Centrifugal casting [28]

## REFERENCES


1. Inoue, T., Nagaki, S., Kishino, T., and Monkawa, M. (1981). *Ingenieur-Archives* **50**(5): 315–327.
2. Inoue, T. (1988). *Thermal Stresses*, pp. 192–278, Hetnarski, Richard B., ed., North-Holland.
3. Inoue, T. (1990). *Computational Plasticity: Current Japanese Materials Research*, pp. 73–96, vol. 7, Inoue, T., et al., eds., Elsevier Applied Science.
4. Inoue, T., and Raniecki, B. (1978). *J. Mech. Phys. Solids* **3**: 187–212.
5. Inoue, T. (1987). *Berg- und Huttenmannische Monatshefte* **3**: 63–71.
6. Bowen, R. M. (1976). *Continuum Physics*, Vol. 3, New York: Academic Press.
7. Malvern, L. E. (1969). *Introduction to the Mechanics of Continuous Medium*, Prentice-Hall.
8. Perzyna, P. (1979). *Advances in Applied Mechanics*, pp. 315–354, vol. 9.
9. Wang, Z.-G., and Inoue, T. (1985). *Material Science and Technology* **1**: 899–903.
10. Johnson, A. W., and Mehl, R. F. (1939). *AIME* **135**: 416–458.
11. Magee, C. L. (1979). *Phase Transformations*, p. 115, London: Chapman Hall.
12. Bhattacharyya, S., and Kel, G. L. (1955). *ASM* **47**: 351–379.
13. Onodera, H., Gotoh, H., and Tamura, I. (1976). *Proc. First JIM Int. Symp. New Aspects of Martesitic Transformation*, pp. 327–332.
14. Inoue, T., Arimoto, K., and Ju, D.-Y. (1992). *Proc. First Int. Conf. Quenching and Control of Distortion*, Chicago, pp. 205–212.
15. Inoue, T., Arimoto, K., and Ju, D.-Y. (1992). *Proc. 8th Int. Congr. on Heat Treatment of Materials*, Kyoto, pp. 569–572.
16. Inoue, T. (1998). *Mathematical Modelling of Weld Phenomena*, pp. 547–575, vol. 4, Cerjak, H., ed.
17. Inoue, T., and Arimoto, K. (1997). *J. Materials Engineering and Performance, ASM* **6**(1): 51–60.
18. Yamanaka, S., Sakanoue, T., Yosii, S., Kozuka, T., and Inoue, T. (1998). *Proc. 18th Conf. Heat Treating, ASM International*, pp. 657–664.
19. Inoue, T., Inoue, H., Uehara, T., Ikuta, F., Arimoto, K., and Igari, T. (1996). *Proc. 2nd Int. Conf. Quenching and Control of Distortion*, Cleveland, pp. 55–62.
20. Inoue, T., Inoue, H., Ikuta, F., and Horino, T. (1999). *Proc. 3rd Int. Conf. Quenching and Control of Distortion*, Prague, pp. 243–250.
21. Inoue, T., and Uehara, T. (1995). *Proc. Int. Symp. Phase Transformations during the Thermal/Mechanical Processing of Steel*, The Metallurgical Society of the Canadian Institute of Mining, Metallurgy and Petroleum, pp. 521–524.
22. Inoue, T. (1997). *Materials Science Research International* **3**(4): 193–203.
23. Inoue, T. (1999). *Proc. 8th Int. Conf. on Mechanical Behaviour of Materials*, Victoria, vol. 2, 458–468.
24. Wang, Z.-G., and Inoue, T. (1985). *Material Science and Technology* **1**(1): 899–903.
25. Sakuma, A., and Inoue, T. (1995). *Proc. 5th Int. Symp. on Plasticity and Its Current Applications*, Sakai-Osaka, pp. 721–724.
26. Inoue, T., and Wang, Z.-G. (1988). *Ingenieur-Archives* **58**(4): 265–275.
27. Inoue, T., and Ju, D. Y. (1991). *Advances in Continuum Mechanics*, pp. 389–406. Bruller, O., Mannl., V., and Najar, J., eds., Springer-Verlag.
28. Du, D. Y., and Inoue, T. (1996). *Materials Science Research International* **2**(1): 18–25.

SECTION **9.12**

# Models for Stress-Phase Transformation Couplings in Metallic Alloys

S. Denis, P. Archambault and E. Gautier
*Laboratoire de Science et Génie des Matériaux et de Métallurgie, UMR 7584 CNRS/INPL, Ecole des Mines de Nancy, Parc de Saurupt, 54042 Nancy Cedex, France*

Contents


## 9.12.1 DOMAIN OF APPLICATION

The models described in this article, which concern phase transformation kinetics and thermomechanical behavior, can be used for many steels and high-strength aluminium alloys submitted to various thermal histories (heating, cooling, isothermal holding) and to stress states that lead to small plastic strains (a few percent). Typical applications are the modeling of heat treatment processes (quenching, surface hardening [induction, laser, electron beam hardening], case hardening) and also welding processes (for the heat-affected zone behavior).


*Handbook of Materials Behavior Models.* ISBN 0-12-443341-3.

## 9.12.2 THEORY USED TO DERIVE THE MODELS

The kinetics models rely on the general theory of solid–solid phase transformations [1]. In order to predict kinetics for various thermal histories, we have developed either a global approach based on the Johnson Mehl Avrami formalism that assumes an additivity principle (for diffusion-dependent transformations in steels or titanium alloys), or an approach that explicitly includes nucleation and growth laws (for precipitation in aluminium alloys).

The thermomechanical behavior of the material during the phase change is described by phenomenological behavior laws that include the thermoelastic and plastic-viscoplastic behavior of the stable multiphase material (mixture rules) and the effect of the phase transformations (volume changes, transformation plasticity) as proposed by several authors (for example, [2]). For precipitation at very small volume fractions, the approach through mixture laws is no longer valid, and a model describing both mechanical softening due to heterogeneous precipitation (according to the literature about solute hardening [3]) and hardening due to homogeneous precipitation (by including dislocation–precipitate interactions [4]) has been proposed.

## 9.12.3 DESCRIPTION OF THE MODELS

### 9.12.3.1 Kinetics Models

#### 9.12.3.1.1 Global Description

The model uses isothermal transformation kinetics (IT diagrams) to calculate the anisothermal kinetics on heating and cooling by applying an additivity principle [5]. Thus, for steels, the austenitization kinetics on heating as well as the kinetics of decomposition of austenite (into ferrite, pearlite, bainite, martensite) during cooling can be described. We recall here only the model on cooling. For diffusion-dependent transformations, incubation and growth periods are treated separately. The incubation period is determined according to Scheil's method: the transformation during continuous cooling begins when the sum S becomes equal to unity:

$$S = f\Sigma_i \ \Delta t_i/\tau(T_i)$$

where $f$ is a heredity factor that has been introduced to take into account nonadditivity of incubation at the transition from the pearlitic to the bainitic domain, $\tau(T_i)$ is the incubation time at temperature $T_i$, and $\Delta t_i$ is the time step.

The diffusion-dependent transformations (in steels, the formation of proeutectoïd ferrite or cementite, pearlite, or bainite) are modeled according to the law developed by Johnson-Mehl Avrami and Kolmogorov:

$$y_k = y_k^{\max}[1 - \exp(-b_k(T)t^{nk(T)})]$$

where $y_k$ is the volume fraction of constituent $k$, $y_k^{\max}$ is the maximum that can be formed at a given temperature, and $n_k$ and $b_k$ are coefficients that are determined for each temperature from the IT diagram ($t = 0$ corresponds to the end of the incubation period). The transition from isothermal to nonisothermal kinetics is done through the fictitious time method [6].

For martensitic transformation, the progress of the transformation is calculated using the relation established by Koistinen and Marburger:

$$y_M = y_\gamma[1 - \exp(-a_M(M_s - T))]$$

where $y_M$ is the volume fraction of martensite, $y_\gamma$ is the remaining austenite fraction at $M_s$, and $a_M$ is a constant.

This model takes into account the effect of austenitic grain size on the transformation kinetics on cooling as the effect of a carbon enrichment of austenite due to a previous ferritic transformation on the subsequent bainitic and martensitic transformations. Moreover, the influence of a stress state has been included: for diffusion-dependent transformations, the acceleration of the transformations due to a stress state is taken into account through phenomenological laws relating $\tau(T_i)$, $n_k$, and $b_k$ to the von Mises stress; for martensitic transformation, Ms is related to the hydrostatic stress and to the von Mises stress. The effects of carbon content heterogeneities (such as those introduced by carburizing [5] or inheritated from solidification segregations [7]) are included as well. A hardness calculation is associated with this phase transformation model, which implicitly takes into account the effect of the morphology of the different constituents.

Recently, the description of tempering kinetics on heating and self-tempering kinetics on cooling has also been added for low alloyed steels [8]. A similar modeling approach has been used for the prediction of the precipitation of the $\alpha$ phase from the $\beta$ phase in a titanium alloy during continuous cooling [9].

This global approach is efficient in predicting the volume fractions of the different constituents during heat treatments of industrial alloys. But, since it does not take into account the nucleation and growth rates explicitly, it does not allow one to properly take into account the effects of hot forming on the subsequent transformations on cooling (enhancement of nucleation).

#### 9.12.3.1.2 Nucleation and Growth Modeling

Considering the precipitation, as in aluminium alloys, the nucleation and growth of the precipitates have been modeled in order to predict the volume fractions of precipitates during a cooling process and also their mean radius and their distribution as well as the chemical composition of the remaining solid solution.

The model developed [10, 11] includes both heterogeneous and homogeneous precipitation, which occur, respectively, at higher and lower temperatures during cooling. The equilibrium conditions are described by the solubility product. The heterogeneous nucleation rate is thus expressed as:

$$\dot{\mathrm{N}} = \dot{\mathrm{N}}_0 \lambda (N_{\mathrm{tot}} - N_{\mathrm{act}}) \exp\left[-(\Delta G^* + Q_d)/RT\right]$$

where $\dot{\mathrm{N}}_0$ is a constant, $\lambda$ is a vacancy acceleration ratio, $N_{\mathrm{tot}}$ and $N_{\mathrm{act}}$ are the total and active density of heterogeneous nucleation sites, and $Q_d$ is the diffusion activation energy. $\Delta G^*$ is the critical free enthalpy variation that depends on the precipitate surface energy $\gamma$ and on the variation of the nucleation free enthalpy variation $\Delta Gv$. For cylindrical precipitates, we have:

$$\Delta G^* = 224/81\pi\alpha N f \gamma^3/\Delta G_v^2$$

where $N$ is the Avogadro number, $\alpha$ is a geometrical parameter, and $f$ is a wetting function representing easier nucleation on heterogeneous sites. Homogeneous nucleation differs through the morphology of the precipitates (spheres), the interfacial energy, the wetting function, which disappears, and the elastic energy, which has to be taken into account.

Postulating quasi-stationarity, the growth rate can be expressed as a function of the precipitate–matrix interface concentration $X_i^{int}$ of element $i$:

$$dr/dt = Di(X_i^{ss} - X_i^{int})/(X_i^p - X_i^{int})$$

where $X_i^{ss}$ and $X_i^p$ are, respectively, the concentration of element $i$ in the solid solution and in the precipitate, and, $D_i$ is the diffusion coefficient of element $i$. Local equilibrium is considered at the interface. Because of the relative large size of precipitates, the Gibbs-Thomson effect has been neglected. This model has been applied to an Al-Zn-Mg-Cu alloy (7000 series) considering the precipitation of the $\eta$ phase.

### 9.12.3.2 Thermomechanical Model

The modeling of the thermomechanical behavior of a material undergoing a phase transformation must include the thermoelastic and plastic-viscoplastic behavior of the stable multiphase material and the effect of the phase transformations.

In order to be able to describe the great complexity of real materials like steels, mainly macroscopic phenomenological behavior laws have been developed [2] (see also review paper [12] for other references). It is generally assumed that the total strain rate is an addition of different contributions:

$$\dot{\varepsilon}_{ij}^{t} = \dot{\varepsilon}_{ij}^{e} + \dot{\varepsilon}_{ij}^{th} + \dot{\varepsilon}_{ij}^{tr} + \dot{\varepsilon}_{ij}^{tp} + \dot{\varepsilon}_{ij}^{in} \quad (1)$$

- $\dot{\varepsilon}_{ij}^{e}$ is the elastic strain rate, which is related to the stress rate by Hooke's law. Young's modulus and Poisson's ratio have to be temperature and microstructure dependent. (Here, *microstructure* means "volume fractions of the different phases").
- $\dot{\varepsilon}_{ij}^{th}$ is the thermal strain rate, which takes into account the thermal expansion coefficients $\alpha_k$ of the different phases and their dependence on temperature: $\varepsilon_{ij}^{th} = \Sigma_k y_k \int \alpha_k(T)dT$.
- $\dot{\varepsilon}_{ij}^{in}$ is the inelastic strain rate: either the plastic strain rate when no viscous effects are considered or the viscoplastic strain rate. It is calculated using the classical theory of plasticity or viscoplasticity with the associated hardening rules (isotropic and/or kinematic) or is obtained from a micro–macro approach. All material parameters (yield stress, hardening parameters, strain rate sensitivity) are to be considered as temperature and microstructure dependent. Mixture rules are generally assumed. In the case of chemical composition heterogeneities, material properties depend also on local composition. Presently, a dependency with carbon content can be taken into account. In addition, it should be mentioned that taking hardening into account is quite complex when a phase transformation occurs. Models have been proposed to account for some possible recovery of strain hardening during a phase transformation; i.e., the new phase "remembers" or does not remember part of the previous hardening.$\dot{\varepsilon}_{ij}^{tr}$ is the strain rate due to the volume change associated with the different phase transformations: $\varepsilon_{ij}^{tr} = \Sigma_k y_k \varepsilon_k^{tr}$.
- $\dot{\varepsilon}_{ij}^{tp}$ is the transformation plasticity strain rate. It is generally assumed that it is proportional to the stress deviator (discussion about this assumption can be found in [12]):

  $$\dot{\varepsilon}_{ij}^{tp} = 3/2 \sum K_k f'(y_k)\dot{y}_k s_{ij}$$

  where $K_k$ and $f$ are experimental parameter and function, and $s_{ij}$ are the components of the stress deviator.

When considering aluminium alloys, the mechanical behavior of the supersaturated solid solution during cooling can be described through Eq. 1. But, because of the very low precipitate volume fractions, the deformation associated with the volume change is negligible as transformation plasticity. A thermo-elasto-viscoplastic behavior has been considered, and the flow stress is written:

$$\sigma_{ss} = \sigma_0 + H(\varepsilon^{vp})^n + K(\dot{\varepsilon}^{vp})^m$$

where $\sigma_0$ represents a threshold stress, $H(\varepsilon^{vp})^n$ is the hardening due to the deformation, $K(\dot{\varepsilon}^{vp})^m$ is the viscous stress, $\varepsilon^{vp}$ and $\dot{\varepsilon}^{vp}$ are, respectively, the viscoplastic strain and strain rate, $H$ and $n$ are hardening parameters, and $K$ and $m$ are the consistence and the strain rate sensitivity. All parameters depend on temperature.

For considering the effects of precipitation on the mechanical behavior, we consider that the flow stress of the alloy can be written: $\sigma = \sigma_{ss} + \Delta\sigma$. This assumes that the precipitation process only influences the yield stress and has no influence on the hardening behavior. $\Delta\sigma$ accounts for solute and precipitation hardening mechanisms in the following way.

For heterogeneous precipitation, precipitates are non coherent and do not lead to hardening. Thus the mechanical behavior is mainly sensitive to the depletion of the solid solution (decrease of the solute content). According to literature about solute hardening, the mechanical softening can be expressed as [11, 13, 14]

$$\Delta\sigma_{he}(T,t) = (\Sigma_i a_i X_i^{ss})^n - (\Sigma_i a_i X_i^{ss0})^n$$

where $X_i^{ss0}$ is the concentration of element $i$ in the supersaturated solid solution, $a_i$ are temperature-dependent coefficients, and $n$ is a temperature-independent coefficient.

Homogeneous precipitation (which is coherent or semicoherent) increases the yield stress of the alloy. The modeling of this hardening accounts for the dislocation–precipitate interaction which leads to [11]: $\Delta\sigma_{ho}(T,t) = g(f_v, r)$, where $f_v$ and $r$ are, respectively, the volume fraction and the radius of the precipitates. The resulting precipitation-induced mechanical effect is then: $\Delta\sigma(T,t) = \Delta\sigma_{he} + \Delta\sigma_{ho}$, where $\Delta\sigma_{he}$ is negative and $\Delta\sigma_{ho}$ is positive.

## 9.12.4 IDENTIFICATION OF THE MATERIAL PARAMETERS

In general, we perform material characterizations (metallurgical and thermomechanical) at constant temperatures after rapid cooling from the austenitization temperature for steels or from the solutionizing temperature for aluminium or titanium alloys. Anisothermal tests are mainly used to validate the models. (In some particular cases, they are used to get parameters that cannot be obtained through isothermal tests.)

### 9.12.4.1 Metallurgical Characterization

The global isothermal transformation kinetics are measured at different temperatures by dilatometry and/or *in situ* resistivity measurements (in titanium and aluminium alloys resistivity is much more sensitive to phase transformation than dilatometry). In addition, microstructural analysis and quantitative image analysis of the isothermally transformed specimen allow one to obtain the equilibrium volume fractions at the different temperatures. (They can also be obtained through thermodynamic calculations of multi-component systems.) The hardness of the different phases depending on the transformation temperature are also obtained.

For the nucleation and growth model, identification of nucleation sites and precipitated phases (chemical composition) and determination of mean precipitate size evolutions (by transmission electron microscopy) are done after isothermal interrupted quenching tests. Different parameters of the model can be taken from the literature (equilibrium diagram, diffusion coefficient), but some other parameters (like nucleation density, wetting function, or surface energy) are unknown. Thus they are determined through comparisons between calculated and experimental results.

Kinetics measurements during continuous cooling tests (at constant cooling rates) and microstructural analysis of the specimen are also performed in order to validate the metallurgical calculations (comparison between measured and calculated kinetics, final microstructures and hardnesses). These tests also give access to the thermal expansion coefficients of the different phases and to the volume changes associated with the phase transformations.

Of course, additional experiments are necessary to quantify the effect of the parent phase grain size on the transformation kinetics or the incidence of chemical composition variations of the alloys.

### 9.12.4.2 Thermomechanical Characterization

On one hand, the mechanical behavior of the individual phases (austenite, ferrite, pearlite, bainite in steels or supersaturated solid solution in aluminium alloys) is obtained through tensile tests (at constant deformation rates) followed by relaxation tests at constant temperatures after formation of the required phases. These tests are performed on our thermomechanical testing apparatus (DITHEM) and allow us to identify the evolutions of the yield stress, the hardening parameters, and the parameters describing the viscous effects versus temperature for the different phases (i.e., by using an optimization software like SIDOLO).

On the other hand, the effects of the phase transformations on the material behavior must be quantified. The changes in mechanical properties due to precipitation in aluminium alloys have been quantified by performing tensile tests for different holding times at a constant temperature. Thus, for example, the decrease of the yield stress related to heterogeneous precipitation and consequently to the depletion of the solid solution has been obtained [13].

In the case of alloys that, in addition to the change in mechanical properties induced by the new phase, exhibit transformation plasticity (like steels, titanium alloys), experimental studies aim to quantify this additional deformation. Typical tests consist in cooling down rapidly a specimen at the transformation temperature and in applying a constant stress before transformation starts and maintaining it all along the transformation. From the dilatometry and/or resistivity measurements, we get both the transformation plasticity amplitudes versus the applied stress and the effect of the applied stress on the transformation kinetics (for ferritic, pearlitic, bainitic transformations in steels, for example). Moreover, experiments in which the material is submitted to a tensile test either during its isothermal transformation or during continuous cooling transformations allow us to validate Eq. 1 [12].

Let us mention that numerous steel characterizations have been performed, but for confidentiality reasons only few of them can be found in publications (see, for instance, [12]).

## 9.12.5 IMPLEMENTATION IN CODES

Our global approach for describing phase transformation kinetics during heating and cooling in steels has been implemented in different finite element codes (in house software [5] and different industrial softwares like LAGAMINE, REFPROF, SYSWELD, FORGE) in order to predict the coupled thermal-metallurgical-mechanical processes during heat treatments of steels (quenching, surface hardening, case hardening).The kinetics model for precipitation and the associated thermomechanical model have been introduced in the finite element code MARC for calculating temperature fields, metallurgical fields, and residual stresses and deformations of quenched high-strength aluminium work pieces [14].

## REFERENCES

1. Christian, J. W. (1975). *The Theory of Transformations in Metals and Alloys*. Pergamon Press.
2. Sjöström, S. (1994). Physical, mathematical and numerical modelling for calculation of residual stress: Fundamentals and applications, *Proc. 4th Int. Conf. on Residual Stresses ICRS4* pp. 484–497, James, M. R., ed., Bethel: Society of Experimental Mechanics.

3. Sigli, C., Vichery, H., and Grange, B. (1996). *Proc. ICAA-5, 5th Int. Conf. on Al Alloys*, Vol. 1, pp. 391–396, Transtech Publications, Switzerland.
4. Ardell, J. A. (1985). Precipitation hardening. *Metall. Trans. A*, **16A**: 2131–2165.
5. Denis, S., Archambault, P., Aubry, C., Mey, A., Louin, J. Ch., and Simon, A. (1999). Modelling of phase transformation kinetics in steels and coupling with heat treatment residual stress predictions. *Journal de Physique IV France* **9**: 323–332.
6. Pumphrey, W. I., and Jones, F. W. (1948). Inter-relation of hardenability and isothermal transformation data. *JISI* **159**: 137–144.
7. Louin, J. Ch., Denis, S., Combeau, H., Lesoult, G., Simon, A., Aliaga, Ch., and Massoni, E. (2000). Effect of solidification segregations on phase transformation kinetics and on the development of internal stresses and deformations during cooling of steels, in *Proc. 5th European Conference on Residual Stresses*, pp. 205–210, Böttger, A. J., Delhez, R., and Mittemeijer, E. J., eds., ECRS5, Trans Tech Publications Ltd, Switzerland.
8. Aubry, C., Denis, S., Archambault, P., Simon, A., and Ruckstuhl, F. (1998). Modelling of tempering kinetics for the calculation of heat treatment residual stresses, *Proc. ICRS5 (Int. Conf. on Residual Stresses)*, pp. 412–417, Ericsson, T., Oden, M., and Andersson, A., eds., Linköping University, Sweden.
9. Laude, E., Gautier, E., Archambault, P., and Denis, S. (1996). Cinétique de transformation des alliages de titane en fonction du traitement thermomécanique. Etude expérimentale et calcul. *La Revue de Métallurgie-CIT/Science et Génie des Matériaux*, 1067–1078.
10. Godard, D., Gautier, E., and Archambault, P. (1999). Modelling heterogeneous precipitation kinetics in a Al-Zn-Mg-Cu aluminum alloy, in *Proc. PTM'99, International Conference on Solid-Solid Phase Transformations*, pp. 145–148, Koiwa, M., Otsuka, K., and Miyasaki, T., eds., The Japan Institute of Metals, Kyoto.
11. Godard, D. (1999). Influences de la précipitation sur le comportement thermomécanique lors de la trempe d'un alliage Al- Zn-Mg-Cu. Thèse de Doctorat de l'INPL, Nancy, France.
12. Denis, S., Archambault, P., Gautier, E., Simon, A., and Beck, G. (1999). Prediction of residual stress and distortion of ferrous and nonferrous metals: Current status and future developments, *Proc. 3rd Int. Conf. on Quenching and Control of Distortion*, pp. 263–276, Totten, G. E., Liscie, B., and Tensi, H. M., eds., ASM International; Journal of Materials Engineering and Performance (accepted October 2000).
13. Godard, D., Archambault, P., Houin, J. P., Gautier, E., and Heymes, F. (1998). Mechanical softening kinetics at high temperatures in an AlMgZnCu alloy: Experimental characterization and microstructural interpretation, *Proc. ICAA-6* pp. 1033–1038, Vol. 2, Sato, T., *et al.*, eds., The Japan Institute of Light Metals.
14. Archambault, P., Godard, D., Denis, S., Gautier, E., and Heymes, F. (1999). Prediction of heat treatment residual stresses: Application to quenching of high strength aluminium alloys including precipitation effects, in *Proc. 7th Int. Seminar on Heat Treatment and Surface Engineering of Light Alloys*, pp. 249–258, Lendva, J., and Reti, T., eds., Hungarian Scientific Society of Mechanical Engineering.

SECTION **9.13**

# Elastoplasticity Coupled with Phase Changes

F. D. FISCHER
*Montanuniversität Leoben, Franz-Josef-Strasse 18, A-8700 Leoben, Austria*

Contents



## 9.13.1 VALIDITY

The transformation–induced plasticity (TRIP) term as an additional strain rate term to the "classical" plastic strain rate term is valid for ductile metals like steel under diffusional or displacive (e.g., martensitic) phase transformations subjected to a monotonic loading path. The transformation kinetics described herein is valid for many technically relevant metals subjected to nearly any paths of heating or cooling. The selection of mechanical material data is typical of heat treatment simulations, welding simulations, or surface treatment simulations by induction hardening, carburizing, nitriding, etc.

## 9.13.2 BACKGROUND

In a recent review paper, Fischer *et al.* [1] defined TRIP as "significantly increased plasticity during a phase change. For an externally applied load for

*Handbook of Materials Behavior Models.* ISBN 0-12-443341-3.

which the corresponding equivalent stress is even small compared to the normal yield stress of the material, plastic deformation occurs. . . .” This softening has its origin in the fact that during a phase transformation a certain part of the material (say, a microregion) may change its volume and, occasionally, its shape, too. An internal stress state is necessary to achieve compatibility between the neighboring material and the transforming microregion. This internal stress state is superimposed by the load stress. The total stress state, e.g., in the case of a volume change of several percentage points, leads to plastification of the neighboring material and often of the microregion itself. One may speak about a plastic accommodation of the transformation eigenstrain. Finally, a macroscopic plastic deformation of the specimen can be observed.

Among several authors, Leblond *et al.* [2] suggested a splitting of the plastic strain rate $\dot{\varepsilon}_{ij}^{P}$ (the dot represents a time derivative) into a classical plastic strain rate $\dot{\varepsilon}_{ij}^{PC}$ representing plastification without any phase change and an extra term $\dot{\varepsilon}_{ij}^{TP}$ corresponding to plastification without any changes of external loading and only effected by the phase change process.

Assuming only two phases, the parent phase and the product phase, the volume fraction of the product phase is denoted as $\xi$. Based on some elementary research by Greenwood and Johnson as well as Mitter (see Fischer *et al.* [1]), the following relation for $\dot{\varepsilon}_{ij}^{TP}$ has eventually been suggested:

$$\dot{\varepsilon}_{ij}^{TP} = \frac{3}{2} K \frac{d\varphi(\xi)}{d\xi} S_{ij} \dot{\xi} \tag{1}$$

where $K$ is a coefficient depending on the transformation volume change $\delta$, $\varphi(\xi)$ is a function describing the evolution of $\dot{\varepsilon}_{ij}^{TP}$ with $\xi$ for a constant load-stress state, and $S_{ij}$ is the global (mesoscopic) stress deviator and corresponds to the stress deviator in the standard analysis of a work piece. The original derivation of the TRIP strain is based on $S_{ij}$ being constant in time. However, in the current formulation of Eq. 1, $S_{ij}$ is allowed to vary with time. Furthermore, $\dot{\varepsilon}_{ij}^{PC}$ is usually denominated as $\dot{\varepsilon}_{ij}^{P}$. Equation 1 is considered to be valid for both a diffusional (time-controlled) and a martensitic (temperature- and stress-controlled) transformation. However, this cannot be held for the martensitic transformation, especially under nonmonotonic loading and temperature paths, since the martensitic transformation is characterized by a volume change $\delta$ and a shape change (transformation shear) $\gamma$ of a single transforming microregion. Therefore, in addition to the previously mentioned elastic-plastic accommodation process, often referred to as the “Greenwood–Johnson effect,” an orientation process takes place due to the formation of preferred variants in the martensitic microregions that may arrange themselves in some (partially) “self-accommodating” groups. This orientation process, often referred to as the "Magee effect," may contribute in the same magnitude to the irreversible length change of elastic plastic material as the "Greenwood–Johnson effect." Thus shape memory alloys "thrive" exclusively on the orientation process! However, if monotonic loading paths are considered, a relation like Eq. 1 can be applied, if the coefficient $K$ is taken from a "Greenwood–Johnson" diagram. Concerning an improved TRIP term for martensitic transformation, the reader is referred to a recent paper by Fischer *et al.* [3]. However, it must be said that this research is not yet finished. Finally, the book edited by Berveiller and Fischer [4] is recommended; it presents an extensive treatment of both types of transformations in various kinds of metals.

## 9.13.3 DESCRIPTION OF THE MODEL

The whole model must consist of two more or less connected parts:

- the transformation process itself, i.e., the metal-physical part described in following text as "transformation kinetics"; and
- the deformation process described as the "deformation process."

External variables to the model are both the load-stress state and the temperature field $T(x, y, z, t)$. The evolution of the temperature field is obviously coupled with the phase change process by the production or extraction of heat (the "latent" heat or "transformation" heat or "recalescence" heat). This coupling must not be omitted. Here, the reader is referred to the chapter on heat treatment and corresponding data, e.g, the specific heat (heat capacity). Usually the coupling of the temperature field with plastic dissipation is weak and consequently need not be taken into account.

### 9.13.3.1 Transformation Kinetics

In practice, a picture on the transformation kinetics is reflected by the time (t)-temperature (T)-transformation (T-T-T) diagram (continuous cooling diagram [CC], cooling transformation diagram [CT], continuous cooling transformation diagram [CCT]). This diagram is based on t–T curves for standard specimens. A typical T-T-T diagram for a certain steel grade is depicted in Figure 9.13.1. The cooling velocity $\dot{T} = dT/dt$ may vary along the curve. Several points are marked describing the type of the phase and the corresponding volume fraction of the phase under consideration. Typically,

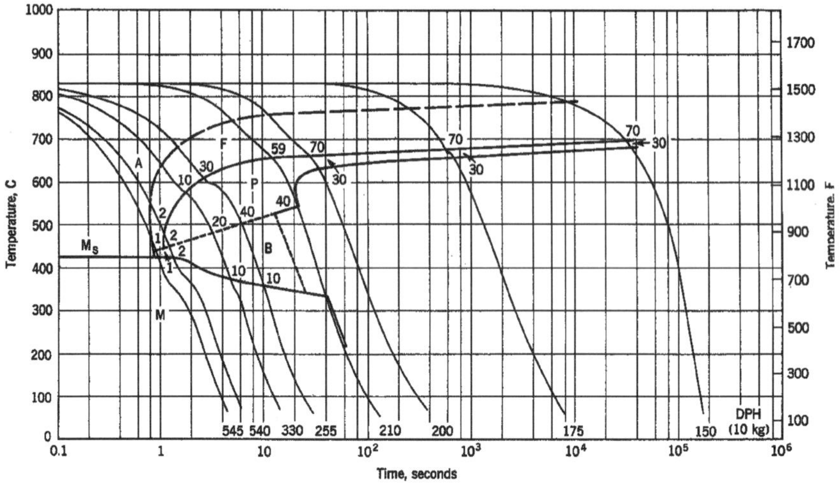


FIGURE 9.13.1 Typical CCT diagram for a certain steel grade with $AC_3$=830°C. Abbreviations: A austenite, B bainite, F ferrite, M martensite, P pearlite. The numbers in the individual areas (as 1, 2, 10, 30, etc.) stand for the volume fraction of the corresponding formed phases. The numbers at the termination of the curves stand for a certain hardness, e.g., DPH.

several phases can be formed. The reader is referred to textbooks such as References [5, 6].

The T-T-T diagram must be transferred into the calculational procedure. Leblond *et al.* [7] made a proposal for how to perform this. For sake of simplicity this proposal is briefly explained by assuming only one phase transformation with the transformed volume fraction $\xi$. Details can be taken from Leblond *et al.* [7] as well as from the SYSWELD-Manuals [8].

The transformed volume fraction $\xi$ is described by the following differential equation:

$$\frac{d\xi}{dt} = \frac{\xi_{eq}(T) - \xi}{\tau(T)} f(\dot{T}) \tag{2}$$

where $\xi_{eq}(T)$ is the equilibrium volume fraction of the phase corresponding to a given temperature $T$ and can only be estimated from the T-T-T diagram. $\xi_{eq}(T)$ could be found in the "isothermal" T-T-T diagram (isothermal transformation diagram [IT]), which, however, is often not available. The functions $\tau(T)$ and $f(\dot{T})$ must be selected such that the curves in the T-T-T diagram are approximated in a satisfying way. For example, SYSWELD [8] provides a specific routine for constructing $\tau(T)$ and $f(\dot{T})$.

Equation 2 can also be applied to the martensitic transformation, whose kinetics are usually described by the Koistinen–Marburger relation (see

Reference [9]) in total quantities and not in rate form as

$$\xi(T) = 1 - \exp[-b(M_s - T)] \quad \text{for } T < M_s \tag{3}$$

It follows that $\xi_{eq} = 1, \tau = 1/b, f(\dot{T}) = -\dot{T}$. It should be mentioned that $\xi(T)$ in Eq. 3 should also include a stress-dependent term. A constant load stress shifts the martensite start temperature $M_s$ to a higher temperature $M_\sigma$; $(M_\sigma - M_s)$ is of the order of magnitude of some degrees centigrade. The influence of the stress state on $\xi(T)$ plays a significant role in the case of shape memory alloys (SMAs) and a minor role in the case of steels.

The kinetics explained previously deal with a cooling path. In the case of steel and many other alloys, the situation is totally different with respect to a heating path from room temperature to a high temperature level, say, ca. 1000°C.

The transformation is totally "unsymmetric" with respect to a $t - T$ cycle. All the phases mentioned remain more or less unchanged until the $AC_1$ temperature, the austenite start temperature, is reached. At this stage a diffusional transformation to austenite starts, which stops at $AC_3$. Of course, $AC_1$ and $AC_3$ depend on the heating rate. Very high heating rates shift $AC_1$ and $AC_3$ to remarkably higher temperatures. The T-T-T diagram offers mostly $AC_3$ and sometimes $AC_1$ for slow heating and no further information on the heating transformation kinetics. It is interesting to note that some recent experiments give clear evidence that also a martensite-to-austenite displacive "back transformation" may occur in steels at higher temperatures, say, $\geq 600$°C; see Reference [3].

Finally, it is trivial to state that in the case of several phases the volume fraction $\xi_i$ of each individual phase, $i = 1, \ldots, n$, must add up to 1!

The reader is also referred to some further papers [10–12] dealing with the mathematical modeling of phase transformations and temperature fields. These papers also deliver some relevant data for the different phases.

### 9.13.3.2 Deformation Process

The deformation process is modeled according to the standard concept of elastic- (visco-) plastic materials. Therefore, only some specific details on the material data with respect to phase change phenomena are given. The following data are needed for a stress analysis:

- Young's modulus $E(T)$. For steel $E(T)$ is mostly the same for all phases. In the case of shape memory alloys, Young's modulus may differ significantly for the austenitic ($\gamma$-) and martensitic ($\alpha'$-) phase.
- Poisson's ratio $\nu$, which is only weakly dependent on the temperature.

- Integral coefficient of thermal expansion (CTE), which differs significantly between the austenitic ($\gamma$-) phase and the ferritic-pearlitic-bainitic-martensitic ($\alpha$- or $\alpha'$-) phases. Usually the CTE depends weakly on the temperature and increases slightly with increasing temperature. The corresponding volume change is usually assumed to be isotropic and is 3 CTE. $(T-T_{ref})$, $T_{ref}$ being the thermal strain-free reference temperature.
- Volume change due to phase transformation, $\delta$. As mentioned in Section 9.13.2, a phase transformation is accompanied by a volume change. The corresponding standard test is the dilatometer test. A typical value for $\delta$ for the transformation of Fe-based alloys from the $\gamma$- to $\alpha$- (or $\alpha'$-) phase is $\leq 0.04$. In the case of shape memory alloys, the value $\delta$ is nearly 0! If several phases are present, a "mixed" value for $\delta$ is applied as $\delta = \sum_{i=1}^{n} \xi_i \delta_i$, $i$ being number of phase, $\delta_i$ being volume change of phase $i$, $\xi_i$ see previous text.
- Yield stress $\sigma_y^i(T, \varepsilon)$, $i$ being number of phase; for details see Section 9.13.4.
- In the case of several phases present, a mixture rule for the yield stress must be installed. For two phases, e.g., $\alpha$ and $\gamma$, nowadays a nonlinear mixture rule is applied,

$$\sigma_y^{\max}(T) = [1 - f(\xi_\alpha)]\sigma_y^\gamma(T) + f(\xi_\alpha)\sigma_y^\alpha(T), \quad \xi_\alpha + \xi_\gamma = 1 \tag{4}$$

  $f(\xi)$ is given from numerical studies. A representative data set can be found in Table 9.13.1.
- Type of hardening; for details, see Section 9.13.4.
- Transformation-induced plasticity (TRIP) term $\dot{\varepsilon}_{ij}^{TP}$. For the sake of completeness, a current formulation of the TRIP strain rate is presented which is implemented in user-supplied material law formulations in several programs or is obligatory in the program SYSWELD [8]. The theoretical background has been explained in Section 9.13.2. The term $\dot{\varepsilon}_{ij}^{TP}$ adds up to the other strain rate terms as

$$\dot{\varepsilon}_{ij} = \dot{\varepsilon}_{ij}^{e} + \dot{\varepsilon}_{ij}^{VP} + \dot{\varepsilon}_{ij}^{th} + \dot{\varepsilon}_{ij}^{TP} \tag{5}$$

  where $\dot{\varepsilon}_{ij}^{e}$ is the elastic strain rate, $\dot{\varepsilon}_{ij}^{P}$ is the conventional plastic strain rate or $\dot{\varepsilon}_{ij}^{VP}$ is the conventional viscoplastic strain rate, $\dot{\varepsilon}_{ij}^{th}$ is the thermal strain rate, and $\dot{\varepsilon}_{ij}$ is the total strain rate.

**TABLE 9.13.1 Weighting Function $f(\xi_\alpha)$ for Nonlinear Mixture Rule**

| $\xi_\alpha$ | 0 | 12.5% | 25% | 50% | 75% | 100% |
|---|---|---|---|---|---|---|
| $f(\xi_\alpha)$ | 0 | 0.044 | 0.124 | 0.391 | 0.668 | 1 |

In accordance with the notation used in Eq. 1, the following data and function are proposed:

$$K = 2\delta/\sigma_y^\gamma h(\sigma_{eq}\sigma_y^\gamma) \tag{6a}$$

$$\varphi(\xi) = \xi(1 - \ln \xi) \tag{6b}$$

where $\xi$ corresponds to the fraction of the $\alpha$- or $\alpha'$-phase, $h(\sigma_{eq}/\sigma_y^\gamma)$ is a correction factor, which is 1.0 for $\sigma_{eq}/\sigma_y^\gamma < 1/2$, and $\sigma_{eq}$ is the equivalent stress to $S_{ij}$.

## 9.13.4 IDENTIFICATION

The kinetics data can partially be identified from a *T-T-T* diagram. However, $\xi_{eq}$ to a certain phase is often difficult to establish. A respresentative data set for various steel phases is given in Table 9.13.2. Usually the steel specimen is austenitized at 850 to 1000°C and held for some time (e.g., half an hour). Then the cooling process is started and plotted in a *T-T-T* diagram.

A typical data set for martensitic transformation is $M_s = 420°C$, $b = 0.03$.

A typical austenitization interval ranks between $AC_1 = 704°C$ and $AC_3 = 830°C$.

A typical data set for E, $\nu$, CTE is E(20°C) ~ 210 Gpa, E(1000°C) ~ 105 Gpa, $\nu = 0.3$, $CTE_\alpha \sim 12.10^{-6} C^{\circ -1}$ for 20°C, $CTE_\gamma \sim 18.10^{-6} C^{\circ -1}$ for 20°C.

The transformation volume change $\delta$ is obtained from a standard dilatometer test. Please check for the presence of a texture (orthotropic plastic anisotropy) in the specimen. A sure indication will be that $\delta$ will not be distributed isotropically.

With respect to the yield stress $\sigma_y^i(T, \varepsilon)$ and the type of hardening, usually uniaxial stress-strain curves for monotonic loading are used as input data to the programs. Of course, data for all phases in the relevant temperature range are needed. It should be mentioned that very often the material data for a specific phase cannot easily be found, since only a given mixture of the phases is thermodynamically stable (or metastable). This forces us to extrapolate a

**TABLE 9.13.2 Phase Transformation Intervals for a High-Strength Steel**

| Phase | Start T(°)C | End T(°)C | *dT/dt* | $\xi$ (total) % |
|---|---|---|---|---|
| Ferrite | 704 | 675 | −2.1 | 5 |
| Pearlite | 675 | 537 | −2.1 | 25 |
| Bainite | 537 | 279 | −1.9 | 55 |
| Martensite | 279 | 20 | −2.6 | 100 |

large amount of necessary data. It is highly recommended to consult a metallurgist for this problem. Usually $\sigma_y^i(T, \varepsilon)$ decreases with increasing temperature and fixed strain. However, there exist metals like Intermetallics, where $\sigma_y^i(T, \varepsilon)$ may increase with increasing temperature. Furthermore, the yield stress for the austenite $\sigma_y^\gamma(20°C, \varepsilon)$ is usually smaller than that of ferritic, pearlitic, bainitic, or martensitic ($\alpha$- or $\alpha'$-) phase $\sigma_y^\alpha(20°C, \varepsilon)$ or $\sigma_y^{\alpha'}$ at room temperature. At higher temperatures, say >600°C, the situation may change, and the $\alpha$- or $\alpha'$-phase becomes the softer phase. If a temperature cycle is applied for a temperature interval including the "crossing" point of the yield stresses of both phases, ratcheting will occur even in free specimens without any load stress; see details in Siegmund *et al.* [13] or Silberschmidt *et al.* [14]. That means that the specimen will change its size monotonically during a temperature cycle. Obviously, this could be extremely dangerous for devices. However, modern alloying techniques allow us to control the temperature dependence of $\sigma_y^\alpha$ or $\sigma_y^{\alpha'}$ and $\sigma_y^\gamma$, for example, by a certain amount of nitrogen. Both cases are now possible, a $\gamma$-phase with a yield stress higher than that of the $\alpha$- or $\alpha'$-phase and vice versa for a temperature range between 20°C and 1000°C. Here the reader is referred to modern duplex steels; see also Silberschmidt *et al.* [14].

Since, for example, a heat treatment may lead to several changes of the sign and orientation of the principal stress components in a material point, one has to consider, at least for some of the possibly developing phases, how they behave under cyclic stressing. Of course, standard hardening mechanisms like isotropic hardening or kinematic hardening are available in most of the programs. However, one must keep in mind that especially during heating the dislocation density may be reset by recrystallization to nearly its original amount. That means that very careful control over the amount and the kind of hardening must be exercised in the simulation of thermal cycles. Specifically, almost no hardening occurs for martensite if the alloy contains some carbon. In addition, the existence of a very high yield strength ($\sigma_y^{\alpha'} > 1500$ MPa) is usually accompanied by a small fracture strain (only a few percentage points).

The only additional mechanical quantity, the coefficient $K$, in the TRIP term (see Eqs. 1), 6a, needs some comments. Classical TRIP tests (sometimes also called "creep" tests) must be performed to find $K$ by applying a constant load $\Sigma$ on a specimen (e.g., weight on a wire) and then performing the transformation by cooling. The irreversible length change $\varepsilon^{TP}$ ($\varepsilon^{TP} = \varepsilon - \delta/3 - \varepsilon^e$) must now be represented in relation to the load stress. The data points should lie on a straight line for $|\Sigma| < \kappa\sigma_y^\gamma, \kappa \sim 2/3$, according to Greenwood and Johnson; see Fischer *et al.* [1]. K now is the inclination of this line. A typical value for $K$ is $4.5.10^{-4}$ to $10.10^{-4}$ ($\text{MPa}^{-1}$). Finally, it should be mentioned that several proposals for $\varphi(\xi)$, Eq. 6b, exist in the open literature which usually do not lead to significantly different results.

## 9.13.5 HOW TO USE THE MODEL

The TRIP term $\dot{\varepsilon}_{ij}^{TP}$, Eq. 1, must be added to the other contributions of the total strain rate $\dot{\varepsilon}_{ij}^{TP}$. Furthermore, a control over the transformation kinetics must be implemented. Only few general purpose programs allow for both; one is the program SYSWELD [8]. The proper implementation is by no means an easy piece of work. At least an option for a user-supplied material description, e.g., a UMAT in ABAQUS [15], must be available. The computational algorithms are the same as those for elasto-(visco)plastic behavior, but with an additional strain rate $\dot{\varepsilon}_{ij}^{TP}$, which is not calculated via a consistency relation but provided externally like a creep strain rate.

## 9.13.6 TABLE OF PARAMETERS

Several data are given in Section 9.13.4, where only two tables are mentioned.

## REFERENCES


1. Fischer, F. D., Sun, Q.-P., and Tanaka, K. (1996). Transformation-induced plasticity (TRIP). *Appl. Mech. Rev.* **46**: 317–364.
2. Leblond, J. B., Mottet, G., and Devaux, J. C. (1986). A theoretical and numerical approach to the plastic behaviour of steels during phase transformations. I. Derivation of general relations; II. Study of classical plasticity for ideal-plastic phases. *J. Mech. Phys. Solids* **34**: 395-410, 411–432.
3. Fischer, F. D., Reisner, G., Werner, E., Tanaka, K., Cailletaud, G., and Antretter, T. (2000). A new view on transformation induced plasticity (TRIP). *Int. J. Plasticity* **16**: 723–748.
4. Berveiller, M., and Fischer, F. D. (1997). *Mechanics of Solids with Phase Changes,* CISM Courses and Lectures No 368, New York, Springer.
5. Krauss, G. (1980). *Principles of Heat Treatment of Steel*, Metals Park, Ohio: American Society for Metals.
6. *Transformation and Hardenability in Steels* (1967). No editors, Symp. Climax Molybdenum Comp. and Univ. Michigan, The Univ. Michigan Extension Service.
7. Leblond, J. B., and Devaux, J. (1984). A new kinetic model for anisothermal metallurgical transformations in steels including effect of austenite grain size. *Acta Metall.* **32**: 137–146.
8. SYSWELD + 2.0: Reference Manual (1997), SYSTUS International, ESI Group: Simulation of Welding and Heat Treatment Processes, Framasoft + CSI.
9. Koistinen, D. P., and Marburger, R. E. (1959). A general equation describing the extent of the austenite-martensite transformation in pure iron-carbon alloys and plain carbon steels. *Acta Metall.* 7: 59–60.
10. Denis, S., Farias, D., and Simon, A. (1992). Mathematical model coupling phase transformations and temperature evolutions in steels. *ISIJ International* **32**: 316–325.
11. Hunkel, M., Lübben, T., Hoffmann, F., and Mayr, P. (1999). Modellierung bei bainitischen und perlitischen Umwandlung bei Stählen. *HTM (Härterei-Technische Mitteilungen—Zeitschrift für Wärmebehandlung und Werkstofftechnik)* **54**: 365–372.

12. Centinel, H., Toparli, M., and Özsoyeller, L. (2000). A finite element based prediction of the microstructural evolution of steels subjected to the Tempcore process. *Mechanics of Materials* **32**: 339–347.
13. Siegmund, T., Werner, E., and Fischer, F. D. (1995). On the thermomechanical deformation behavior of duplex-type materials. *J. Mech. Phys. Solids* **43**: 495–532.
14. Silberschmidt, V. V., Rammerstorfer, F. G., Werner, E. A., Fischer, F. D., and Uggowitzer, P. J. (1999). On material immanent ratchetting of two-phase materials under cyclic purely thermal loading. *Arch. Appl. Mech.* **69**: 727–750.
15. ABAQUS, www.hks.com

SECTION **9.14**

# Mechanical Behavior of Steels during Solid–Solid Phase Transformations

JEAN-BAPTISTE LEBLOND

*Laboratoire de Modélisation en Mécanique, Université Pierre et Marie Curie, 8 rue du Capitaine Scott, 75015 Paris, France*

Contents



## 9.14.1 VALIDITY OF THE MODELS PROPOSED

The models proposed herein for the plastic behavior of solids during solid–solid transformations are valid for many steels; see examples provided in Reference [1]. They are also applicable to other, nonferrous metals, as shown by Greenwood and Johnson [2] in their pioneering work involving a first, simple version of the more refined model described in following text for ideal plasticity. *They are not applicable to shape memory alloys*, for which the transformation is a quasi–elastic phenomenon.

*Handbook of Materials Behavior Models.* ISBN 0-12-443341-3.

## 9.14.2 THEORY USED TO DERIVE THE MODELS

The models presented here do not involve any *ad hoc*, adjustable parameter, but simply the usual thermomechanical characteristics of the material. They are not derived from heuristic considerations but from (approximate) homogenization [3–7]. It is supposed that the anomalous plastic behavior during phase transformations, and especially the so-called *transformation plasticity phenomenon*, arise from the sole microscopic mechanism proposed by Greenwood and Johnson [2], with that suggested by Magee [8] being disregarded. In spite of this limiting assumption, the models proposed are often experimentally found to be sufficient, as already mentioned.

## 9.14.3 DESCRIPTION OF THE MODELS

### 9.14.3.1 Notations

- s: stress deviator
- $\sigma_\gamma^Y$: yield stress of the parent ($\gamma$) phase
- $\sigma_\alpha^Y$: yield stress of the product ($\alpha$) phase
- $\sigma^Y$: "global" yield stress
- $z$: volume fraction of the product ($\alpha$) phase
- $T$: temperature
- $f(z), 2ptg(z)$: functions given in Table 9.14.1 below
- $h(X)$: function given by $h(X) = \begin{cases} 1 & \text{if} \quad 0 \leq X \leq 0.7 \\ 1 + 5(X - 0.7) & \text{if} \quad 0.7 \leq X \leq 1 \end{cases}$.
- $\dot{\boldsymbol{\varepsilon}}^p = \dot{\boldsymbol{\varepsilon}}^{tp} + \dot{\boldsymbol{\varepsilon}}_\sigma^{cp} + \dot{\boldsymbol{\varepsilon}}_T^{cp}$: total plastic strain rate
- $\dot{\boldsymbol{\varepsilon}}^{tp}$: transformation plastic strain rate
- $\dot{\boldsymbol{\varepsilon}}_\sigma^{cp}$: classical plastic strain rate arising from variations of the stresses
- $\dot{\boldsymbol{\varepsilon}}_T^{cp}$: classical plastic strain rate arising from variations of the temperature
- $\Delta V/V$: difference of specific volume between the two phases (function of temperature)
- $E$: Young's modulus (assumed to be the same for both phases)
- $\lambda_i$: thermal expansion coefficient of phase (i)

TABLE 9.14.1 The functions $f(z)$ and $g(z)$

| $z$ | 0 | 0.125 | 0.25 | 0.50 | 0.75 | 1 |
|---|---|---|---|---|---|---|
| $f(z)$ | 0 | 0.0186 | 0.101 | 0.392 | 0.672 | 1 |
| $g(z)$ | 0 | 2 | 3 | 1.75 | 1.75 | 1 |

- $\varepsilon_i^{eff}$: hardening parameter of phase (i) (isotropic hardening)
- $\mathbf{a}_i$: center of yield locus of phase (i) (kinematic hardening)
- $\theta$: transformation "memory coefficient" for hardening (= 0 for ferritic, pearlitic, and bainitic transformations; = 1 for martensitic transformations)

### 9.14.3.2 Case of Ideal Plastic Phases

With

$$\sigma^{eq} \equiv \left(\frac{3}{2}\mathbf{s} : \mathbf{s}\right)^{1/2} \quad \text{and } \sigma^Y = [1 - f(z)]\sigma_\gamma^Y + f(z)\sigma_\alpha^Y :$$

If $\sigma^{eq} < \sigma^Y$:

$$\dot{\boldsymbol{\varepsilon}}^{tp} = -\frac{1}{\sigma_\gamma^Y}\frac{\Delta V}{V} h\left(\frac{\sigma^{eq}}{\sigma^Y}\right)\mathbf{s}(\ln z)\dot{z};$$

$$\dot{\boldsymbol{\varepsilon}}_\sigma^{cp} = \frac{3}{2\sigma_\gamma^Y}\frac{(1-z)g(z)}{E}\mathbf{s}\dot{\sigma}^{eq};$$

$$\dot{\boldsymbol{\varepsilon}}_T^{cp} = \frac{3}{\sigma_\gamma^Y}(\lambda_\gamma - \lambda_\alpha)z(\ln z)\mathbf{s}\dot{T}$$

If $\sigma^{eq} = \sigma^Y$:

$$\dot{\boldsymbol{\varepsilon}}^p = \dot{\lambda}\,\mathbf{s} \quad \text{where} \quad \dot{\lambda} \geq 0 \text{ is indeterminate.}$$

### 9.14.3.3 Case of Isotropically Hardenable Phases

With

$$\sigma^{eq} \equiv \left(\frac{3}{2}\mathbf{s} : \mathbf{s}\right)^{1/2}, \quad \dot{\boldsymbol{\varepsilon}}^{eq} \equiv \left(\frac{2}{3}\dot{\boldsymbol{\varepsilon}}^p : \dot{\boldsymbol{\varepsilon}}^p\right)^{1/2}$$

and

$$\sigma^Y = [1 - f(z)]\sigma_\gamma^Y\left(\varepsilon_\gamma^{eff}\right) + f(z)\sigma_\alpha^Y\left(\varepsilon_\alpha^{eff}\right) :$$

If $\sigma^{eq} < \sigma^{Y}$:

$$\dot{\boldsymbol{\varepsilon}}^{tp} = -\frac{1}{\sigma_{\gamma}^{Y}\left(\varepsilon_{\gamma}^{eff}\right)} \frac{\Delta V}{V} h\left(\frac{\sigma^{eq}}{\sigma^{Y}}\right) \mathbf{s}(\ln z)\dot{z};$$

$$\dot{\boldsymbol{\varepsilon}}_{\sigma}^{cp} = \frac{3}{2\sigma_{\gamma}^{Y}\left(\varepsilon_{\gamma}^{eff}\right)} \frac{(1-z)g(z)}{E} \mathbf{s}\, \dot{\sigma}^{eq};$$

$$\dot{\boldsymbol{\varepsilon}}_{T}^{cp} = \frac{3}{\sigma_{\gamma}^{Y}\left(\varepsilon_{\gamma}^{eff}\right)} (\lambda_{\gamma} - \lambda_{\alpha}) z(\ln z)\, \mathbf{s}\, \dot{T};$$

$$\dot{\varepsilon}_{\gamma}^{eff} = -\frac{2}{3(1-z)} \frac{\Delta V}{V} h\left(\frac{\sigma^{eq}}{\sigma^{Y}}\right)(\ln z)\dot{z} + \frac{g(z)}{E}\dot{\sigma}^{eq} + 2(\lambda_{\gamma} - \lambda_{\alpha})\frac{z \ln z}{1-z}\dot{T};$$

$$\dot{\varepsilon}_{\alpha}^{eff} = -\frac{\dot{z}}{z}\varepsilon_{\alpha}^{eff} + \theta \frac{\dot{z}}{z}\varepsilon_{\gamma}^{eff}.$$

If $\sigma^{eq} = \sigma^{Y}$:

$$\dot{\boldsymbol{\varepsilon}}^{p} = \frac{3}{2}\frac{\dot{\varepsilon}^{eq}}{\sigma^{eq}}\mathbf{s};$$

$$\dot{\varepsilon}_{\gamma}^{eff} = \dot{\varepsilon}^{eq}; \quad \dot{\varepsilon}_{\alpha}^{eff} = \dot{\varepsilon}^{eq} - \frac{\dot{z}}{z}\varepsilon_{\alpha}^{eff} + \theta\frac{\dot{z}}{z}\varepsilon_{\gamma}^{eff}.$$

### 9.14.3.4 Case of (Linearly) Kinematically Hardenable Phases

With

$$\mathbf{a}_{\gamma} \equiv h_{\gamma}(T)\mathbf{b}_{\gamma}, \; \mathbf{a}_{\alpha} \equiv h_{\alpha}(T)\mathbf{b}_{\alpha}, \; \mathbf{a} = (1-z)\mathbf{a}_{\gamma} + z\mathbf{a}_{\alpha},$$

$$\sigma^{eq} \equiv \left[\frac{3}{2}(\mathbf{s}-\mathbf{a}) : (\mathbf{s}-\mathbf{a})\right]^{1/2}, \quad \dot{\sigma}_{s}^{eq} \equiv \frac{3}{2\sigma'^{eq}}(\mathbf{s}-\mathbf{a}_{\gamma}) : \dot{\mathbf{s}},$$

$$\sigma'^{eq} \equiv \left[\frac{3}{2}(\mathbf{s}-\mathbf{a}_{\gamma}) : (\mathbf{s}-\mathbf{a}_{\gamma})\right]^{1/2}, \quad \dot{\varepsilon}^{eq} \equiv \left(\frac{2}{3}\dot{\boldsymbol{\varepsilon}}^{p} : \dot{\boldsymbol{\varepsilon}}^{p}\right)^{1/2},$$

$$\text{and} \quad \sigma^{Y} = [1 - f(z)]\sigma_{\gamma}^{Y} + f(z)\sigma_{\alpha}^{Y} :$$

If $\sigma^{eq} < \sigma^{Y}$:

$$\dot{\boldsymbol{\varepsilon}}^{tp} = -\frac{1}{\sigma_{\gamma}^{Y}}\frac{\Delta V}{V}h\left(\frac{\sigma^{eq}}{\sigma^{Y}}\right)(\mathbf{s}-\mathbf{a}_{\gamma})(\ln z)\dot{z};$$

$$\dot{\boldsymbol{\varepsilon}}_{\sigma}^{cp} = \frac{3}{2\sigma_{\gamma}^{Y}}\frac{(1-z)g(z)}{E}(\mathbf{s}-\mathbf{a}_{\gamma})\dot{\sigma}_{s}^{eq};$$

$$\dot{\boldsymbol{\varepsilon}}_{T}^{cp} = \frac{3}{\sigma_{\gamma}^{Y}}(\lambda_{\gamma}-\lambda_{\alpha})z(\ln z)(\mathbf{s}-\mathbf{a}_{\gamma})\dot{T};$$

$$\dot{\mathbf{b}}_{\gamma} = \frac{1}{1-z}\left(\dot{\boldsymbol{\varepsilon}}^{tp}+\dot{\boldsymbol{\varepsilon}}_{\sigma}^{cp}+\dot{\boldsymbol{\varepsilon}}_{T}^{cp}\right); \quad \dot{\mathbf{b}}_{\alpha} = -\frac{\dot{z}}{z}\mathbf{b}_{\alpha}+\theta\frac{\dot{z}}{z}\mathbf{b}_{\gamma}.$$

If $\sigma^{eq} = \sigma^{Y}$:

$$\dot{\boldsymbol{\varepsilon}}^{p} = \frac{3}{2}\frac{\dot{\varepsilon}^{eq}}{\sigma^{eq}}(\mathbf{s}-\mathbf{a});$$

$$\dot{\mathbf{b}}_{\gamma} = \dot{\boldsymbol{\varepsilon}}^{p}; \quad \dot{\mathbf{b}}_{\alpha} = \dot{\boldsymbol{\varepsilon}}^{p} - \frac{\dot{z}}{z}\mathbf{b}_{\alpha}+\theta\frac{\dot{z}}{z}\mathbf{b}_{\gamma}.$$

## 9.14.4 IDENTIFICATION OF MATERIAL PARAMETERS

Purely mechanical parameters (Young's modulus, yield limit, and hardening slope of each phase) can be obtained through standard tests at temperatures sufficiently high or low for a single phase to be present. The thermal expansion coefficients of the phases and the difference of specific volume between them can be obtained through (stress-free) dilatometry tests. If one wishes to assess the accuracy of the formulae given previously for the transformation plastic strain rate, one must perform in addition dilatometry tests with some external stress applied, which requires more sophisticated equipment.

## 9.14.5 NUMERICAL IMPLEMENTATION

The numerical implementation of the models described in preceding text is quite easy. For each time step, one should first correct the "elastically computed" (that is, assuming the total minus thermal strain increment to be purely elastic) stresses to account for the values of the transformation plastic strain increment and the classical plastic one given previously (implicit or explicit algorithm). If the equivalent von Mises stress calculated in that way is smaller than the global yield stress, the treatment is finished. If it is greater, then the previous corrections should be abandoned and a classical projection method employed to ensure satisfaction of the criterion at the final instant of the time interval considered and of the flow rule during this interval (implicit algorithm).

## REFERENCES

1. Leblond, J.-B. (1990). *Qualification expérimentale du modèle de plasticité de transformation*, FRAMASOFT+CSI Internal Report no. CSS/L/NT.90/4022.
2. Greenwood, G. W., and Johnson, R. H. (1965). *The deformation of metals under small stresses during phase transformations. Proc. Roy. Soc.* **A 283**: 403–422.
3. Leblond, J.-B., Mottet, G., and Devaux, J.-C. (1986). A theoretical and numerical approach to the plastic behaviour of steels during phase transformations. I: Derivation of general relations. *J. Mech. Phys. Solids* **34**: 395–409.
4. Leblond, J.-B., Mottet, G., and Devaux, J.-C. (1986). A theoretical and numerical approach to the plastic behaviour of steels during phase transformations. II: Study of classical plasticity for ideal-plastic phases. *J. Mech. Phys. Solids* **34**: 411–432.
5. Leblond, J.-B., Devaux, J., and Devaux, J.-C. (1989). Mathematical modelling of transformation plasticity in steels. I: Case of ideal-plastic phases. *Int. J. Plast.* **5**: 551–572.
6. Leblond, J.-B. (1989). Mathematical modelling of transformation plasticity in steels. II: Coupling with strain hardening phenomena. *Int. J. Plast.* **5**: 573–591.
7. Devaux, J., Leblond, J.-B., and Bergheau, J.-M. (2000). Numerical study of the plastic behaviour of a low alloy steel during phase transformation, *Proceedings of the "First International Conference on Thermal Process Modelling and Computer Simulation"*, Shanghai, China.
8. Magee, C. L. (1966). *Transformation Kinetics, Microplasticity and Ageing of Martensite in Fe-Ni.* Ph.D. Thesis, Carnegie–Mellon University, Pittsburgh.

SECTION 9.15

# Constitutive Equations of a Shape Memory Alloy under Complex Loading Conditions

MASATAKA TOKUDA
*Department of Mechanical Engineering, Mie University, Kamihama 1515 Tsu 514-8507, Japan*

Contents



## 9.15.1 INTRODUCTION

The deformation mechanism of shape memory is the martensitic phase transformation induced by applied stress and temperature. Quite unique, complicated, and interesting mechanical properties can be obtained by applying the simultaneous change of temperature and stress, which has six independent components (i.e., by applying the complex loading conditions).

*Handbook of Materials Behavior Models*. ISBN 0-12-443341-3.

The proposed constitutive equations can predict the complicated nonlinear deformation behavior of a shape memory alloy.

## 9.15.2 VALIDITY (DOMAIN OF APPLICATIONS)

Roughly speaking, the quite complicated deformation behavior observed in the shape memory alloy subjected to the complex loading path can be predicted (at least qualitatively) by the proposed constitutive equations. The reliability was confirmed experimentally by applying the combined loads of axial force and torque to the thin-walled tubular specimen of a Cu-based shape memory alloy. Figure 9.15.1 shows one example of comparisons between the experimental and predicted results. Figure 9.15.1a shows the given thermomechanical (complex) loading path to the thin-walled tubular specimen: 1–2: proportional loading of torsion stress $\tau$, axial stress $\sigma$; 2–3: increase of temperature $T$ while keeping the combined stress state constant; 3–4: decrease of temperature $T$ while keeping the stress state constant; 4–5: decrease of shear stress $\tau$ while keeping the axial stress $\sigma$ and

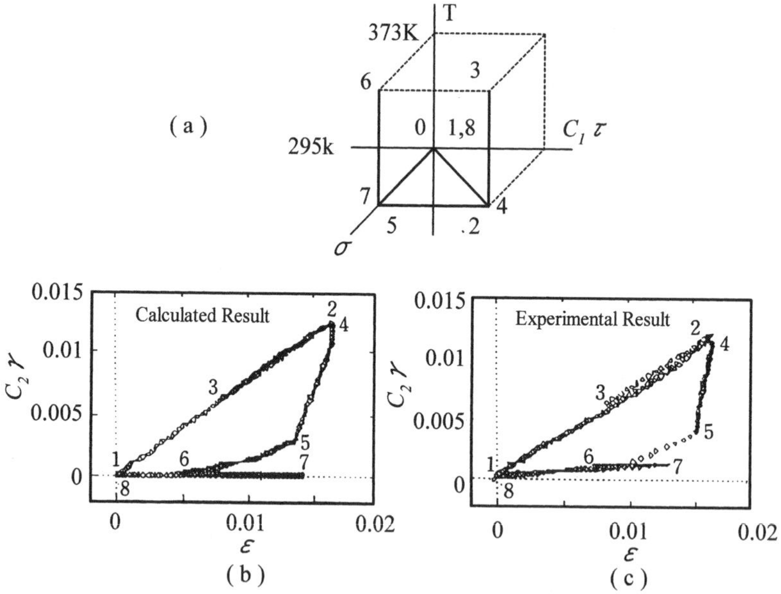


FIGURE 9.15.1 An example of comparisons between experimental and predicted results. (a) Thermo-mechanical loading path. (b) Calculated result of strain response. (c) Experimental result of strain response.

temperature $T$; 5–6 and 6–7: increase and decrease, respectively, of temperature while keeping the stress state constant: and 7–8: proportional unloading to zero stress state. Figure 9.15.1b,c show the computational (predicted) result and experimental result. As the figure shows, the computational result coincides with the experimental results qualitatively very well.

## 9.15.3 MODELING

The formulation of constitutive equations is based on the semimicroscopic mesomechanical approach by considering the multilayered structure of a polycrystalline shape memory alloy (SMA). The formulation procedure is composed of the following stages.

### 9.15.3.1 Modeling of Phase Transformation System

The basic phase transformation mechanisms of SMA are the temperature-induced and stress-induced martensite phase transformations. The deformation induced by these martensite phase transformations is considered to be the twin-type (shear) deformation that has a negligibly small change of volume and that is controlled by the temperature and the stress tensor. This twin-type (shear) deformation occurs only in the crystallographically determined (easy twin) direction on the crystallographically determined (easy twin) planes. This "twin (shear) deformation system," with a specific direction and a specific plane, is named the *phase transformation system*, and it may correspond to the "slip system" in crystal plasticity, whose mechanism is dislocation slip. When SMA is subjected to thermal loading (the change of temperature) only, the martensite phase transformation occurs and produces the (twin-type) shear deformation. However, in this case, no macroscopically significant deformation of crystal grain can be observed. This phenomenon can be explained as follows: in the case of temperature change without any external (applied) or internal stress, the phase transformation can occur simultaneously in every transformation system with different (randomly distributed) orientations, and thus the induced shear (twin) strains cancel each other on average. On the other hand, in the case of the stress-induced phase transformation, the phase transformation occurs in the phase transformation systems with the preferable orientation to the applied stress state. Thus only the shear strains with specific direction are summed up, and consequently, the significant macroscopic inelastic deformation can be observed in the crystal component (as well as in the polycrystal). Our engineering interests are in the macroscopic inelastic deformations, the force produced when the deformation is constrained, and their combination. Therefore, only the case when the material is subjected to some stresses is considered in the proposed constitutive equations. The effect of temperature is incorporated as the temperature effect on the shear stress which is necessary for the phase transformation, i.e., on the critical (resolved) shear stress. Details of the (critical) shear stress $\tau$ and shear strain $\gamma$ relation formulated in the previously mentioned scheme are given in References [1, 2].

## 9.15.3.2 Modeling of Crystal Component

The deformation property (stress-strain-temperature relation) of the crystal grain component with the phase transformation systems described in Section 9.15.3.1 is formulated as follows, according to the crystal plasticity manner.

The number of phase transformation systems in the crystal grain is assumed to be $M$ (for example, $M = 24$ for the martensite phase transformation of Cu-based SMA). When the crystal grain is subjected to the uniform stress $\sigma_{\mathrm{ij}}$ $(\mathrm{l}, j = 1, 2, 3)$, the shear stress $\tau$ resolved on the $m$-th $(m = 1, 2, \ldots, M)$ phase transformation system can be obtained by the following equation:

$$\tau_{(m)} = \sum_{\mathrm{i},k=1}^{3} \alpha_{(m)\mathrm{i}j}\sigma_{\mathrm{i}j} \tag{1}$$

The coefficient $\alpha(m)_{ij}$ in Eq. 1 is the so-called generalized Schmid factor, defined as follows:

$$\alpha_{(m)ij} = [s_{(m)i}n_{(m)j} + s_{(m)j}n_{(m)i}] \tag{2}$$

where $s_{(m)i}$ and $n_{(m)i}$ $(i = 1, 2. 3)$ are the unit vector along the shear direction and the unit vector normal to the transformation plane of the $m$-th phase transformation system, respectively. By using the obtained resolved shear stress $\tau_{(m)}$, the shear strain increment $d\gamma^{PT}_{(m)}$ of the $m$-th phase transformation system can be estimated by using the model of phase transformation system explained in Section 9.15.3.1, when only the $m$-th system becomes active. If some systems become active simultaneously in the crystal grain component, some corrections are necessary in order to incorporate the interactions among the phase transformation systems (see details in Reference [1]).

### 9.15.3.3 Model of SMA Polycrystal

When the stress-strain-temperature relation is derived on the basis of the stress-strain-temperature relation of its crystal grain component explained in Section 9.15.3.2, the interactions among grain components have to be incorporated. That is, each grain component has its own stress and strain, depending on its own orientation in the polycrystal, because each grain has the anisotropy related to the phase transformation system. Thus the complicated interaction among grains appears to satisfy the compatibility condition by strain and the equilibrium condition of stress in the polycrystal. Thus the nonuniform stress and strain distributions appear even if the applied stress is quite simple, for example, the uniaxial tension of a solid bar. The effects of the nonuniform distribution of stress and strain on the mechanical properties of polycrystalline materials are very important from the viewpoint of path dependency, especially when the strain or stress path is complex. This kind of interaction among grains can be taken into account by using one of the well-developed mechanical models of inhomogeneous solids: the self-consistent model. According to the self-consistent model, the following equation can be obtained:

$$s_{ij}^{(k)} - S_{ij} = \alpha G(e_{ij}^{PT(k)} - E_{ij}^{PT}) \tag{3}$$

where $s_{ij}^{(k)}$ is the deviatoric stress component of the $k$-th grain component embedded in the polycrystal, $S_{ij}$ is the averaged (macroscopic) deviatoric stress, $e_{ij}^{PT(k)}$ is the (deviatoric) phase transformation strain components of the $k$-th grain, $E_{ij}^{PT}$ is the averaged (macroscopic) phase transformation strain, and $G$ is the averaged shear modulus of the polycrystal. The coefficient $\alpha$ is still under discussion, and several values of $\alpha$ are proposed, for example,

$\alpha = 0$ (for the stress constant model: the so-called Maxwell model)
$\alpha = 0.2$ (for the modified Kröner–Budiansky–Wu model by Berveiller–Zaoui)
$\alpha = 1.0$ (for the KBW original model)
$\alpha = 2.0$ (for the strain constant model: the so-called Voigt model)
$\alpha = \infty$ (for the inelastic strain constant model: the Taylor model)

## 9.15.4 CONSTITUTIVE EQUATIONS

The constitutive equations based on the modeling mentioned in Section 9.15.3 not generally be expressed in a compact closed form but in the flow chart for computing because some iterative (convergence) computation is necessary. In a special case, for example, the constitutive equations based on the stress constant model can be expressed in a closed form when the strain path can be given. It should be noted that the proposed constitutive equations can be described symbolically in the form of an internal variable theory whose internal variables are $\gamma_{(m)}^{PT(k)}$, in the following manner:

$$\begin{aligned} dE_{ij} = & \sum_{k,l=1}^{3} F_{ijkl}(S_{pq}, \Sigma, T : \gamma_{(m)}^{PT(n)}) dS_{kl} \\ & + F_{ij}\, (S_{ij}, \Sigma, T : \gamma_{(m)}^{PT(n)}) d\Sigma \\ & + F_{ij}\, (S_{pq}, \Sigma, T : \gamma_{(m)}^{PT(k)}) dT \end{aligned} \tag{4}$$

$$\begin{aligned} d\gamma_{(m)}^{PT(k)} = & \sum_{p,q=1}^{3} H_{(m)}^{(k)} pq(S_{ij}, \Sigma, T : \gamma_{(m)}^{PT(k)}) dS_{pq} \\ & + H_{(m)}^{(k)}\ (S_{pq}, T : \gamma_{(m)}^{PT(k)}) d\Sigma \\ & + H_{(m)}^{T}(k)\ (S_{ij}, \Sigma, T : \gamma_{(m)}^{PT(k)}) dT \end{aligned} \tag{5}$$

where $\gamma_{(m)}^{PT(k)}$ $(m = 1, 2,..., M,\ k = 1, 2,..., N)$ is a shear strain of the $m$-th transformation system in the $k$-th grain component, $N$ is the number of grain components of the polycrystal model, and $\Sigma$ is the volumetric part of the macroscopic (averaged) stress of the polycrystal model.

## 9.15.5 IDENTIFICATION OF MATERIAL PARAMETERS

The following material parameters have to be determined experimentally. All of them can be determined by two simple uniaxial tension (or torsion) tests with different temperatures.

$G$: shear modulus
$K$: bulk modulus
$H$: shear strain hardening parameter of critical shear stress
$\beta$: temperature hardening parameter of critical shear stress
$\tau_{OMS}$: shear stress for the martensite phase transformation starting at a reference temperature $T_0$
$\tau_{OAF}$: shear stress of reverse (austenite) phase transformation finishing at a reference temperature $T_0$
$\gamma_{MAX}^{PT}$: maximum shear strain of a phase transformation system.

The number M of phase transformation systems in a single crystal grain component and the generalized Schmid factor $\alpha_{ij}$ are determined by the crystallographic structure of selected materials (can be found in the handbook of metals). The number $N$ of crystal grain components in the polycrystal model is determined from a compromise of accuracy and computational time. From the author's experience, $N \doteqdot 100$ is recommended.

## 9.15.6 HOW TO USE THE MODEL

The proposed constitutive equations were first formulated in order to predict the unique behavior of a shape memory alloy under complex loading conditions which are difficult to conduct experimentally. To this purpose, the proposed constitutive equations are quite successful and give us quite interesting information which cannot be obtained experimentally. On the other hand, the proposed constitutive equations may not yet be convenient for the stress-strain analysis of a shape memory structure by the finite element method or other numerical analyses. In this analysis, we need a much faster computer or some more simplification of constitutive equation which does not need the iterative computation.

## 9.15.7 PARAMETERS

The proposed constitutive equations have been used only for the Cu-based shape memory alloy, whose chemical composition is Cu-10 wt%Al5 wt%Mn5 wt%Zn. The values of the material parameters for this material are as follows:

$$G = 16.62 pt \text{GPa}$$
$$H = 5.0 \times 10^2 \text{ MPa}$$
$$\beta = 5.0 \times 10 \text{ MPa/K}$$
$$\tau_{OMS} = 90.0 \text{ MPa} (T_0 = 25^\circ\text{C})$$
$$\tau_{OAF} = 70.0 \text{ MPa}, (T_0 = 25^\circ\text{C})$$
$$\gamma_{MAX}^{PT} / \sqrt{3} = 2.5\%$$

The sets of parameters for other materials are at present unknown.

## REFERENCES


1. Tokuda, M, Ye, M, Bundara, B, and Sittner, P. (1999) 3D constitutive equations of polycrystalline shape memory alloy. *Archive of Mechanics* 51(6): 847–864.
2. Tokuda, M, Ye, M, Takakjura, M, and Sittner, P. (1998). Thermo-mechanical behavior of shape memory alloy under complex loading conditions. *International Journal of Plasticity* 15(2): 223–239.

SECTION **9.16**

# Elasticity Coupled with Magnetism

RENÉ BILLARDON, LAURENT HIRSINGER and FLORENCE OSSART

*ENS de Cachan/CNRS/Université Paris 6, 61 avenue du Président Wilson, 94235 Cachan Cedex, France*

Contents





*Handbook of Materials Behavior Models*. ISBN 0-12-443341-3.

## 9.16.1 GENERAL VALIDITY

This chapter is devoted to different scalar magnetic hysteresis models that are valid for isotropic soft ferri- and ferromagnetic materials subjected to uniaxial magnetic excitations and uniaxial elastic stresses.

## 9.16.2 HEREDITARY MODEL: PREISACH MODEL

### 9.16.2.1 Background

The Preisach model [1] consists of a collection of bistable elementary hysteresis operators $\gamma_{\alpha\beta}(H)$, whose switching fields are denoted $\alpha$ and $\beta$ (Fig. 9.16.1). This collection is described by the so-called Preisach distribution $\mu(\alpha, \beta)$, which gives the weight of the operator $\gamma_{\alpha\beta}$ and is usually represented by its isovalues in the Preisach plane $(\alpha, \beta)$ (Fig. 9.16.2). The magnetic behavior of the material is modeled by the following formula:

$$M(H) = M_s \int\int_{\alpha \geq \beta} \mu(\alpha, \beta)\gamma_{\alpha\beta}(H)d\alpha d\beta \tag{1}$$

The model has two important properties [2]:

- The extrema of the applied field $H$ are detected and stored until they are erased by an extremum with a larger magnitude.
- Minor loops are closed and congruent (no reptation).

### 9.16.2.2 Description of the Model

The Preisach model is currently recognized as a static scalar hysteresis model with well-defined properties and the more accurate prediction of minor loops. Numerous ideas have been proposed in order to extend the initial scalar model, either by assuming that the distribution $\mu(\alpha, \beta)$ depends on some state

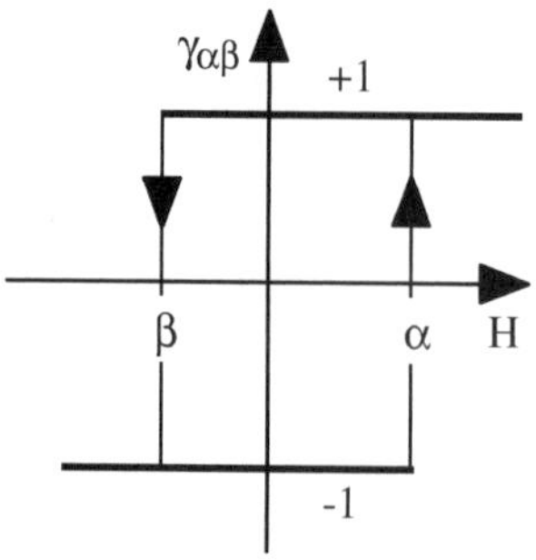


FIGURE 9.16.1 Elementary hysteresis operator definition. $\alpha \geq \beta$ so that hysteresis always causes dissipation. The magnetization of the material $M(H)$ saturates for $H \geq H_{sat}$. Hence, there is no operator such that $\alpha > H_{sat}$ or $\beta < H_{sat}$.

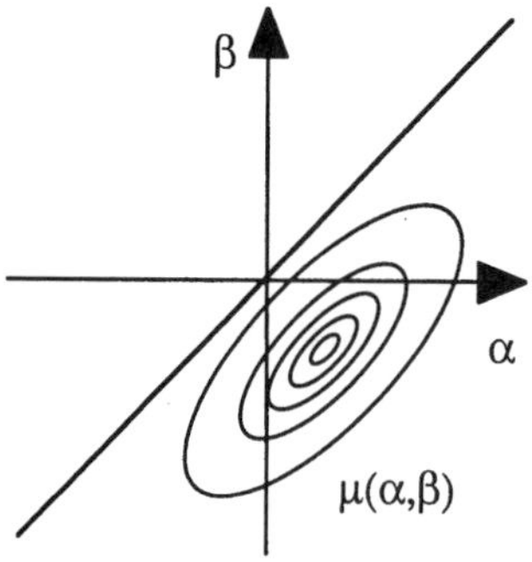


FIGURE 9.16.2 Isovalues of the Preisach distribution. $H_c$ being the coercive field, the distribution has a sharp maximum at $(H_c, -H_c)$ and vanishes for $\alpha > H_{sat}$ or $\beta < -H_{sat}$.

variables of the system, or by changing the behavior of the elementary operator $\gamma_{\alpha\beta}$. Some examples are as follows:

- Moving model [5]: the distribution $\mu(\alpha, \beta)$ depends on the current magnetization $M$, which improves the description of minor loops.
- Dynamic model [6]: $\gamma_{\alpha\beta}$ switchings are damped, just as the actual motions of magnetic domain walls are "damped" by eddy currents.
- Mayergoyz vector model [3]: the classical Preisach model is distributed over the space.
- Charap vector model [7]: $\gamma_{\alpha\beta}$ is an anisotropic vector operator, including rotation of the magnetization $M$.

### 9.16.2.3 Description of the Coupling with Elasticity

The mechanical state and in particular the stress state of the material have a great influence on its magnetic properties. Two different approaches have been proposed to build a scalar Preisach-type hysteresis model that accounts for the influence of a uniaxial stress:

- Identification of the Preisach distribution as a function of stress $\sigma$ so that $\mu(\alpha, \beta, \sigma)$ [8]: results obtained for standard Fe-3%Si laminations do not exhibit a simple evolution law.
- Definition of the elementary operator as a function of stress so that $\gamma_{\alpha\beta}(H, \sigma)$ [9]: experimental identification has to be done from a set of $M(H, \sigma)$ plots.

The second approach can be used to build an anisotropic vector model based on the definition of vector operators $\gamma_{\alpha\beta}(H, \sigma)$ [10]. Since the magnetomechanical couplings are fundamentally anisotropic and different in tension and compression, using a vector elementary operator improves the scalar description of hysteresis even along a fixed direction.

The classical Preisach model is easy to implement and use. All the needed details are found in the literature. However, none of its generalizations preserves the simplicity and well-defined properties of the initial model. Using them requires some rules of thumb and know-how.

### 9.16.2.4 Identification of the Parameters

The identification of the model consists of linking the distribution $\mu(\alpha, \beta)$ to experimental data. Two approaches may be used:

- The distribution $\mu(\alpha, \beta)$ is assumed to be described by an analytical function whose parameters are derived from the minimization of some global criterion (surface or global shape of the major loop) [4].
- The Preisach plane is discretized and the distribution is calculated at each point by using a set of minor loops [3]. The higher the number of loops, the higher the accuracy of the model.

## 9.16.3 INTERNAL VARIABLE MODEL: STATE COUPLING BETWEEN ELASTICITY AND MAGNETISM

### 9.16.3.1 Background

Isothermal situations are considered. The specific enthalpy $\Psi$ is partitioned into four terms [14]. $\Psi_m$ and $\Psi_{\mu m}$, respectively, correspond to the mechanical state of the material and to the magnetoelastic couplings. $\Psi_{\mu an}$ and $\Psi_{\mu hys}$, respectively, correspond to the magnetic state of a comparison ideal material (with the same behavior as the anhysteretic response of the real material) and to the hysteretic part of the magnetic behavior of the real material. Hence,

$$\begin{aligned}\rho\Psi(\mathbf{H}, \mathbf{X}, \sigma) &= \rho\Psi_m(\sigma) + \rho\Psi_{\mu m}(\sigma, \mathbf{H}; \mathbf{X}) + \rho\Psi_{\mu an}(\mathbf{H}) + \rho\Psi_{\mu hys}(\mathbf{H}; \mathbf{X}) \\ &= -\frac{1}{2}\sigma : \mathbb{E}^{-1} : \sigma - \int_0^\sigma \varepsilon^\mu(\sigma, \mathbf{H}; \mathbf{X}) : d\sigma - \int_0^{\mathbf{H}} \mu_0 \mathbf{M}_{an0}(\mathbf{h}) \cdot d\mathbf{h} \\ &\quad + \rho\Psi_{\mu hys}(\mathbf{H}; \mathbf{X}) \end{aligned} \tag{2}$$

where $\mathbb{E}$, $\varepsilon^\mu$, $\mathbf{M}_{an0}$, respectively, correspond to the elastic moduli tensor, the magnetostriction strain tensor, the anhysteretic magnetic response. $\mathbf{H}$, $\sigma$ and $\mathbf{X}$, respectively, correspond to the magnetic field, stress tensor and the internal variable(s) representative of magnetic hysteresis. Subscript 0 refers to zero stress state.

State laws are derived from the specific enthalpy $\Psi$ used as state potential:

$$\varepsilon = -\frac{\partial\rho\Psi}{\partial\sigma} = \mathbb{E}^{-1} : \sigma + \varepsilon^\mu(\sigma, \mathbf{H}, \mathbf{X}) \tag{3}$$

$$\mu_0 \mathbf{M}_1 = -\frac{\partial\rho\Psi}{\partial\mathbf{H}} = \mu_0 \mathbf{M}_{an0}(\mathbf{H}) - \frac{\partial\rho\Psi_{\mu hys}}{\partial\mathbf{H}} + \int_0^\sigma \frac{\partial\varepsilon^\mu}{\partial\mathbf{H}}(\sigma, \mathbf{H}; \mathbf{X}) : d\sigma \tag{4}$$

$$\mu_0 \mathbf{Y} = -\frac{\partial\rho\Psi}{\partial\mathbf{X}} = -\frac{\partial\rho\Psi_{\mu hys}}{\partial\mathbf{X}} + \int_0^\sigma \frac{\partial\varepsilon^\mu}{\partial\mathbf{X}}(\sigma, \mathbf{H}; \mathbf{X}) : d\sigma \tag{5}$$

and Clausius-Duhem inequality becomes:

$$D_i = -\rho\dot{\Psi} - \varepsilon : \dot{\sigma} - \mu_0 \mathbf{M} \cdot \dot{\mathbf{H}} = \mu_0(\mathbf{M}_1 - \mathbf{M}) \cdot \dot{\mathbf{H}} + \mu_0 \mathbf{Y} \cdot \dot{\mathbf{X}} \geq 0 \tag{6}$$

This condition, which must be satisfied for any physically admissible evolution, gives an important limitation to the possible expressions of the evolution laws for internal variables $\mathbf{X}$.

## 9.16.3.2 Validity of the State Coupling

This state coupling has been experimentaly validated for pure nickel [13] and silicon iron alloy [14]. In the case of anhysteretic magnetoelastic evolutions ($\dot{\mathbf{X}} = 0$), the Clausius–Duhem inequality (Eq. 6) simplifies and variable $\mathbf{M}_1$ appears as magnetization $\mathbf{M} = \mathbf{M}_{an}(\sigma, \mathbf{H})$. Hence, the state law (Eq. 4) gives the expression of the anhysteretic magnetic response of the material when it is subjected to elastic loadings:

$$\mu_0 \mathbf{M}_{an}(\sigma, \mathbf{H}) = \mu_0 \mathbf{M}_{an0}(\mathbf{H}) + \int_0^{\sigma} \frac{\partial \varepsilon_{an}^{\mu}}{\partial \mathbf{H}}(\sigma; \mathbf{H}) : d\sigma \tag{7}$$

where $\varepsilon_{an}^{\mu}(\sigma, \mathbf{H}) = \varepsilon^{\mu}(\sigma, \mathbf{H}, \mathbf{X} = 0)$ denotes the anhysteretic magnetostriction strain of the material when it is subjected to elastic loadings. Derivation of this relation with respect to stress $\sigma$ leads to the so-called Maxwell expression:

$$\mu_0 \frac{\partial \mathbf{M}_{an}}{\partial \sigma}(\sigma; \mathbf{H}) = \frac{\partial \varepsilon_{an}^{\mu}}{\partial \mathbf{H}}(\sigma; \mathbf{H}) \quad \forall (\sigma, \mathbf{H}) \tag{8}$$

## 9.16.3.3 Possible Choices for Internal Variable X

Further assumptions must be made to complete the magnetoelastic model:

- Internal variable $\mathbf{X}$ must be defined. In the case of a purely phenomenological approach, this definition may be directly deduced from the modeling of the magnetization response $\mathbf{M}(\sigma, \mathbf{H}, \mathbf{X})$. Different assumptions can be made. For instance, one may introduce
  - a partition of magnetization $\mathbf{M}$ so that:

$$\mathbf{M} = \mathbf{M}_{an}(\sigma, \mathbf{H}) + \mathbf{M}_{hys} \quad \text{and} \quad \mathbf{X} = \mathbf{M}_{hys} \tag{9}$$

  - or a partition of magnetic field H so that:

$$\mathbf{M} = \mathbf{M}_{an}(\sigma, \mathbf{H} - \mathbf{H}_{hys}) \quad \text{and} \quad \mathbf{X} = \mathbf{H}_{hys} \tag{10}$$

These variables are depicted in Figure 9.16.3 in the uniaxial case.

- A kinetic law for internal variable $\mathbf{X}$ that satisfies the Clausius–Duhem inequality (Eq. 6) for any magnetomechanical loading must be introduced:

$$\dot{\mathbf{X}} = \dot{\mathbf{X}}(\sigma, \mathbf{H}, \dot{\mathbf{H}}, \mathbf{X}) \tag{11}$$

- A kinetic law for magnetostriction strains $\varepsilon^{\mu}(\sigma, \mathbf{H}, \mathbf{X})$ must also be introduced.

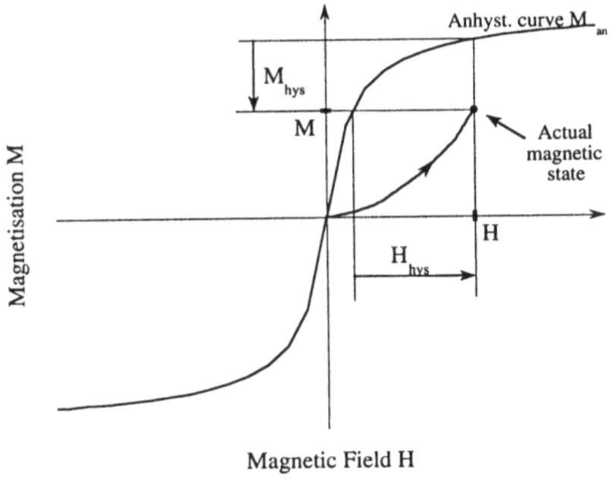


FIGURE 9.16.3 Definition of variables $M_{hys}$ and $H_{hys}$ in the uniaxial case.

# 9.16.4 MAGNETOELASTIC SCALAR REVERSIBLE BEHAVIOR: ANHYSTERETIC BEHAVIOR $X = 0$

## 9.16.4.1 Description of Models for the Anhysteretic Magnetization

The anhysteretic magnetic response of a material free of any mechanical loading has been modeled by Frohlich [19] as

$$M_{an0}(H) = M_s \frac{\chi_{a0} H}{M_s + \chi_{a0} H} \tag{12}$$

where $M_{an0}$ and $H$, respectively, denote $\mathbf{M}_{an0}$ and $\mathbf{H}$ moduli and where material parameters $\chi_{a0}$ and $M_s$ are, respectively, related to the initial slope of the anhysteretic curve and to the saturation magnetization.

By analogy with the behavior of paramagnetic materials, Jiles and Atherton [25] used the Langevin function to model the magnetic anhysteretic curve, such that:

$$M_{an0}(H) = M_s \left( \coth\left(\frac{H_e}{a}\right) - \frac{a}{H_e} \right) \quad \text{with } H_e = H + \alpha M \tag{13}$$

where $H_e$ denotes the effective magnetic field as introduced by Weiss to account for ferromagnetism ($\alpha$ is the mean molecular field constant representing interactions of magnetic moments), whereas $a$ denotes a material parameter that depends on temperature and that controls the anhysteretic curve shape.

From phenomenological considerations [18], the anhysteretic curve has also been modeled by the following relation:

$$M_{an0}(H) = \frac{2M_s}{\pi} \text{Arctan}\left(\frac{\pi}{2}\chi_{a0}\frac{H}{M_s}\right) \tag{14}$$

This latter expression gives a better fit of the experimental response of Fe-3%Si alloys than Eq. 13 within the range of $\pm 3000$ A/m.

### 9.16.4.2 Description of Models of Anhysteretic Magnetostriction Strain

For isotropic materials, it is reasonable to assume that the anhysteretic magnetostriction strain tensor takes the following form [22]:

$$\varepsilon^{\mu}_{an}(\mathbf{H}) = \begin{pmatrix} \varepsilon^{\mu}_{an\|}(\mathbf{H}) & 0 & 0 \\ & \varepsilon^{\mu}_{an\perp}(\mathbf{H}) & 0 \\ sym. & & \varepsilon^{\mu}_{an\perp}(\mathbf{H}) \end{pmatrix}_{(\boldsymbol{u}_{\|},\boldsymbol{v}_{\perp},\boldsymbol{w}_{\perp})} \tag{15}$$

where $\boldsymbol{u}_{\|}$, $\boldsymbol{v}_{\perp}$, and $\boldsymbol{w}_{\perp}$, respectively, denote unit vectors that define a direct frame with direction $\boldsymbol{u}_{\|}$ parallel to magnetization vector $\mathbf{M}$. Besides, $\varepsilon^{\mu}_{an\|}$ and $\varepsilon^{\mu}_{an\perp}$, respectively, denote the magnetostriction strains in the direction of magnetization and in any direction transverse to the direction of magnetization. Since the magnetization process is at quasi-constant volume, the transverse magnetostriction is such that [19]:

$$\varepsilon^{\mu}_{an\|}(\mathbf{H}) = -\frac{1}{2}\varepsilon^{\mu}_{an\perp}(\mathbf{H}) = \frac{3}{2}\lambda_s\left(\frac{M_{an}(\mathbf{H})}{M_s}\right)^2 \tag{16}$$

where $\lambda_s$ denotes the saturated magnetostriction strain along the saturation direction. For a randomly oriented polycrystalline cubic material, an estimation of $\lambda_s$ is obtained by averaging the magnetostriction of grains and neglecting the strain incompatibility between neighboring grains [21,22], such that:

$$\lambda_s = \tfrac{2}{5}\lambda_{100} + \tfrac{3}{5}\lambda_{111} \tag{17}$$

where $\lambda_{100}$ and $\lambda_{111}$, respectively, denote the magnetostriction constant in [100] and [111] crystallographic directions of the cubic single crystal.

### 9.16.4.3 Description of a Scalar Model

When a uniaxial stress $\sigma$ is applied in the direction of the magnetic field, Jiles and Atherton [25] have proposed to modify the effective magnetic field in the anhysteretic curve (Eq. 13) with

$$H_e = H + \alpha M + \frac{\sigma}{\mu_0}\frac{d\,\varepsilon^{\mu}_{an\|\sigma}}{dM} \tag{18}$$

where $\varepsilon^{\mu}_{an\|\sigma}$ denotes the magnetostriction strain in the direction of the applied uniaxial stress. According to Eqs. 15 and 16, its expression is

$$\varepsilon^{\mu}_{an\|\sigma} = \frac{9}{4}\lambda_s\left(\cos^2\theta - \frac{1}{3}\right)\left(\frac{M}{M_s}\right)^2 \tag{19}$$

where $\theta$ denotes the angle between the direction of the applied stress $\sigma$ and the direction of the applied magnetic field **H**.

Eventually, the anhysteretic behavior of a material subjected to a uniaxial stress is predicted by the following expression of the anhysteretic curve $M_{an}$:

$$M = M_{an}(H, M, \sigma) = M_s\left(\coth\left(\frac{H_e}{a}\right) - \frac{a}{H_e}\right) \tag{20}$$

with

$$H_e,\ M, \sigma = H + \alpha M + \frac{27}{4}\frac{\sigma}{\mu_0}\lambda_s\left(\cos^2\theta - \frac{1}{3}\right)\frac{M}{M_s^2} \tag{21}$$

On the other hand, it appears that the inverse of the initial slope of the anhysteretic curve depends linearly on the applied uniaxial stress [18,20]. For sake of simplicity, the magnetic behavior of the material is assumed to remain isotropic. Finally, anhysteretic curve $M_{an}$ is modeled by the following expression:

$$M = M_{an}(H, \sigma) = \frac{2M_s}{\pi}\,\mathrm{Arctan}\left(\frac{\pi}{2}\chi_a(\sigma)\frac{H}{M_s}\right) \tag{22}$$

with

$$\chi_a^{-1}(\sigma) = \chi_{a0}^{-1} + A\cdot\sigma \tag{23}$$

where $\sigma$ denotes the value of the uniaxial stress applied in the direction of the magnetic field. In other words, the magnetoelastic state coupling (derivative of Eq. 8 with respect to H) is approximately a constant, denoted by $-A\chi_a^2$, in the vicinity of zero magnetization:

$$\frac{\partial^2 M_{an}}{\partial H\,\partial\sigma}(\sigma, H=0) = \frac{1}{\mu_0}\frac{\partial^2\varepsilon^{\mu}_{an\|}}{\partial^2 H}(\sigma, H=0) = -A\chi_a^2(\sigma) \tag{24}$$

#### 9.16.4.4 Identification of the Parameters

The model parameters, four for the former ($M_s$, $a$, $\alpha$, $\lambda_s$) and three for the latter ($M_s$, $\chi_{a0}$, $A$), are identified from the anhysteretic curve measured on samples subjected to constant uniaxial stress. See Table 9.16.1.

### 9.16.5 MAGNETOELASTIC HYSTERETIC BEHAVIOR: MODEL PROPOSED BY JILES AND ATHERTON ($X = M_{\text{hys}}$)

#### 9.16.5.1 Validity

The model proposed by Jiles and Atherton [25] is a scalar magnetic hysteresis model. It is coupled with elasticity; i.e., it accounts for the effect of uniaxial stresses on magnetisation of bulk isotropic materials. Whereas good predictions can be obtained for major loops, minor loops are approximated: the monotonic condition $dM/dH$, always positive, is not always fulfilled just after reversal of the applied magnetic field.

**Table 9.16.1 Table of Parameters**

| Materials | $M_s$ ($10^3$ A/m) | $\chi_{a0}$ | $a$ (A/m) | $\alpha$ ($10^{-3}$) | $\lambda_{100}$ ($10^{-6}$) | $\lambda_{111}$ ($10^{-6}$) | $\lambda_s$ ($10^{-6}$) | A ($10^{-6}$ MPa$^{-1}$) |
|---|---|---|---|---|---|---|---|---|
| N.O. Silicon Iron Alloy (M330–50A) | 1190 | 39300 | — | — | 23 | −4.5 | 6.5 | 0.925 |
| N.O. Silicon Iron Alloy (M450–50E) | 1115–1700 | 31000–41000 | 43–426 | 0.091–0.72 | | | | |
| N.O. Silicon Iron Alloy (M600–50A) | 1230–1700 | 34000–70000 | 80–538 | 0.178–0.92 | | | | |
| Low Carbon Steel | 1397–1600 | 2160–10600 | 342–1100 | 0.64–1.6 | 21 | −21 | −4.2 | |
| Carbon Steel (AISI 4130) | 1350 | 375 | 2100 | 2 | 21 | −21 | −4.2 | −5.73 |
| Nickel | 480 | | | | −46 | −24 | −32.8 | |
| Ferrite Core | 380 | 420 | 30 | $10^{-3}$ | | | | |
| Terfenol D $Tb_{0.3}Dy_{0.7}Fe_{1.9}$ | 765–790 | 11 | 7000 | 32–35 | 90 | 1600 | 1000–1200 | 6100 |

### 9.16.5.2 Description of the Evolution Law

This model is based on the following principal hypotheses:

- Even for real materials with defects (which are the source of hysteresis), the anhysteretic curve exists and corresponds to the magnetic behavior of a comparison ideal material free of any defect: the relationship between anhysteretic magnetization $M_{an}$ and magnetic field H is reversible nonlinear, such that:

$$M_{an} = M_{an}(H_e) = M_s\left(\coth\left(\frac{H_e}{a}\right) - \frac{a}{H_e}\right) \quad (25)$$

where $H_e$ and $a$, respectively, denote the effective magnetic field and a material parameter.

- From the molecular field theory proposed by Weiss, the expression of the effective magnetic field $H_e$ is the following (Eq. 18):

$$H_e = H + \alpha M + \frac{\sigma}{\mu_0}\frac{d\varepsilon^{\mu}_{an\|\sigma}}{dM} \quad (26)$$

- To account for reversible magnetization due to reversible magnetic domain wall bowing and reversible rotation, magnetization $M$ is partitioned into reversible, $M_{rev}$, and irreversible, $M_{irr}$, parts such that:

$$M = M_{rev} + M_{irr} \quad (27)$$

- The reversible magnetization component $M_{rev}$ is supposed to take the form:

$$M_{rev} = c(M_{an} - M_{irr}) \quad (28)$$

where $c$ denotes a material parameter.

- Magnetic behavior irreversibility due to pinning–unpinning of magnetic domain walls during the magnetization process is associated with supplementary energy, noted $E_{loss}$, needed to overcome these pinning sites:

$$E_{loss} = \int_0^M k\, dM_{irr} \quad (29)$$

where $k$ denotes a material parameter characterizing the average energy required to break pinning sites. The energy supplied to material $dE$ is either stored in the material in magnetostatic energy form or dissipated by hysteresis. In the case of anhysteretic behavior of the material, this supplied energy is completely stored in the material, such that:

$$dE = M(H)dH - dE_{loss} = M_{an}(H)dH \quad (30)$$

and consequently

$$\frac{dM_{irr}}{dH} = \frac{M_{an} - M_{irr}}{k} \quad (31)$$

According to Eqs. 27 and 28 and to the effective magnetic field equations (Eqs. 19 and 26), the evolution law (Eq. 31) of the internal variable $M_{irr}$, which describes magnetic irreversibility, becomes:

$$\frac{dM_{irr}}{dH} = \frac{M_{an} - M_{irr}}{\delta k - \alpha^* (M_{an} - M_{irr})} \tag{32}$$

with

$$\alpha^* = \alpha + \frac{27}{4} \frac{\sigma}{\mu_0} \frac{\lambda_s}{M_s^2} \left( \cos^2\theta - \frac{1}{3} \right) \tag{33}$$

where $\delta$ denotes the sign function of the rate of magnetic field $\dot{\mathrm{H}} = dH/dt$, that is to say:

$$\delta = 1 \text{ if } \dot{H} > 0 \quad \text{and} \quad \delta = -1 \text{ if } \dot{H} < 0 \tag{34}$$

It can be noticed that, in this model, Eq. 9 takes the following form

$$M_{hys} = (1 - c)(M_{irr(H,\sigma)} - M_{an(H,\sigma)})$$

### 9.16.5.3 Identification of the Parameters

This magnetic hysteresis model coupled with elasticity proposed by Jiles and Atherton requires one to identify six parameters: saturated magnetization $M_s$ and magnetostriction $\lambda_s$, magnetic moments interaction parameter $\alpha$, shape parameter for anhysteretic curve a, average lost energy parameter $k$, and reversibility coefficient $c$ (see Table 9.16.2).

Parameters $M_s$, $\lambda_s$, and $\alpha$ are identified by accurate physical measurements, and their values are given in the literature. From the viewpoint of applications, some parameters can be directly identified from the hysteresis loop. By derivation of Eq. 31 with respect to $H$, parameter $c$ appears as the ratio between initial susceptibilities (at origin) measured on the first magnetization curve $\chi_{in}$ and on the anhysteretic curve $\chi_{an0}$ [26]:

$$c = \frac{\chi_{in}}{\chi_{an0}} \tag{35}$$

The parameter $k$ can also be defined as:

$$k = M_{an}(H_c) \left( \frac{\alpha}{1 - c} + \frac{1}{(1 - c)\chi_{Hc} - c \frac{dM_{an}}{dH}(H_c)} \right) \tag{36}$$

where $H_c$ and $\chi_{H_c}$, respectively, denote the coercivity force and the slope of the hysteresis loop at $H_c$. Lastly, the parameter $a$ can be identified as:

$$a = \frac{\alpha M_s}{3} \frac{k}{k - H_c} \tag{37}$$

Table 9.16.2 Table of Parameters

| Materials | $M_s$ ($10^3$ A/m) | $a$ (A/m) | $\alpha$ ($10^{-3}$) | $c$ | $k$ (A/m) | $\lambda_s$ ($10^{-6}$) |
|---|---|---|---|---|---|---|
| N.O. Silicon Iron Alloy (M330–50A) | 1190 | | | | | 6.5 |
| N.O. Silicon Iron Alloy (M450–50E) | 1115–1700 | 43–426 | 0.091–0.72 | 0.085–0.1 | 40–36 | |
| N.O. Silicon Iron Alloy (M600–50A) | 1230–1700 | 80–538 | 0.178–0.92 | $1.2\,10^{-3}$–0.04 | 114–104 | |
| Low Carbon Steel | 1397–1600 | 342–1100 | 0.64–1.6 | $1.1\,10^{-3}$–0.2 | 247–400 | −4.2 |
| Carbon Steel (AISI 4130) | 1350 | 2100 | 2 | 0.25 | 1800 | 35 |
| Nickel | 480 | | | | | −32.8 |
| Ferrite Core | 380 | 30 | $10^{-3}$ | 0.95 | 10 | |
| Terfenol D $Tb_{0.3}Dy_{0.7}Fe_{1.9}$ | 765–790 | 7000 | 32–35 | 90 | 3280–3950 | 1000–1200 |

## 9.16.6 MAGNETOELASTIC HYSTERETIC BEHAVIOR: MODEL PROPOSED BY LMT-CACHAN ($X = H_{\text{hys}}$)

### 9.16.6.1 Validity

The model proposed by LMT-Cachan is a scalar magnetic hysteresis model. It is coupled with elasticity; i.e., it accounts for the effect of uniaxial stresses (applied in the same direction as the applied magnetic field) on magnetization of bulk isotropic materials.

### 9.16.6.2 Description of the Model

This model is based on the following principal hypotheses:

- Magnetic field $H$ is partitioned into reversible anhysteretic and irreversible parts denoted by $H_{an}$ and $H_{hys}$, respectively, so that:

$$H = H_{an} + H_{hys} \qquad \forall M \tag{38}$$

• The anhysteretic curve exists for a real material and the relation between magnetization $M$ and anhysteretic magnetic field $H_{an}$ is reversible nonlinear, as introduced in Section 9.16.4 by Eqs. 23 and 24:

$$M = M_{an}(H_{an}, \sigma) = \frac{2M_s}{\pi} \text{Arctan}\left(\frac{\pi}{2} \chi a(\sigma) \frac{H}{M_s}\right) \tag{39}$$

with

$$\chi_a^{-1}(\sigma) = \chi_{a0}^{-1} + A \cdot \sigma \tag{40}$$

where $\sigma$ denotes the value of the uniaxial stress applied in the direction of the magnetic field.

• For sake of simplicity and as a first approximation, it is assumed that a uniaxial stress mainly affects the anhysteretic curve and has no significant influence on the hysteretic behavior. The evolution law of internal variable $H_{hys}$ is chosen such that, on one hand, Eq. 6 is satisfied. From phenomenological considerations it has been proposed that:

$$\dot{H}_{hys} = \chi_0 \dot{H} \text{ if } \dot{H} > 0 \quad \text{and} \quad H_{hys} \leq H_y \quad \text{or} \quad \text{if } \dot{H} < 0 \quad \text{and} \quad H_{hys} \geq -H_y \tag{41}$$

$$\dot{H}_{hys} = \frac{H_c - sign(H) H_{hys}}{H_c - H_y} \chi_0 \dot{\mathrm{H}} \quad \text{if } \dot{\mathrm{H}} > 0 \text{ and } H_{hys} > H_y \quad \text{or if } \dot{\mathrm{H}} < 0 \text{ and } H_{hys} < -H_y \tag{42}$$

where constants $\chi_0$, $H_c$, and $H_y$ denote material parameters, respectively, related to the initial slope of the first magnetization curve, the coercive magnetic field, and the so-called yield magnetic field.

### 9.16.6.3 Identification of the Parameters

This magnetic hysteresis model coupled with elasticity proposed by LMT-Cachan requires the identification of seven parameters: saturated magnetization $M_s$ and magnetostriction $\lambda_s$, initial slopes of the anhysteretic curve ($M$ vs. $H_{an}$), $\chi_{a0}$ and of the first magnetisation curve ($M$ vs. $H$), $\chi_0$, coercive force $H_c$, yield magnetic field $H_y$, and stress dependence coefficient of the anhysteretic curve initial slope $A$ (see Table 9.16.3). To identify these parameters, major hysteresis loops with first magnetization curve and anhysteretic response of stressed material are needed.

Table 9.16.3 Table of Parameters

| Materials | $M_s$ ($10^3$ A/m) | $\chi_{a0}$ | $\chi_0$ | $H_c$ (A/m) | $H_y$ (A/m) | $\lambda_s$ ($10^{-6}$) | A ($10^{-6}$MPa)$^{-1}$ |
|---|---|---|---|---|---|---|---|
| N.O. Silicon Iron Alloy (M330–50HA) | 1190 | 39300 | 0.85 | 60 | 10 | 6.5 | 0.925 |
| N.O. Silicon Iron Alloy (M450–50E) | 1115–1700 | 31000–41000 | | | | | |
| N.O. Silicon Iron Alloy (M600–50A) | 1230–1700 | 34000–70000 | | | | | |
| Low Carbon Steel | 1397 | 2160–10600 | | 10–100 | | −4.2 | |
| Carbon Steel (AISI 4130) | 1350 | 375 | | 10–100 | | −4.2 | −5.73 |
| Nickel | 480 | | | | | | |
| Ferrite Core | 380 | 420 | | 10 | | | |
| Terfenol D $Tb_{0.3}Dy_{0.7}Fe_{1.9}$ | 765–790 | 11 | | 3000 | | 1000–1200 | 6100 |

## 9.16.6.4 How to Use the Model

The constitutive equations (Eqs. 41 and 42) could be solved iteratively by applying a $\theta$-method and a pure Newton algorithm [29].

## REFERENCES


1. Preisach, F. (1935). Uber die magnetische nachwirking. *Z. Phys.* **94**: 277–302.
2. Bertotti, G. (1998). *Hysteresis in Magnetism.* Academic Press.
3. Mayergoyz, I. D. (1991) *Mathematical Models of Hysteresis.* New York, Springer-Verlag.
4. Bertotti, G., Fiorillo, F., and Soardo, G. P. (1987). Dependence of power losses on peak magnetization and magnetization frequency in grain-oriented and non-oriented 3% SiFe. *IEEE Trans. Magn.* **23**(5): 3520.
5. Torre, E. Della (1991). Existence of magnetization dependent Preisach models. *IEEE Trans. Magn.* **27**(4): 3697–3699.
6. Bertotti, G. (1992). Dynamic generalization of the scalar Preisach model of hysteresis. *IEEE Trans. Magn.* **28**(5): 2599.
7. Davidson, R., and Charap, S. H. (1996). Combined vector hysteresis models and applications. *IEEE Trans. Magn.* **32**(5): 4198–4203.
8. LoBue, M., Basso, V., Fiorillo, F., and Bertotti, G. (1999). Effect of tensile and compressive stress on dynamic loop shapes and power losses of Fe-Si electrical steels. *J. Magn. Magn. Mat.* **196–197**: 372–374.

9. Berqvist, A., and Engdhal, G. (1991). A stress-dependent magnetic Preisach hysteresis model. *IEEE Trans. Magn.* **27**(6): 4796–4798.
10. Sasso, C. P., Basso, V., LoBue, M., and Bertotti, G. (2000). Vector model for the study of hysteresis under stress. *J. Appl. Phys.* **87**: 4774–4776.
11. Barbier, G. (1995). Proposition d'un modèle de couplage magnéto-mécanique pour les matériaux ferromagnétiques doux, Rapport de DEA de Mécanique, Université Paris 6, LMT-Cachan.
12. Bassiouny, E., *et al.* (1998). Thermodynamical formulation for coupled electromechanical hysteresis effects. *International Journal of Engineering Science* **26** (12): 1279–1295.
13. Gourdin, C., *et al.* (1998). Experimental identification of the coupling between the anhysteretic magnetic and magnetostrictive behaviours. *Journal of Magnetism and Magnetic Materials* **177–181**: 201–202.
14. Hirsinger, L., *et al.* (2000). Application of the internal variable formalism to the modeling of magnetoelasticity, in *Studies Applied Electromagnetics and Mechanics, Vol. 29: Mechanics of Electromagnetic Materials and Structures*, pp. 54–67, Yang, J. S., and Maugin, G. A., eds., IOS Press.
15. Maugin, G. A. (1998). *Continuum Mechanics of Electromagnetic Solids*, North-Holland.
16. Maugin, G. A. (1991). Compatibility of magnetic hysteresis with thermodynamics. *Int. J. Appl. Electromag. Mat.* **2**: 7–19.
17. Maugin, G. A., and Muschik, W. (1994). Thermodynamics with internal variables. *J. Non-Equilib. Thermodyn.* **19**: 217–289.
18. Gourdin, C., *et al.* (1998). Finite element implementation of an isotropic internal variable magneto-elastic hysteresis model, in *Non-Linear Electromagnetic Systems*, pp. 625–628, Kose, V., and Sievert, J., eds., Amsterdam: IOS Press.
19. Jiles, D. C. (1991). *Introduction to Magnetism and Magnetic Materials*, Chapman & Hall.
20. Sablik, M. J. (1989). Modeling stress dependence of magnetic properties for NDE of steels. *Nondestr. Test. Eval.* **5**: 49–65.
21. du Tremolet de Lacheisserie, E. (1990). *Magnetostriction: Theory and Applications of Magnetoelasticity*, CRC Press.
22. du Tremolet de Lacheisserie, E. (1999). *Magnetisme, 1: Fondements*, Presses Universitaires de Grenoble.
23. Calkins, F. T., Smith, R. C., and Flateau, A. B. (2000). Energy-based hysteresis model for magnetostrictive transducers. *IEEE Trans. Magn.* **36**(2): 429–439.
24. Hernandez, E. Del Moral, Muranaka, C. S., and Cardoso, J. R. (2000). Identification of the Jiles–Altherton model parameters using random and deterministic searches. *Physica* B **275**: 212–215.
25. Jiles, D. C., and Atherton, D. L. (1986). Theory of ferromagnetic hysteresis. *Journal of Magnetism and Magnetic Materials* **61**: 48–60.
26. Jiles, D. C., Thoelke, J. B., and Devine, M. K. (1992). Numerical determination of hysteresis parameters for the modeling of magnetic properties using the theory of ferromagnetic hysteresis. *IEEE Trans. Magn.* **28**: 27–35.
27. Jiles, D. C. (1995). Theory of the magnetomechanical effect. *Journal of Physics D: Applied Physics* **28**: 1537–1546.
28. Sablik, M. J., and Jiles, D. C. Coupled magnetoelastic theory of magnetic and magnetostrictive hysteresis. *IEEE Trans. Magn.* **29**: 2113–2123.
29. Gourdin, C., Hirsinger, L., and Billardon, R. (1998). Finite element implementation of an internal variable magneto-elastic hysteresis model, in *Studies in Applied Electromagnetics and Mechanics, Vol. 13: Non-linear Electromagnetic Systems*, pp. 625–628, Kose, V., and Sievert, J., eds., IOS Press.

SECTION **9.17**

# Physical Aging and Glass Transition of Polymers

RACHID RAHOUADJ and CHRISTIAN CUNAT
*LEMTA, UMR CNRS 7563,ENSEM INPL 2, avenue de la Forêt-de-Haye, 54500 Vandoeuvre-lès-Nancy, France1*

Contents





*Handbook of Materials Behavior Models*. ISBN 0-12-443341-3.

## 9.17.1 VALIDITY

In the case of glassy materials near the glass transition, the aging transformation is usually associated with the recovery of a thermodynamic equilibrium state. It concerns various physical properties, such as the mechanical compliance, the refractive index, the apparent volume expansion, and the enthalpy related to the apparent specific heat. The last-mentioned is measured during continuous heat treatment.

Examples of experimental results of isothermal volume recovery and apparent specific heat of polymers are analyzed near the glass transition as functions of heat treatment and temperature, respectively. The present theoretical modeling is based on an irreversible thermodynamics approach called "the distribution of nonlinear relaxations" (DNLR).

## 9.17.2 BACKGROUND

The theoretical background [1, 2], has been briefly described in the article entitled "A Nonlinear Viscoelastic Model Based on Fluctuating Modes", Section 2.6. This modeling is in accordance with the fundamental works on internal variables by Coleman and Gurtin [3]. In the frame of a modal analysis of the dissipation, we have shown that the constitutive relationship can be written as

$$\dot{\beta}_m = \sum_{j=1}^{n} (\dot{\beta}_j)_m = \sum_{j=1}^{n} \left( p_j^0 a_{mq}^u \dot{\gamma}_q - \frac{(\beta_j)_m - p_j^0 \beta_m^r}{\tau_j(T, (\beta_m - \beta_m^r))} \right) \quad (1)$$

where $\gamma_q$ and $\beta_m$ refer to the perturbation and to the corresponding response components, respectively, and $j$ to a normal dissipation mode. The indexes $r$ and $u$ denote the relaxed (or equilibrium) and the unrelaxed states, respectively. The parameter $p_j^0$ represents the relative weight of the process $j, j = 1, \ldots, n$, in the overall relaxation spectrum. The parameter $a^u$ is the symmetrical matrix of Tisza. Referring to the theorem of equipartition of the entropy production, [2], the initial relaxation spectrum near equilibrium can be defined as

$$p_j^0 = \sqrt{\tau_j^r} / \sum_{k=1}^{n} \sqrt{\tau_k^r} \quad (2)$$

Note that this distribution of relaxation times, $\tau_j^r$, is also assumed to be operational for completely frozen states as for the beginning of the aging transformation.

The data for volume recovery are shown in Figure 9.17.1a–c. The representation in Figure 9.17.1c is preferred to others for comparing theoretical simulations and experiments by means of the effective relaxation time $\tau_{eff} = -\delta/\dot{\delta}$ and the volume variation $\delta$ due to nonequilibrium, as defined by Kovacs [4, 5].

Some nonlinearities come obviously from the statistical nature of the relaxation times distribution, whereas other nonlinearities may find their origin in the dependence of relaxation times with the departure from equilibrium states. This major effect can be taken into account by introducing a shift function, $a_\delta$, into the relaxation spectrum. Then, the relaxation times may be defined as inversely proportional to the jump frequency and to the probability of occurrence of an activated state (denoted by the symbol $+$). Thus one has

$$\Delta F_j^+ = \Delta F_j^{+,r}(T) + K_\delta \delta = (\Delta E^{+,r} - T\Delta S_j^{+,r}) + K_\delta \delta \qquad (3)$$

assuming that $\Delta E_j^{+,r} = \Delta E^{+,r}, \quad \forall j = 1, \ldots, n$.

From this relation, the relaxation times become

$$\tau_j = \frac{h}{kT} \exp(\Delta F_j^+/RT) = \tau_j^r \, a(t) \qquad (4)$$

where $\tau_j^r = \exp\left[(\Delta E^{+,r} - T\Delta S_j^{+,r})/RT\right]$ and $a(t) = a_\delta = \exp(K_\delta \delta/RT)$.

To simplify the present analysis, we will consider the simplest case where $\delta = \mathrm{trace}(\boldsymbol{\beta} - \boldsymbol{\beta}^r)$. Equation 1 can be rewritten more explicitly as

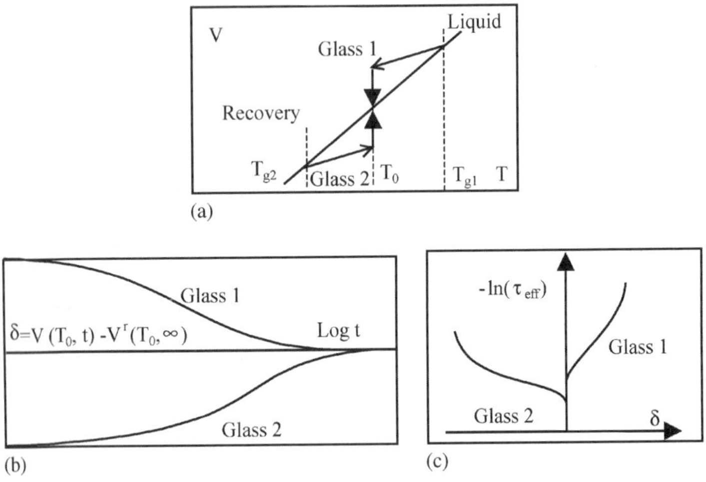


FIGURE 9.17.1 Asymmetry of the recovery response with respect to the sign of deviation from equilibrium. a: Thermal sequences; b. and c Two different representations of the same responses.

follows:

$$\begin{bmatrix} -\dot{S} \\ \dot{\boldsymbol{\varepsilon}} \end{bmatrix} = \begin{bmatrix} -CT^{-1} & \boldsymbol{\alpha} \\ \boldsymbol{\alpha} & J \end{bmatrix}^{u} \begin{bmatrix} \dot{T} \\ \dot{\boldsymbol{\sigma}} \end{bmatrix} - \begin{bmatrix} -(S_1 - S_1^r) \ldots -(S_j - S_j^r) \ldots -(S_n - S_n^r) \\ (\boldsymbol{\varepsilon}_1 - \boldsymbol{\varepsilon}_1^r) \ldots (\boldsymbol{\varepsilon}_j - \boldsymbol{\varepsilon}_j^r) \ldots (\boldsymbol{\varepsilon}_n - \boldsymbol{\varepsilon}_n^r) \end{bmatrix} \begin{bmatrix} 1/\tau_1 \\ \vdots \\ 1/\tau_j \\ \vdots \\ 1/\tau_n \end{bmatrix} \tag{5}$$

when the system is submitted only to a restrictive thermal or thermomechanical solicitation, $\dot{T}$ and $\dot{\sigma}$ being independent. The parameters C, $\boldsymbol{\alpha}$, and J represent the specific heat, the tensor of expansion coefficients, and the tensor of compliance, respectively.

## 9.17.3 DESCRIPTION OF THE MODEL

In this section, two specific applications of the DNLR are briefly presented, the volume recovery and the apparent specific heat. Further details are given elsewhere by Marceron and Cunat [6].

### 9.17.3.1 MODEL FOR SIMULATION OF THE VOLUME RECOVERY

In the experiments of Kovacs [4, 5], used to illustrate our purpose, the perturbation $\gamma_n$ (or solicitation) corresponds to the absolute temperature $T$, and the response $\beta_m$ to each component of the strain tensor. Under these conditions, the thermomechanical coupling leads to a specific tensorial relation (instead of Eq. 5):

$$\dot{\boldsymbol{\varepsilon}} = \boldsymbol{\alpha}_L^u \dot{T} - \sum_{j=1}^{n} \frac{\boldsymbol{\varepsilon}_j - \boldsymbol{\varepsilon}_j^r}{\tau_j(T, \delta)}, \quad \text{with} \quad \boldsymbol{\varepsilon} = \sum_{j=1}^{n} \boldsymbol{\varepsilon}_j \tag{6}$$

For an isotropic medium, the tensor of one-dimensional expansion verifies $\boldsymbol{\alpha}_L^u = \alpha_L^u \mathbf{1}$, the trace of the strain tensor being coupled to the temperature variations. In other words, the volume strain may also be introduced by

$$\frac{\dot{V}}{V^r} = \alpha^u \dot{T} - \frac{1}{V^r} \sum_{j=1}^{n} \frac{V_j - V_j^r}{\tau_j} \tag{7}$$

where $V = \sum_{j=1}^{n} V_j$, and the volume $V^r$ corresponds to the overall representative volume element at the relaxed state. Indeed, the experimental data by Kovacs for volume recovery of PVAc suggest the approximation $V \approx V^r$ [4, 5]. The variable $\alpha^u$ represents the coefficient of isotropic expansion of the frozen amorphous phase. The previous relation (Eq. 7) has been actually used in our simulations of volume variations during recovery tests. The time integration of this equation has been performed in a numerical way using our spectral distribution ( $(\tau_j^r, j = 1, \ldots, n)$ and the specific spectral shift function $a(t)$. The actual relaxation spectrum is supposed to be continuous. Its modeling near the equilibrium has been approximated using a discrete distribution of 50 modes equidistributed over six decades of the time scale [2].

Two versions (called versions 1 and 2) corresponding to different levels of approximation of the DNLR formalism have been developed to describe the volume recovery under various thermal histories.

#### 9.17.3.1.1 Version 1: A First Approximation without Entropic Coupling

Version 1 is the simplest application of the DNLR approach. The coupling between the thermodynamic variables has been neglected. This version has led to the expression of the volume recovery as a function of the thermal history. Thus, referring to Eq. 7, the equilibrium condition may be written:

$$\frac{\dot{V}}{V^r} = \alpha^r \dot{T} \tag{8}$$

Combining Eqs. 7 and 8 and introducing the volume variation $(\Delta V = V - V^r)$ and the relaxation times $[\tau_j = \tau_j^r a(\Delta V / V^r)]$ gives:

$$\frac{\dot{V} - \dot{V}^r}{V^r} = \frac{\Delta \dot{V}}{V^r} = (\alpha^u - \alpha^r)\dot{T} - \sum_{j=1}^{n} \left[ \frac{(\Delta V / V^r)_j}{\tau_j^r a(\Delta V / V^r)} \right] \tag{9}$$

Thus, by considering the definition:

$$\Delta V / V^r = \mathrm{trace}\,(\boldsymbol{\varepsilon}) = \delta \tag{10}$$

Equation 9 becomes in its final form

$$\dot{\delta} = \Delta\alpha \dot{T} - \sum_{j=1}^{n} \frac{\delta_j}{\left\{ \frac{h}{kT} \exp\left( \frac{\Delta E^{+,r}}{RT} - \frac{\Delta S_j^{+,r}}{R} \right) \right\} \exp(K_\delta \delta / RT)} \tag{11}$$

where the factor of nonlinearity due to the deviation from equilibrium, $K_\delta = K_\delta(T)$, can be temperature-dependent.

#### 9.17.3.1.2 Version 2: A Second Approximation with Entropic Coupling

To establish a more complete modeling, we need to consider the entropic coupling effects explicitly contained in the general relation (Eq. 5). Indeed, this relation shows that both volume and entropy simultaneously relax during aging transformations. Consequently, the activation entropy $\Delta S_j^+ = S_j^+ - S_j$ involves another type of nonlinearity by means of the evolving relaxation times, $\tau_j^*(T, S)$. Thus we will present a more accurate version of the DNLR, called Version 2, which consists in solving a new complete differential system involving simultaneously the entropy and the volume variations.

As a first example, each isothermal evolutions ($\dot{T} = 0$), during volume recovery, will be characterized by

$$\dot{S} = -\sum_{j=1}^{n} \frac{S_j - S_j^r}{\tau_j^*(T, S)\, a(\Delta S)} \quad \text{with} \quad S = \sum_{j=1}^{n} S_j, \text{ and } S^r = \sum_{j=1}^{n} S_j^r \tag{12a}$$

The adequate shift function and relaxation time are given by

$$a(\Delta S) = \exp\left(\frac{K_s(S - S^r)}{RT}\right) = \exp\left(\frac{K_s \Delta S}{RT}\right) \tag{12b}$$

$$\tau_j^*(T, S) = \frac{h}{k_B T} \exp\left[\frac{\Delta E^{+,r} - T(\Delta S^{+,r} + \int_{t_0}^{t} \dot{S}\, dt')}{RT}\right] \tag{12c}$$

The unknown initial value $\Delta S_0 = S^u - S^r$ becomes a new adjustable parameter calculated for each isothermal experiment in the integration of Eq. 12a. In the case of PVAc, this parameter is not temperature-dependent, in the explored range. Thus Eq. 7 becomes

$$\dot{\delta} = \alpha^u \dot{T} - \sum_{j=1}^{n} \frac{\delta_j}{\tau_j^*(T, S)\, a(\Delta V/V^r)} \tag{12d}$$

where the shift function, $a(\Delta V/V^r) = \exp(K_\delta \delta/RT)$, corresponds to the definition (Eq. 11) of version 1.

Integrating numerically the set of equations (Eqs. 12a–12d) provides the fitting of the isothermal curves of volume recovery.

The second example on PVAc is concerned with the variation of the apparent specific heat during heating or quenching processes. From Eq. 5, it appears that the apparent specific heat may be expressed as a function of the entropy:

$$C_p^{app.} = \frac{T\dot{S}}{\dot{T}} = C_p^u - \frac{T}{\dot{T}} \sum_{j=1}^{n} \frac{(S_j - S_j^r)}{\tau_j^*\, a(\Delta S)} \tag{13a}$$

An alternative formulation, with $C_p^{app.} = \dot{H}/\dot{T}$, is based on the enthapy $H$ obtained by the Legendre transformation of the Helmholtz free energy instead of the entropy. This is the choice of Aharoune [7], who analyzed the DSC response (differential scanning calorimetry) of various glasses with the following relation:

$$\Delta C_p^{app.} = \Delta C_p - \frac{1}{\dot{T}} \sum_{j=1}^{q} \frac{(H_j - H_j^r)}{\tau_j^* \, a(\Delta H)} \tag{13b}$$

where $\Delta C_p = C_p^u - C_p^r$ and

$$a(\Delta H) = \exp\left(\frac{K_h(H - H_r)}{RT}\right) \tag{13c}$$

## 9.17.4 IDENTIFICATION OF THE PARAMETERS

For all examples presented here, we have verified that a discrete distribution of 50 relaxation modes, equidistributed over six decades of time respecting Eqs. 1 and 2, is operational.

### 9.17.4.1 Identification of the Parameters of Version 1 for the Volume Recovery

From Eq. 11, in the case of volume recovery, four parameters must be adjusted, i.e., $\Delta\alpha = \alpha^u - \alpha^r$, $\Delta E^{+,r}$, $\Delta S_{j=n}^{+,r}$ corresponding to the longest relaxation time, and the factor $K_\delta$. The identification of $\Delta\alpha$ is obtained directly from the Kovacs experimental data by estimating the initial deviation from equilibrium, $\delta_0 = V(T_0, t = 0) - V^r(T_0, t = \infty)$, for a given temperature jump and with the approximate relation $\delta_0 = \Delta\alpha(T_g - T_0)$; see Figure 9.17.1b and 9.17.2. The three other parameters, $\Delta E^{+,r}$, $\Delta S_{j=n}^{+,r}$, and $K_\delta(T)$, are estimated for each isothermal by minimizing the deviations between theoretical simulations and experimental results, according to the classical algorithm of Gauss–Newton (Fig. 9.17.2).

### 9.17.4.2 Identification of the Parameters of Version 2 for the Volume Recovery and Apparent Specific Heat

Version 2 involves six parameters in Eqs. 12a–12d, i.e., the four previous ones, $\Delta\alpha(T)$, $\Delta S_{j=n}^{+,r}$, $\Delta E^{+,r}$, and $K_\delta(T)$, completed by two others which may depend on temperature:

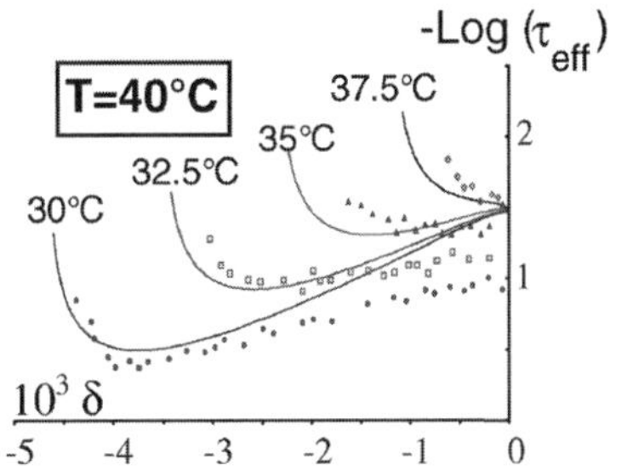


FIGURE 9.17.2 Experimental results (symbols, Kovacs) and theoretical predictions of the model, version 1.

- $\Delta S_o(T) = S^u - S^r$, representing the difference of entropy between the unrelaxed and relaxed states;
- $K_s(T)$, giving the nonlinearity due to the entropic recovery.

To analyze the observations by Kovacs giving $\delta = \delta(T, t)$, we assume that $\Delta\alpha$ is not temperature-dependent. The other five parameters, i.e., $\Delta S_n^{+,r}$, $\Delta E^{+,r}$, $K_\delta(T)$, $\Delta S_o(T)$, and $K_s(T)$, are optimized for all isothermal evolutions using the numerical method mentioned previously.

In addition, in order to simulate the apparent specific heat on the base of Eqs. 13b and 13c, one needs to identify four parameters using the same method, i.e., $\Delta C_p$, $\Delta S$, $\Delta E$, $K_h$, and $\Delta S_o$. In spite of the case of the volume recovery, the correction due to the entropic coupling has no significant effect on the specific heat.

## 9.17.5 HOW TO USE THE MODEL

### 9.17.5.1 Isothermal Behavior: The Case of the Volume Recovery

Version 1 provides a simple and fast calculation of the theoretical response by numerical integration of the set of equations without any entropic coupling. These predictions are similar to those from the KAHR model developed by Kovacs and coworkers [8] and from the model suggested by Moynihan *et al.* [9]. But, as shown in Figure 9.17.2, such a simplified version failed in the cases of expansion due to the aging process, especially near the equilibrium, when $\delta \to 0$.

Version 2, which accounts for the entropic coupling and which is formally contained in the DNLR framework, leads to more accurate results of volume recovery (Fig. 9.17.3).

The most interesting aspect of such a model is, of course, its ability to give some precise predictions for various thermal histories. Figure 9.17.4 gives us a comparison between theory and experiments for complex thermal loading, i.e., sequences of quenching, preannealing, and heating treatments. One can

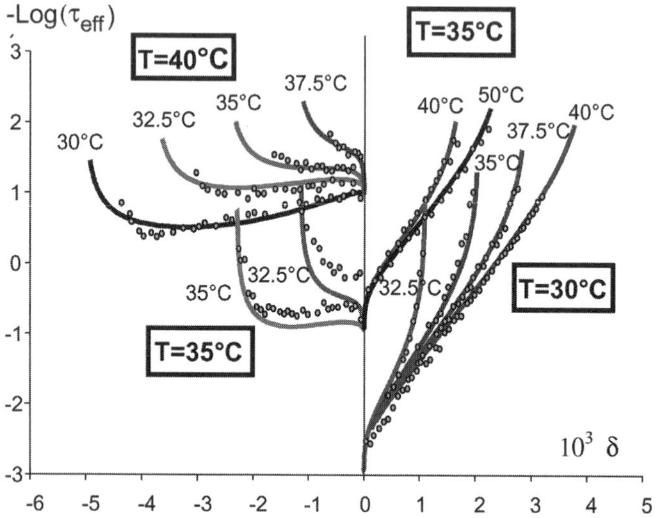


FIGURE 9.17.3 Theoretical curves (version 2) and experimental results for the effective relaxation time at different temperatures $T$ and $T_o$, PVAc.

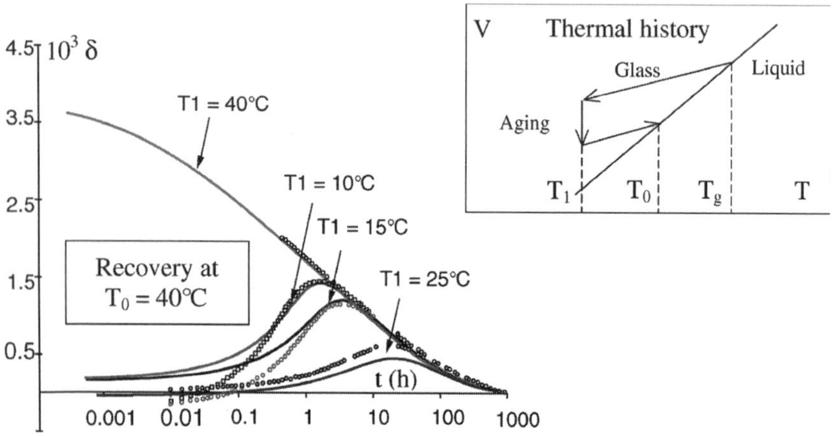


FIGURE 9.17.4 Memory effects induced by preannealing treatments: prediction of version 2 and experimental results (symbols).

see the good compatibility between version 2 and the isothermal volume recovery data from Kovacs.

### 9.17.5.2 Nonisothermal Behavior: The Case of the Specific Heat

We have chosen here to describe the behavior of PVAc, whose specific heat is given by Volkenstein and Sharonov [10].

Figure 9.17.5 gives an example of simulations of heating combined with preannealing. The parameters are completely consistent with those found for the volume recovery experiments.

## 9.17.6 LIST OF PARAMETERS

### 9.17.6.1 Isothermal Volume Recovery of PVAc

#### 9.17.6.1.1 Version 1: Study at 40°C

$\Delta\alpha = 4.65 \times 10^{-4}\,\mathrm{K}^{-1}$
$\Delta E^{+,r} = 712,700\,\mathrm{J/mol.at};$
$\Delta S^{+,r}_{j=n} = 1991\,\mathrm{J/mol.\ K};$
$K_\delta = -2850 \times 10^3\,\mathrm{J/mol.at}.$

#### 9.17.6.1.2 Version 2: Study at Various Temperatures

$\Delta\alpha = 6.65 \times 10^{-4}\,\mathrm{K}^{-1}$
$\Delta E^{+,r} = 712,000\,\mathrm{J/mol};\ \Delta S^{+,r}_{j=n} = 1980\,\mathrm{J/mol.K};\ \Delta S_0 = 11\,\mathrm{J/mol.K};$
$K_\delta = 43,797 + 2141\,\mathrm{T} - 27.7\,\mathrm{T}^2;\ K_s = 82 + 22\,\mathrm{T}.$

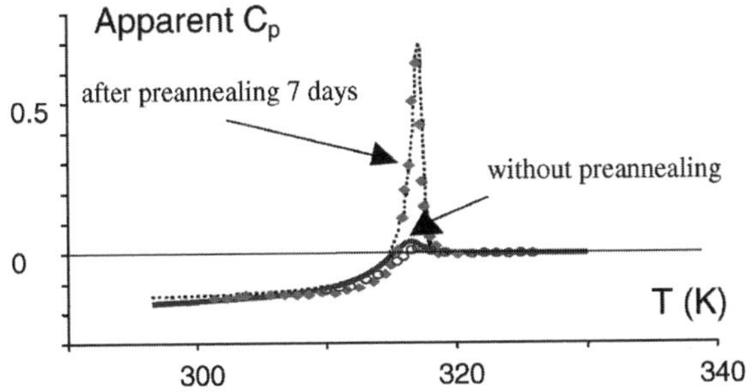


FIGURE 9.17.5 Experimental results (symbols) and theoretical simulation (model version 2) for the apparent specific heat of PVAc.

### 9.17.6.2 Anisothermal Aging and Specific Heat

$\Delta E^{+,r} = 715.2\,\text{kJ/mol}$ (compare with $712.7\,\text{kJ/mol}$ for version 1 and $712\,\text{kJ/mol}$ for version 2); and $K_h = -11870$, referring to the enthalpy–dissipation coupling. $\Delta S_0$ is identical to the previous values and has no significant effect on the responses.

### 9.17.6.3 Comments

It can be concluded from Figure 9.17.3 that the coupling between the entropy variation and the volume recovery leads to a better agreement between experiments and theory. Furthermore, we have examined the predictive ability of this version to traduce the complex thermal histories of volume relaxation as proposed by Kovacs. The obtained agreement, illustrated in Figure 9.17.4, shows a promising ability in spite of the observed small gaps, which are certainly due to the ignorance of the actual thermal rates.

Figure 9.17.5 compares DSC experimental curves from Volkenstein and Sharonov [10] with theoretical simulations. Note also that the activation parameters governing the kinetics of isothermal recovery near the equilibrium are very close to those of the apparent specific heat.

## REFERENCES

1. Cunat, C. (1985). Thèse de Doctorat d'Etat, INPL, Nancy, France.
2. Cunat, C. (1996). *Rev. Gén. Therm.* **35**: 680–685, Elsevier, Paris.
3. Coleman, B. D., and Gurtin, M. (1967). *J. Chem. Phys.* **47**(2): 597.
4. Kovacs, A. J. (1954). Ph.D. thesis, Paris.
5. Kovacs, A. J. (1963). *Fortschr. Hochpol. Forsch.* **3**: 394.
6. Marceron, P., and Cunat, C. (1999). Submitted to *J. Mech. Tim. Dep. Mat.*
7. Aharoune, A. (1991). Ph.D. thesis, INPL, Nancy, France.
8. Kovacs, A. J., Aklonis, J. J., Hutchinson, J. M., and Ramos, A. R. (1979). *J. Polym. Sci., Polym. Phys. Ed.* **17**: 1097.
9. Moynihan, C. T., *et al.* (1976). *Ann. N.Y. Acad. Sci.* **279**: 15.
10. Volkenstein, M. V., and Sharonov, Y. (1961). *Vysokomol. Soed.* **3**: 1739.

CHAPTER 10

# *Composite Media, Biomaterials*

SECTION **10.1**

# Introduction to Composite Media

JEAN LEMAITRE
*Université Paris 6, LMT-Cachan, 61 avenue du Président Wilson, 94235 Cachan Cedex, France*

All previous chapters have been concerned with phenomenological models which apply irrespective of the precise material. Here a special chapter is devoted to formalisms by which the *overall behavior* is obtained by *homogenization* of the properties of different constituents. The class of materials includes *metals* considered as mixtures of different phases, *fibers reinforced composites, laminates, concrete* considered as an assembly of rocks, sand, and cement, reinforced polymers, etc. Also included are *biomaterials* such as bones or soft tissues. Geomaterials are treated in another special chapter because of the particular formalism imposed by porosity, dilatancy, etc. A second reason is that, as for geomaterials, it is a composite community active in its own field.

The objective of the approaches in this chapter is to express the parameters in a constitutive law for the overall deformation behavior in terms of the properties of the constituents, as well as their distribution (random, organised in layers, etc.) and shapes (spherical particles, fibers, etc.), i.e., the *microstructure*. Assuming these are known, there is a rich toolbox of techniques in the literature for *linear* elastic constituents, which are summarized in Section 10.2. For composite materials where one or more of the phases is nonlinear, the methods are less well developed and more difficult. Section 10.3 summarizes some recent developments for nonlinear materials. Almost all available approaches assume that the size of the component made out of the composite is much larger than the typical length scale of the microstructure. As miniaturization continues, one may approach the limits of this assumption and it may be necessary to use nonlocal models, as explained in Section 10.4. In several applications, not only is the material subjected to mechanical loading, but also thermal strains, phase transitions, etc., may take place simultaneously. One of the methods discussed in Section 10.2, based on eigenstrains or transformations, is ideally suited to incorporate such effects; this is discussed in Section 10.5.

*Handbook of Materials Behavior Models.* ISBN 0-12-443341-3.

The *elementary mechanisms* may be observed by microscopes of different kinds, from which it is possible to deduce simple laws, such as the Schmitt slip law in plasticity, the friction law, or the decohesion law. There quantitative identification is difficult because direct measurements at microscale are often impossible to perform. Microhardness and nano-indentation tests on metallic crystals or pullout tests of fibers in composites or microtension tests are used but accuracy is poor. Often the only way is an indirect identification from tests at mesoscale.

It is inherent to *homogenization* that it applies to deformation behavior but not to fracture since this is a localization phenomenon. However, homogenization techniques have been developed to describe the influence of damage inside the material, treated as an evolving additional constituent. For laminate composites, this is demonstrated in Section 10.6, while Section 10.7 is concerned with the behavior of ceramic matrix composites. Section 10.8 is an example of an actual failure analysis, rather than a damage approach, to materials with a particular microstructure.

The special case of reinforced polymers at their viscous state during extrusion is treated in Section 10.9. Finally, *biomechanics* using similar tools is described in Section 10.10 for bones and in Section 10.11 for *soft tissues*.

SECTION **10.2**

# Background on Micromechanics

ERIK VAN DER GIESSEN
*University of Groningen, Applied Physics, Micromechanics of Materials, Nyenborgh 4, 9747 AG Groningen, The Netherlands*

Contents



## 10.2.1 INTRODUCTION

Although the precise origin is not clear, the term *micromechanics* is usually associated with the description of the overall behavior of heterogeneous materials such as composites.

This section summarizes some basic notions and results taken from reference books such as Nemat-Nasser and Horii [9] and Suquet [10], which will serve as background for the more detailed contributions in the rest of this chapter.

Many of the articles in this chapter use the direct notation of tensors, denoted by boldface letters. For example, a second-order tensor is denoted by $\boldsymbol{a}$, with components $a_{ij}$ on an appropriate Cartesion basis $\{\boldsymbol{e}_i\}$. The inverse is denoted by a superscript $-1$, while a superscript $T$ or $t$ denotes the transpose. The dot product is usually used for operations like $\boldsymbol{\sigma} \cdot \boldsymbol{n} = \sigma_{ij} n_j \boldsymbol{e}_i$, but for the product of a second-order and a fourth-order tensor we simply write $\boldsymbol{L\varepsilon} = L_{ijkl}\varepsilon_{kl}\boldsymbol{e}_i\boldsymbol{e}_j$ (although some authors write the same as $\boldsymbol{L}:\boldsymbol{\varepsilon}$). The dyadic or tensorial product of two vectors, $\boldsymbol{ab}$, is a second-order tensor with components $a_i b_j$.

*Handbook of Materials Behavior Models.* ISBN 0-12-443341-3.

## 10.2.2 BASIC CONCEPTS

Almost all engineering materials are *inhomogeneous* at some length scale; i.e., they consist of different *phases* or constituents. Depending on the material and on the scale of observation, the material's microstructure can be granular, such as for a polycrystalline metal, or can be viewed as a composite with one phase serving as a *matrix* in which the other phases are embedded as *inclusions*: particles, fibres, voids, etc. Composite modeling, as discussed in this chapter, becomes relevant when the typical dimension $d$ of the phases is much smaller than $L$, the characteristic length of the body under consideration or the "wavelength" of its loading (Fig. 10.2.1). This then leads to a decoupling of the two scales so that:

- spatial variations in the stress and strain fields at the microstructural scale, denoted by $\boldsymbol{\sigma}(\boldsymbol{x})$ and $\boldsymbol{\varepsilon}(\boldsymbol{x})$, respectively, influence the macroscopic response only through their averages;
- gradients of macroscopic fields, such as $\Sigma(X)$ and $E(X)$, and composition are not significant for the response at the microstructural scale.

The coupling between the continuum descriptions at the two scales is provided by the appropriate averaging, denoted by brackets:

$$\Sigma = \langle \boldsymbol{\sigma} \rangle, \quad E = \langle \boldsymbol{\varepsilon} \rangle \tag{1}$$

The averaging is performed over a *representative volume element* (RVE) (Fig. 10.2.1). This RVE has to be (i) large enough so that statistically meaningful averaging over the microstructure is possible, i.e., $l \gg d$, but (ii) small enough that macroscopic gradients can be neglected, i.e., $l \ll L$. Once the RVE has been identified (region $\Omega$), averaging can be carried out in terms of volume averages; for example,

$$\langle q \rangle = \frac{1}{\Omega} \int_{\Omega} q(x) dV \tag{2}$$

Since the stress and strain fields $\boldsymbol{\sigma}(\boldsymbol{x})$ and $\boldsymbol{\varepsilon}(\boldsymbol{x})$ satisfy equilibrium and compatibility, respectively,

$$\operatorname{div} \boldsymbol{\sigma} = 0, \quad \boldsymbol{\varepsilon} = \tfrac{1}{2}(\nabla \boldsymbol{u} + \nabla \boldsymbol{u}^T) \tag{3}$$

the macroscopic stress and strain can be expressed in terms of the traction fields $\boldsymbol{t}(\boldsymbol{x})$ and the displacement fields $\boldsymbol{u}(\boldsymbol{x})$ over the boundary $\partial\Omega$ of the RVE (with unit outer normal $\boldsymbol{n}$):

$$\Sigma = \langle \boldsymbol{\sigma} \rangle = \frac{1}{2\Omega} \int_{\partial\Omega} (\boldsymbol{xt} + \boldsymbol{tx}) dA, \quad E = \langle \boldsymbol{\varepsilon} \rangle = \frac{1}{2\Omega} \int_{\partial\Omega} (\boldsymbol{un} + \boldsymbol{nu}) dA \tag{4}$$

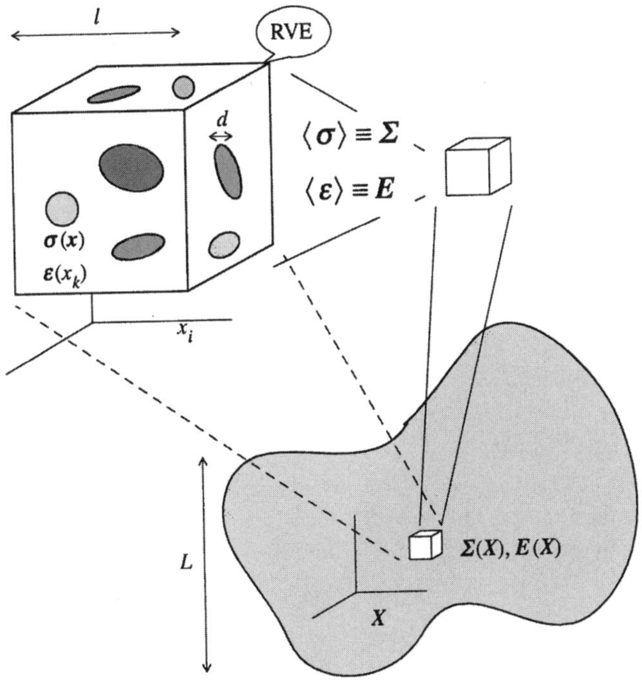


FIGURE 10.2.1 A body made of an inhomogeneous material can be represented as its macroscopic scale $X$ as a continuum when its characteristic length scale $L$ is much larger than the characteristic dimension of the microstructure $d$. The macroscopic continuum then describes the average behavior of representative volume elements (RVEs) of the inhomogeneous material at the microscopic scale $x$.

It is sometimes convenient to decompose the total volume of the RVE into the volumes $\Omega^{(r)}$ (sometimes denoted $\Omega_r$) of each phase $r$. Averages can then be computed also over each phase separately; for example,

$$\langle q\rangle_r = \frac{1}{\Omega^{(r)}} \int_{\Omega^{(r)}} q(x)dV \tag{5}$$

With the aid of such phase averages, the total average $\langle q\rangle$ according to Eq. 2 can be rewritten as

$$\langle q\rangle = \sum_r c^{(r)} \langle q\rangle_r \tag{6}$$

where $c^{(r)}$ is the volume fraction of phase $r$.

## 10.2.3 HOMOGENIZATION TECHNIQUES

Once the averaging is defined, the major challenge is to compute the microscopic fields for a given microstructure. We summarize a few of the most well-known approaches for linear materials; Section 10.3 will discuss some approaches for nonlinear materials.

For linear elastic materials, the microscopic fields depend linearly on the corresponding macroscopic fields. This is expressed as

$$\boldsymbol{\varepsilon}(\boldsymbol{x}) = \boldsymbol{A}(\boldsymbol{x})\boldsymbol{E}, \quad \boldsymbol{\sigma}(\boldsymbol{x}) = \boldsymbol{B}(\boldsymbol{x})\boldsymbol{\Sigma} \tag{7}$$

in terms of the fourth-order *localization* or *concentration* tensors $\boldsymbol{A}$ and $\boldsymbol{B}$ for strain and stress. From Eq. 4, it follows that

$$\langle \boldsymbol{A}(\boldsymbol{x})\rangle = \langle \boldsymbol{B}(\boldsymbol{x})\rangle = \boldsymbol{I\!I} \tag{8}$$

with $\boldsymbol{I\!I}$ the fourth-order unit tensor with Cartesian components $\delta_{ij}\delta_{kl}$. The problem thus is to find these localization tensors for the given microstructure. When these are known, the overall constitutive law can be readily obtained. When the microscopic elasticity (Hooke's law) is written as

$$\boldsymbol{\sigma}(\boldsymbol{x}) = \boldsymbol{L}(\boldsymbol{x})\boldsymbol{\varepsilon}(\boldsymbol{x}) \tag{9}$$

with the moduli $\boldsymbol{L}(\boldsymbol{x})$ being piecewise equal to one of the phase moduli $\boldsymbol{L}^{(r)}$, the overall constitutive law is

$$\boldsymbol{\Sigma} = \boldsymbol{L}^{*}\boldsymbol{E} \tag{10}$$

where the overall or *effective* stiffness tensor $\boldsymbol{L}^{*}$ is given by

$$\boldsymbol{L}^{*} = \langle \boldsymbol{L}\boldsymbol{A}\rangle = \sum_{r} c^{(r)}\boldsymbol{L}^{(r)}\boldsymbol{A}^{(r)} \tag{11}$$

with $\boldsymbol{A}^{(r)} = \langle \boldsymbol{A}\rangle_r$. Equivalently, when the microscopic constitutive equation is given in terms of the compliance tensor $\boldsymbol{M}$,

$$\boldsymbol{\varepsilon}(\boldsymbol{x}) = \boldsymbol{M}(\boldsymbol{x})\boldsymbol{\sigma}(\boldsymbol{x}) \tag{12}$$

the macroscopic equations become

$$\boldsymbol{E} = \boldsymbol{M}^{*}\boldsymbol{\Sigma} \tag{13}$$

with the overall, effective compliance tensor $\boldsymbol{M}^{*}$ given by

$$\boldsymbol{M}^{*} = \langle \boldsymbol{M}\boldsymbol{B}\rangle = \sum_{r} c^{(r)}\boldsymbol{M}^{(r)}\boldsymbol{B}^{(r)} \tag{14}$$

Thus the remaining challenge is to find the localization tensors $\boldsymbol{A}$ or $\boldsymbol{B}$. Since realistic microstructures tend to be so complex that closed-form solutions are beyond the present capabilities, numerous approximate techniques have been proposed. Some of these, usually referred to as mean-field methods, treat only the phase-average localization tensors $\boldsymbol{A}^{(r)}$ or $\boldsymbol{B}^{(r)}$, since these are sufficient to compute the effective moduli; cf. Eqs. 11–14. Methods which actually compute $A(x)$ or $B(x)$ evidently provide more details about the microscopic fields, but are practically always numerical. We mention a few of the most popular ones.

### 10.2.3.1 Mean-Field Approaches

There are many practical examples of composite materials in which the volume fraction of inclusions in the matrix is so small that the interaction between the inclusions can be neglected. Hence, in such so-called *dilute* cases, the inclusion can be considered to live in an infinite matrix. This allows us to make use of some central results by Eshelby [3] for inclusions in a matrix with the same elastic properties but with an eigenstrain in the inclusions through the concept of equivalent homogeneous inclusions. This involves replacing the inhomogeneous inclusion of the composite with a homogeneous inclusion having the proper *eigenstrain* or *transformation strain* (see further in Section 10.5). When the inclusions are modeled as ellipsoids (including spheres as well as discs and needles as limiting cases), Eshelby's [3] key result is that the strains inside such inclusions are uniform and related to the unconstrained eigenstrain through a fourth-order tensor $S$. Hence, the localization tensors $A^{(i)}$ and $B^{(i)}$ in the inclusions ($r = i$; $r = m$ for the matrix) are uniform. It can be shown [6] that they are given by

$$A^{(i)} = [I + SM^{(m)}(L^{(i)} - L^{(m)})]^{-1}, \quad B^{(i)} = L^{(i)}[I + SM^{(m)}(L^{(i)} - L^{(m)})]^{-1}M^{(m)} \tag{15}$$

This is enough for two-phase systems to determine the overall moduli as

$$L^* = L^{(m)} + c^i(L^{(i)} - L^{(m)})A^{(i)} \tag{16}$$

since the phase-average localization tensors in the matrix are correlated because of Eq. 8.

The Eshelby tensor $S$ here depends only on the properties of the matrix and on the shape of the inclusions. Explicit expressions can be found in, e.g., Reference [8]. For instance, the Cartesian components $S_{ijkl}$ for spherical inclusions in an isotropic matrix with bulk modulus $k^{(m)}$ and shear modulus $\mu^{(m)}$ are

$$S_{ijkl} = \alpha\frac{1}{3}\delta_{ij}\delta_{kl} + \beta\left[\frac{1}{2}\left(\delta_{ik}\delta_{jl} + \delta_{il}\delta_{jk}\right) - \frac{1}{3}\delta_{ij}\delta_{kl}\right] \tag{17}$$

with

$$\alpha = \frac{3k^{(m)}}{3k^{(m)} + 4\mu^{(m)}}, \quad \beta = \frac{6\left(k^{(m)} + 2\mu^{(m)}\right)}{5(3k^{(m)} + 4\mu^{(m)})} \tag{18}$$

With these, the overall bulk and shear moduli, $k^*$ and $\mu^*$, are obtained from Eq. 16 as

$$k^* = k^{(m)} + \frac{c^{(i)}\left(k^{(i)} - k^{(m)}\right)}{1 + \alpha(k^{(i)}/k^{(m)} - 1)}, \quad \mu^* = \mu^{(m)} + \frac{c^{(i)}\left(\mu^{(i)} - \mu^{(m)}\right)}{1 + \beta(\mu^{(i)}/\mu^{(m)} - 1)} \tag{19}$$

The dilute approximation loses accuracy when the inclusion volume fraction is larger than several percentages (depending on the contrast, i.e., the difference between $\boldsymbol{L}^{(i)}$ and $\boldsymbol{L}^{(m)}$). Then, the interaction between inclusions must be accounted for. The *Mori-Tanaka method* [7] uses an approximation for this by using the Eshelby equivalent inclusion concept but replacing the actual stress on the inclusion by the matrix average stress. The most straightforward way of proceeding [1] is to replace Eq. 7 by

$$\boldsymbol{\varepsilon}^{(i)} = \boldsymbol{A}^{(i)}_{dil}\boldsymbol{\varepsilon}^{(m)}, \quad \boldsymbol{\sigma}^{(i)} = \boldsymbol{B}^{(i)}_{dil}\boldsymbol{\sigma}^{(m)} \tag{20}$$

with $\boldsymbol{A}^{i}_{dil}$ and $\boldsymbol{B}^{i}_{dil}$ the dilute localization tensors according to Eq. 15. Straightforward algebra to eliminate the matrix averages from these expressions leads to the Mori-Tanaka localization tensors

$$\boldsymbol{A}^{(i)} = \left[(1 - c^{(i)})\boldsymbol{I} + c^{(i)}\boldsymbol{A}^{(i)}_{dil}\right]^{-1}\boldsymbol{A}^{(i)}_{dil}, \quad \boldsymbol{B}^{(i)} = \left[(1 - c^{(i)})\boldsymbol{I} + c^{(i)}\boldsymbol{B}^{(i)}_{dil}\right]^{-1}\boldsymbol{B}^{(i)}_{dil} \tag{21}$$

Estimates of the overall properties of porous elastic materials, i.e., those with a relatively large contrast in properties of the "inclusions" compared to the matrix, are useful up to void volume fractions of $c^{(i)} = 0.25$.

So-called *self-consistent methods* were initially devised for composites, such as polycrystals, in which it is not obvious which phase is the matrix and which is the inclusion [6]. However, it was shown later that they deliver useful estimates for a much wider range of materials. The key idea is to account for inclusion interaction in an approximate manner by embedding the inclusion not in the matrix but in a medium with the overall elastic moduli. Thus one uses, for example, the dilute estimate (Eq. 16) rephrased as

$$\boldsymbol{L}^*_{SC} = \boldsymbol{L}^{(m)} + c^{i}(\boldsymbol{L}^{(i)} - \boldsymbol{L}^{(m)})\boldsymbol{A}^* \tag{22}$$

with $\boldsymbol{A}^*$ now being a function of the elastic moduli $\boldsymbol{L}^*_{SC}$ of the composite. Hence, Eq. 22 is implicit and has to be solved iteratively. Alternative more elaborate versions have been derived subsequently, see for example [9, 2].

### 10.2.3.2 Bounds

According to Eq. 11, the overall modulus $\boldsymbol{L}^*$ is not simply the (Voigt or Taylor) average $\langle \boldsymbol{L}(\boldsymbol{x}) \rangle$ of the microscopic modulus tensor. A similar conclusion holds for the (Reuss or Sachs) average $\langle \boldsymbol{M}(\boldsymbol{x}) \rangle$ in relation to $\boldsymbol{M}^* = (\boldsymbol{L}^*)^{-1}$. However, it follows from minimum potential energy and minimum complementary energy considerations that the two averages do provide bounds on the actual overall modulus:

$$\langle \boldsymbol{M} \rangle^{-1} = \left( \sum_r c^{(r)} \boldsymbol{M}^{(r)} \right)^{-1} \leq \boldsymbol{L}^* \leq \langle \boldsymbol{L} \rangle = \sum_r c^{(r)} \boldsymbol{L}^{(r)} \tag{23}$$

Much tighter bounds on the actual overall modulus can be obtained from a variational principle due to Hashin and Shtrikman [4]. The details are beyond the scope of this section, but the key idea is (i) to estimate the average strain in each phase $r$ by treating it as an ellipsoidal inclusion in a reference matrix with elastic modulus $\boldsymbol{L}^0$ as discussed previously; (ii) to use these as trial solutions in variational theorems. The classical Hashin-Shtrikman bounds apply to composites with a statistically isotropic distribution of particles, and Willis [11] has generalized the approach to more general cases.

### 10.2.3.3 Cell Methods

The actual calculation of the microfields is frequently carried out by cell methods. The starting point is the idea that the fields at the microscale can be computed if the RVE with the distribution of phases is modeled as a body subjected to boundary conditions that are consistent with the macroscopic state. There are two distinct approaches: either one prescribes the boundary displacements $\boldsymbol{u}(\boldsymbol{x})$ in accordance with the macroscopic strain $\boldsymbol{E}$ as

$$\boldsymbol{u}(\boldsymbol{x}) = \boldsymbol{E} \cdot \boldsymbol{x} \quad \text{on } \partial\Omega \tag{24}$$

so that $\langle \boldsymbol{\varepsilon}(\boldsymbol{u}) \rangle = \boldsymbol{E}$, or one prescribes uniform boundary tractions,

$$\boldsymbol{t}(\boldsymbol{x}) = \Sigma \cdot \boldsymbol{n}(\boldsymbol{x}) \quad \text{on } \partial\Omega \tag{25}$$

so that $\langle \boldsymbol{\sigma}(\boldsymbol{t}) \rangle = \Sigma$. In either one of these boundary conditions, Hill's [5] lemma holds:

$$\langle \boldsymbol{\sigma\varepsilon} \rangle = \Sigma \boldsymbol{E} \tag{26}$$

But Eqs. 24 and 25 are not equivalent (except in the limit $d/l \to 0$) and will only give the same fields away from a boundary layer (more on this in Section 10.4).

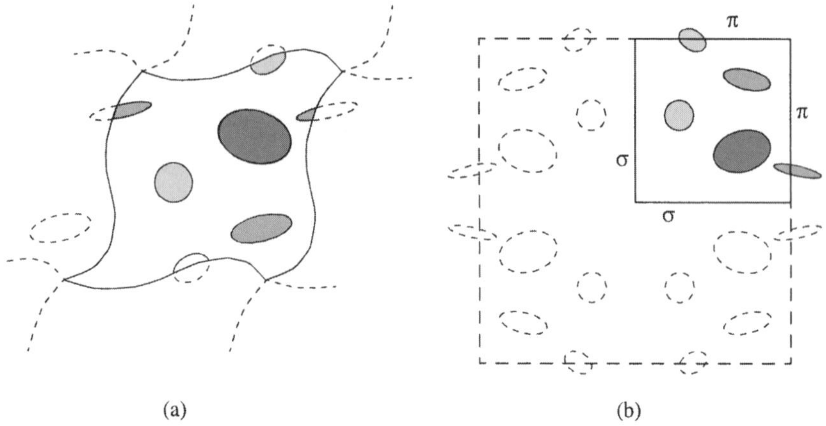


FIGURE 10.2.2 a. A unit cell in a material with a periodic distribution of phases. b. A unit cell (dashed square) that has two reflection symmetries. Only a quarter (solid square) needs to be analyzed, by imposing periodic and symmetric boundary conditions on the periodic ($\pi$) and symmetric ($\sigma$) boundaries.

By definition, the RVE contains many phases, which makes the computation usually impractical. If the distribution of phases exhibits symmetries, these can be used to define a much smaller *unit cell*. Numerical solution of the microscopic fields in such cells may be quite feasible with current computing facilities to very high degrees of accuracy. In case of periodic microstructures, the microfields are periodic, and the proper boundary conditions are

$$\boldsymbol{u}(\boldsymbol{x}) = \boldsymbol{E} \cdot \boldsymbol{x} + \boldsymbol{u}^* \quad \text{on } \partial\Omega \tag{27}$$

with the fluctuations $\boldsymbol{u}^*$ prescribed periodic (see Fig. 10.2.2a). For these boundary conditions, there are no boundary layer artifacts, and Hill's lemma (Eq. 26) continues to hold.

In particular cases, even simpler boundary conditions can be applied. The most well-known ones are the symmetry boundary conditions. These apply to (i) unit cells that have reflection symmetries (ii) under macroscopic states that respect the same symmetry. A typical example is shown in Figure 10.2.2b for a composite with a square packing of particles (in two dimensions; fibers in three dimensions) subject to a macroscopic strain with principal directions coinciding with the directions of packing. Periodic boundary conditions in such a case imply that the cell boundaries remain straight; the symmetry boundaries remain straight because of symmetry.

## REFERENCES


1. Benveniste, Y. (1987). A new approach to the application of Mori-Tanaka's theory in composite materials. *Mech. Mater.* **6**: 147–157.
2. Christensen, R. M. (1979). *Mechanics of Composite Materials*, New York: John Wiley & Sons.
3. Eshelby, J. D. (1957). The determination of the elastic field of an ellipsoidal inclusion and related problems. *Proc. Roy. Soc. Lond.* **A241**: 376–396.
4. Hashin, Z., and Shtrikman, S. (1963). A variational approach to the theory of the elastic behavior of multiphase materials. *J. Mech. Phys. Solids* **11**: 127–140.
5. Hill, R. (1963). Elastic properties of reinforced solids: Some theoretical principles. *J. Mech. Phys. Solids* **11**: 357–372.
6. Hill, R. (1965). A self-consistent mechanics of composite materials. *J. Mech. Phys. Solids* **13**: 213–222.
7. Mori, T., and Tanaka, K. (1973). Average stress in the matrix and average elastic energy of materials with misfitting inclusions. *Acta Metall.* **21**: 571–574.
8. Mura, T. (1987). *Micromechanics of Defects in Solids*. Dordrecht: Martinus Nijhoff.
9. Nemat-Nasser, S., and Hori, M. (1993). *Micromechanics: Overall Properties of Heterogeneous Materials*, Amsterdam: North-Holland.
10. Suquet, P. (1997). *Continuum Micromechanics*. CISM Lecture Notes 377.
11. Willis, J. R. (1977). Bounds and self-consistent estimates for the overall properties of anisotropic composites. *J. Mech. Phys. Solids* **25**: 185–202.

SECTION **10.3**

# Nonlinear Composites: Secant Methods and Variational Bounds

PIERRE M. SUQUET
*LMA/CNRS 31 chemin Joseph Aiguier, 13402, Marseille, Cedex 20, France*

Contents




*Handbook of Materials Behavior Models.* ISBN 0-12-443341-3.

## 10.3.1 INTRODUCTION

The problem addressed here is that of the effective behavior of *nonlinear composites*. By composites we understand not only man-made materials but also all types of inhomogeneous materials which are ubiquitous in nature (polycrystals, wood, rocks, bone, porous materials).

Consider a finite volume element $V$ of such a composite material, large enough to be representative of the composite microstructure and nevertheless small enough for the different phases (grains or different mechanical phases) to be clearly distinguished. This representative volume element (RVE) $V$ is composed of $N$ distinct homogeneous phases $V_r$, $r = 1, \ldots, N$, the behavior of which is characterized by a nonlinear relation between the (infinitesimal) strain and stress fields, $\boldsymbol{\varepsilon}$ and $\boldsymbol{\sigma}$,

$$\boldsymbol{\varepsilon}(x) = \mathscr{G}^{(r)}(\boldsymbol{\sigma}(x)) \quad \text{when } \boldsymbol{x} \text{ is in phase } r \tag{1}$$

This constitutive relation corresponds either to *nonlinear elastic behavior* within the context of small strains, or to *finite viscous deformations* when $\boldsymbol{\varepsilon}$ and $\boldsymbol{\sigma}$ are interpreted as the Eulerian strain rate and Cauchy stress, respectively.

When the RVE $V$ is subjected to an average strain $\bar{\boldsymbol{\varepsilon}}$, it reacts to this strain by an average stress $\bar{\boldsymbol{\sigma}}$. The relation between $\bar{\boldsymbol{\sigma}}$ and $\bar{\boldsymbol{\varepsilon}}$ is the effective constitutive relation of the composite.

The question addressed here is: *Can predictions be made regarding this effective constitutive relation, given the constitutive relations of the phases and some (often limited) information about the composite microstructure?*

When the phases are linear, there exists a large body of literature partially answering this question. However, when the phases are nonlinear, there are very few *really* nonlinear schemes to analyze the global, as well as the local, response of nonlinear composites. Most methods are heuristic and are extensions or modifications of the secant method (which we will discuss in Section 10.3.2) and of the incremental method (which we will not discuss here). Most of these schemes proceed in three successive steps:

1. First, the constitutive relations of each individual phase are linearized in an appropriate manner. This is done pointwisely and serves to define a *linear comparison solid* with local elastic moduli which, in general, vary from point to point.
2. Then, the problem is reduced to that of estimating the effective properties of a *linear comparison solid* with a *finite* number of phases. To this aim, an approximation is introduced by assuming that the local moduli are piecewise uniform. In most cases (but not all), the regions

where the moduli are uniform are precisely the domains occupied by the material phases.

3. Finally, the effective linear properties of the linear comparison solid are estimated or bounded by a scheme which is relevant for the type of microstructure exhibited by the linear comparison solid. These linear effective properties are used to estimate the nonlinear effective properties of the actual nonlinear composite.

Following the seminal work of J. Willis [1], more rigorous results, namely, bounds for the nonlinear effective properties of composites, have been developed in the past ten years by Ponte Castañeda [2], Willis [3], and Suquet [4], among others. Ponte Castañeda's variational procedure [5], which will be briefly recalled in Section 10.3.3, is probably the most rigorous bounding theory available to date. Other methods, less rigorous but sometimes more accurate — for instance, the second-order procedure of Ponte Castañeda [6] or the affine procedure of Masson and Zaoui [7] — will not be discussed here. Some connections do exist between the heuristic secant methods and the more elaborate bounding techniques, and we will briefly outline them in Section 10.3.3.3, following Suquet [8, 9]. More details can be found in the review papers by Ponte Castañeda and Suquet [10] and Willis [11] (see also several contributions in Reference [12]).

## 10.3.2 SECANT METHODS

### 10.3.2.1 Nonlinear Local Problem

The local stress and strain fields within $V$ are solutions of the *local problem* consisting of the constitutive equations (Eq. 1), the compatibility conditions satisfied by $\boldsymbol{\varepsilon}$, and the equilibrium equations satisfied by $\boldsymbol{\sigma}$:

$$\boldsymbol{\varepsilon}(x) = \mathscr{G}^{(r)}(\boldsymbol{\sigma}(x)) \text{ in phase } r, \quad \boldsymbol{\varepsilon} = \tfrac{1}{2}(\nabla \boldsymbol{u} + \nabla \boldsymbol{u}^t), \quad \operatorname{div}(\boldsymbol{\sigma}) = 0 \quad \text{in } V \tag{2}$$

Several classes of boundary conditions can be considered on $\partial V$ (all conditions being equivalent in the limit of a large RVE under appropriate growth conditions on the functions $\mathscr{G}^{(r)}$). For definiteness, we will assume uniform tractions on the boundary $\boldsymbol{\sigma}(x).\boldsymbol{n}(x) = \bar{\boldsymbol{\sigma}}.\boldsymbol{n}(x)$ on $\partial V$. Once Eq. 2 is solved (at least theoretically), the spatial average $\bar{\boldsymbol{\varepsilon}}$ of the strain field $\boldsymbol{\varepsilon}$ can be taken:

$$\bar{\boldsymbol{\varepsilon}} = \langle \boldsymbol{\varepsilon} \rangle, \quad \text{where} \quad \langle . \rangle = \frac{1}{|V|}\int_V . \, dx$$

Then the relation between this average strain and the imposed average stress defines the *effective constitutive relation* of the composite:

$$\bar{\boldsymbol{\varepsilon}} = \tilde{\mathcal{G}}(\bar{\boldsymbol{\sigma}}) \tag{3}$$

## 10.3.2.2 Linearization

It is in general impossible to solve the nonlinear local problem (Eq. 2) exactly. Therefore, a first step in most nonlinear schemes is to write the constitutive relations (Eq. 1) in the form (the reference to the phase is implicit here):

$$\boldsymbol{\varepsilon}(x) = \boldsymbol{M}(\boldsymbol{\sigma}(x)) : \boldsymbol{\sigma}(x) + \boldsymbol{\eta}(x) \tag{4}$$

where $\boldsymbol{M}$ and $\boldsymbol{\eta}$ have to be specified. Two particular choices, giving rise to two broad classes of models referred to as *secant* and *tangent* models and schematically depicted in Figure 10.3.1 (several different choices are reviewed in Gilormini [13]), correspond respectively to the following linearlizations:

$$secant : \boldsymbol{M}(\boldsymbol{\sigma}(x)) = \boldsymbol{M}_{sct}(\boldsymbol{\sigma}(x)), \quad \boldsymbol{\eta} = 0 \tag{5}$$

and

$$tangent : \boldsymbol{M}(\boldsymbol{\sigma}(x)) = \boldsymbol{M}_{tgt}(\boldsymbol{\sigma}(x)), \quad \boldsymbol{\eta} = e_0 \tag{6}$$

Note that there are several choices for $\boldsymbol{M}_{sct}$ in (Eq. 5) for which $\boldsymbol{M}_{sct}(\boldsymbol{\sigma}) : \boldsymbol{\sigma} = \mathcal{G}(\boldsymbol{\sigma})$. Consequently, the secant moduli are not uniquely defined

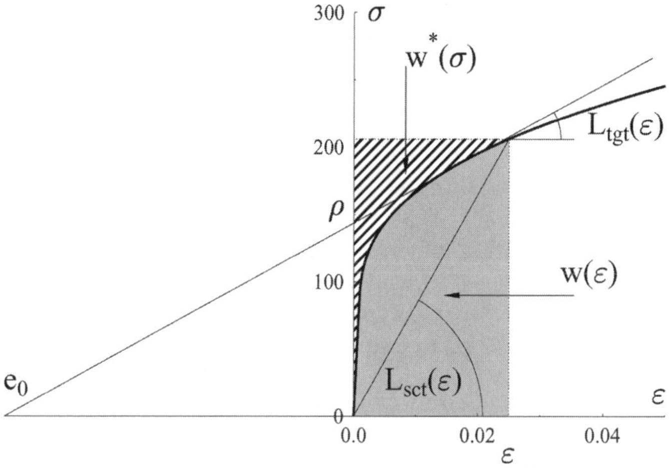


FIGURE 10.3.1 Secant and tangent moduli in a tensile uniaxial test. The secant and the tangent compliance are the inverse of the secant and tangent stiffness $L_{sct}$ and $L_{tgt}$, respectively.

(this observation was made by Gilormini [13]). The tangent moduli are defined (uniquely in general) as $M_{tgt}(\boldsymbol{\sigma}) = d\mathcal{G}(\boldsymbol{\sigma})/d\boldsymbol{\sigma}$.

We will not discuss the second choice (Eq. 6), namely, the class of tangent methods (the interested reader is referred to References [6, 13, 14] for additional details).

### 10.3.2.3 Secant Methods in General

Using the equivalent writing (Eq. 5) of the constitutive relations (Eq. 1), the local problem (Eq. 2) can be reformulated as

$$\boldsymbol{\varepsilon}(x) = \mathbf{M}^{(r)}_{sct}(\boldsymbol{\sigma}(x)) : \boldsymbol{\sigma}(x) \text{ in phase } r, \quad \boldsymbol{\varepsilon} = \tfrac{1}{2}(\nabla\mathbf{u} + \nabla\mathbf{u}^t), \quad \mathrm{div}(\boldsymbol{\sigma}) = 0 \text{ in } V \tag{7}$$

This problem can be considered a linear elasticity problem for a composite with infinitely many phases, since the moduli $M^{(r)}_{sct}(\boldsymbol{\sigma}(x))$ may vary from point to point. Equation 7 is therefore not simpler than the original (Eq. 2). A simplifying assumption is introduced by assuming that the secant moduli take piecewise uniform values $M^{(r)}$ on subdomains $W_r, r = 1, \ldots, M$, which coincide with, or are contained in, the physical domains $V_r$. For simplicity we will assume in the following that the domains $W_r$ and $V_r$ coincide.

In addition, the uniform moduli are assumed to be evaluated at some "effective stress" $\tilde{\boldsymbol{\sigma}}^{(r)}$ for the phase $r$:

$$\mathbf{M}^{(r)} = \mathbf{M}^{(r)}_{sct}\left(\tilde{\boldsymbol{\sigma}}^{(r)}\right) \tag{8}$$

The simplified secant problem now consists of the following:

$$\boldsymbol{\varepsilon}(x) = \mathbf{M}^{(r)} : \boldsymbol{\sigma}(x) \text{ in phase } r, \quad \boldsymbol{\varepsilon} = \tfrac{1}{2}(\nabla\mathbf{u} + \nabla\mathbf{u}^t), \quad \mathrm{div}(\boldsymbol{\sigma}) = 0 \quad \text{in } V \tag{9}$$

Assuming for a moment that the $M^{(r)}$'s are given, Eq. 9 is a problem for a linear $N$-phase composite called the *linear comparison composite*. The determination of the $\mathbf{M}^{(r)}$'s is made possible through Eq. 8. These relations involve the effective stress $\tilde{\boldsymbol{\sigma}}^{(r)}$ (which has not been specified yet, but this will be done in the next two paragraphs), which itself depends on the average stress $\bar{\boldsymbol{\sigma}}$. Therefore, Eq. 8 is a closure condition which renders the problem nonlinear.

In summary, any secant method involves three steps:

1. A linear theory providing an expression for $\tilde{\mathbf{M}}$ as a function of the moduli $M^{(r)}$ of the individual phases in the linear comparison solid.

2. The resolution of $N$ nonlinear tensorial problems for the $N$ unknown tensors $\boldsymbol{M}^{(r)}$:

$$\mathbf{M}^{(r)} = \mathbf{M}^{(r)}_{sct}\left(\tilde{\boldsymbol{\sigma}}^{(r)}\right), \quad \tilde{\boldsymbol{\sigma}}^{(r)} = \text{function of } \left\{\mathbf{M}^{(r)}\right\}_{r=1,\dots,N} \text{ and } \bar{\boldsymbol{\sigma}} \tag{10}$$

   Note that these equations depend on the particular expression ("function of") chosen for the effective stress $\tilde{\boldsymbol{\sigma}}^{(r)}$. Different choices give rise to different secant methods.
3. Once the $N$ nonlinear tensorial problems (Eq. 10) are solved, the overall stress-strain relation is given by

$$\bar{\boldsymbol{\varepsilon}} = \tilde{\boldsymbol{M}} : \bar{\boldsymbol{\sigma}} \tag{11}$$

   where $\tilde{\boldsymbol{M}}$ is the effective compliance derived by means of the linear theory of step 1 from the individual moduli $\boldsymbol{M}^{(r)}$ as determined in step 2.

### 10.3.2.4 A Secant Method Based on First-Order Moments (Classical Secant Method)

It remains to define the effective stress $\tilde{\boldsymbol{\sigma}}^{(r)}$. In the classical secant method, this effective stress is set equal to the average stress over phase $r$ (it should be emphasized that the stress field under consideration is now that in the linear comparison solid):

$$\tilde{\boldsymbol{\sigma}}^{(r)} = \bar{\boldsymbol{\sigma}}^{(r)} = \langle \boldsymbol{\sigma} \rangle_r, \quad \text{where} \quad \langle . \rangle_r = \frac{1}{|V_r|} \int_{V_r} . \, dx \tag{12}$$

This "first order moment" of the stress over phase $r$ can be expressed in terms of the overall stress $\bar{\boldsymbol{\sigma}}$ by means of the "stress-localization" tensor $\boldsymbol{B}^{(r)}$:

$$\bar{\boldsymbol{\sigma}}^{(r)} = \langle \boldsymbol{B} \rangle_r : \bar{\boldsymbol{\sigma}} \tag{13}$$

Most linear theories provide (more or less) explicit expressions for the "stress-localization" tensors $\boldsymbol{B}^{(r)}$ as functions of the individual compliances $\boldsymbol{M}^{(r)}$. A typical example will be given in following text. Equation 13 completes Eq. 10.

### 10.3.2.5 A Secant Method Based on Second-Order Moments (Modified Secant Method)

The classical secant method has several serious limitations. One of them is illustrated by its unphysical prediction for the response of nonlinear porous materials under hydrostatic loadings. Consider an RVE composed of an incompressible matrix with voids and subject to an hydrostatic stress. The average stress in the matrix is hydrostatic (the average stress in the voids is 0). Since the matrix is incompressible, it is insensitive to hydrostatic stresses. Therefore, the secant compliance associated with a purely hydrostatic stress by Eq. 12 always coincides with the initial compliance (under zero stress) of the material. The secant method applied with Eq. 12 predicts a linear overall response of the porous material. However, the actual response of the porous material is nonlinear, since the local stress state in the matrix is not hydrostatic (analytic calculations can be carried out on the hollow sphere model to prove this point explicitly), even if the average stress is hydrostatic. The occurrence of shear stresses in some regions of the RVE introduces nonlinearities both in the local and overall responses of the RVE which are not taken into account by Eq. 12.

This observation has motivated the introduction of theories based on *the second moment of the stress field*, in particular form by Buryachenko [15], in approximate form by Qiu and Weng [16], or in general and rigorous form by Suquet [8] and Hu [17]. It is indeed observed that in many cases of interest the secant compliance $\boldsymbol{M}_{sct}$ depends on the stress through the "quadratic stress"

$$\boldsymbol{M}_{sct}(\boldsymbol{\sigma}) = \boldsymbol{M}_{sct}(\mathcal{S}), \quad \text{where} \quad \mathcal{S} = \tfrac{1}{2}\,\boldsymbol{\sigma} \otimes \boldsymbol{\sigma} \tag{14}$$

Therefore, rather than expressing $\boldsymbol{M}^{(r)}$ in terms of an "effective stress" $\tilde{\boldsymbol{\sigma}}^{(r)}$, one can express $\boldsymbol{M}^{(r)}$ in terms of an effective "quadratic stress" $\tilde{\mathcal{S}}^{(r)}$. A very natural choice for this effective quadratic stress is

$$\tilde{\mathcal{S}}^{(r)} = \left\langle \tilde{\mathcal{S}} \right\rangle_r = \tfrac{1}{2}\langle \boldsymbol{\sigma} \otimes \boldsymbol{\sigma} \rangle_r \tag{15}$$

This effective "second-order moment" of the stress over phase $r$ has definite advantages over the first-order moment used in the classical secant method. For instance, it better accounts for local fluctuation of the stress. To see this, note that $\tilde{\mathcal{S}}^{(r)}_{ijij} = \left\langle \sigma_{ij}\sigma_{ij} \right\rangle_r$. Therefore, as soon as $\boldsymbol{\sigma}$ is nonzero in a (non-negligible) region of phase $r$, the second-moment $\tilde{\mathcal{S}}^{(r)}$ of the stress does not vanish. In particular, the overall response of porous materials under hydrostatic loading, as predicted by the secant method based on the "second-order moment," is nonlinear (as it should be) and close to the exact solution [16]. The modified secant theory consists in solving Eq. 10 together with the definition (Eq. 15) of the effective stress of phase $r$.

In practice, one has to compute $\tilde{\mathcal{S}}^{(r)}$ for the linear comparison solid. This can be done analytically by means of a result previously used in different contexts by several authors (see, for instance, Kreher [18]).

*Consider a linear composite composed of N homogeneous phases with elastic compliance* $\mathbf{M}^{(r)}$. *Let* $\tilde{\mathbf{M}}(\mathbf{M}^{(1)},\ldots,\mathbf{M}^{(r)},\ldots,\mathbf{M}^{(N)})$ *be the overall compliance tensor of this composite, and let* $\boldsymbol{\sigma}$ *denote the stress field in this linear composite. Then:*

$$\langle \boldsymbol{\sigma} \otimes \boldsymbol{\sigma} \rangle_r = \frac{1}{c^{(r)}} \bar{\boldsymbol{\sigma}} : \frac{\partial \tilde{\mathbf{M}}}{\partial \mathbf{M}^{(r)}} : \bar{\boldsymbol{\sigma}} \tag{16}$$

A detailed proof of this result can be found in References [9, 18] (among others).

In conclusion, the nonlinear systems of equations to be solved to complete step 2 of the secant method read as:

$$\tilde{\mathcal{S}}^{(r)} = \frac{1}{2c^{(r)}} \bar{\boldsymbol{\sigma}} : \frac{\partial \tilde{\mathbf{M}}}{\partial \mathbf{M}^{(r)}} : \bar{\sigma}, \quad \mathbf{M}^{(r)} = \mathbf{M}^{(r)}_{sct}\left(\tilde{\mathcal{S}}^{(r)}\right) \tag{17}$$

## 10.3.2.6 Example: Deformation Theory of Plasticity

A rather general form of stress-strain relations for isotropic elastic-plastic materials in the context of a deformation theory is given by the following:

$$\boldsymbol{\varepsilon} = \boldsymbol{\varepsilon}^e + \boldsymbol{\varepsilon}^p, \quad \boldsymbol{\varepsilon}^e = \frac{\boldsymbol{\sigma}_m}{k}\,\mathrm{i} + \frac{\mathbf{s}}{2\mu_0}, \quad \boldsymbol{\varepsilon}^p = \frac{3}{2}\frac{p(\sigma_{eq})}{\sigma_{eq}}\mathbf{s} \tag{18}$$

where $\sigma_m = (1/3)\,\mathrm{tr}\,(\boldsymbol{\sigma})$ is the hydrostatic stress, $\sigma_{eq} = [(3/2)\mathbf{s} : \mathbf{s}]^{1/2}$ is the von Mises equivalent stress ($\mathbf{s}$ being the stress deviator), and $p(\sigma_{eq})$ is the inverse of the plastic hardening curve $\sigma_{eq}(p)$ of the material. An alternate writing of Eq. 18 is

$$\boldsymbol{\varepsilon} = \mathbf{M}_{sct}(\boldsymbol{\sigma}) : \boldsymbol{\sigma}, \quad \text{with } \mathbf{M}_{sct}(\boldsymbol{\sigma}) = \frac{m}{3}J + \frac{\theta_{sct}(\sigma_{eq})}{2}\,\mathbf{K} \tag{19}$$

where

$$m = \frac{1}{k}, \quad \theta_{sct}(\sigma_{eq}) = \frac{1}{\mu_{sct}(\sigma_{eq})} = \frac{1}{\mu_0} + \frac{3p(\sigma_{eq})}{\sigma_{eq}} \tag{20}$$

and where $J$ and $K$ are the fourth-order tensors which project any second-order tensor on its hydrostatic and deviatoric parts:

$$J = \tfrac{1}{3}\,\mathrm{i} \otimes \mathrm{i}, \quad K = I - J$$

$i$ and $\boldsymbol{I}$ being the identity for symmetric second-order and fourth-order tensors, respectively.

Since we will be manipulating isotropic tensors throughout this section, simplified (and classical) notations will be helpful. Any isotropic fourth-order tensor $\boldsymbol{B}$ with minor and major symmetries can be decomposed as $\boldsymbol{B} = b_m \boldsymbol{J} + b_{dev}\boldsymbol{K}$, where $b_m = \boldsymbol{B} :: \boldsymbol{J}$ and $b_{dev} = (1/5)\boldsymbol{B} :: \boldsymbol{K}$. We use the compact notation $\boldsymbol{B} = \{b_m,\ b_{dev}\}$. The algebra over isotropic (and symmetric) tensors is then very simple: for two such tensors $\boldsymbol{B}$ and $\boldsymbol{C}$, one has $\boldsymbol{B} :: \boldsymbol{C} = \{b_m c_m,\ b_{dev} c_{dev}\}$ and $(\boldsymbol{B})^{-1} = \{(1/b_m),\ (1/b_{dev})\}$.

The response of the phases as described by Eq. 19 is linear for purely hydrostatic loadings (characterized by a constant bulk modulus $k$), and nonlinear in shear (characterized by a stress-dependent secant shear modulus $\mu_{sct}$). Note that $\boldsymbol{M}_{sct}$, which depends on $\boldsymbol{\sigma}$ through $\sigma_{eq}$ only, is indeed a function of the quadratic stress $\mathscr{S}$ defined by Eq. 14. Indeed, $\sigma_{eq}$ is itself a function of $\mathscr{S}$: $\sigma_{eq} = (\frac{3}{2}\,\boldsymbol{s} : \boldsymbol{s})^{1/2} = (3\boldsymbol{K} :: \mathscr{S})^{1/2}$.

Consider now a two-phase material, where both phases are elastic-plastic and obey the constitutive relations (Eq. 19) with different material constants. We further consider the case where the phases in the composite are arranged isotropically in a particle-matrix configuration, corresponding, for instance, to the case of spherical particles of phase 1 randomly distributed in phase 2 (which is the matrix). The linear comparison composite is a two-phase linear composite with the same microstructure as the initial nonlinear composite and with isotropic phases characterized by a compliance tensor $\boldsymbol{M}^{(r)} = \{m^{(r)}/3,\ \theta^{(r)}/2\}$.

Regarding the first step of the three-step procedure outlined previously, namely, a theory for the effective properties of the linear comparison solid, the Hashin-Shtrikman formalism is known to provide, in most cases, an accurate estimate of the effective properties of isotropic linear composites with particle-matrix microstructure. The corresponding estimate reads as

$$\tilde{\boldsymbol{M}} = \boldsymbol{M}^{(2)} + c^{(1)}\left(\boldsymbol{M}^{(1)} - \boldsymbol{M}^{(2)}\right) : \boldsymbol{B}^{(1)},\quad \boldsymbol{B}^{(1)} = \left(\boldsymbol{I} + c^{(2)}\boldsymbol{Q} : \left(\boldsymbol{M}^{(1)} - \boldsymbol{M}^{(2)}\right)\right)^{-1} \tag{21}$$

where $c^{(1)}$ and $c^{(2)}$ are the phase volume fractions and where $\boldsymbol{Q} = \{3q_m,\ 2q_{dev}\}$ with

$$q_m = \frac{1-\alpha}{m^{(2)}},\quad q_{dev} = \frac{1-\beta}{\theta^{(2)}},\quad \alpha = \frac{3k^{(2)}}{3k^{(2)} + 4\mu^{(2)}},\quad \beta = \frac{2}{5}\,\frac{3k^{(2)} + 6\mu^{(2)}}{3k^{(2)} + (4/3)\mu^{(2)}}$$

Then, the effective compliance $\tilde{\boldsymbol{M}} = \{\tilde{m}/3,\ \tilde{\theta}/2\}$ and the stress-localization tensors $\boldsymbol{B}^{(r)} = \{b_m^{(r)},\ b_{dev}^{(r)}\}$ read, respectively:

$$\tilde{m} = m^{(2)} + c^{(1)}\left(m^{(1)} - m^{(2)}\right)b_m^{(1)},\quad \tilde{\theta} = \theta^{(2)} + c^{(1)}\left(\theta^{(1)} - \theta^{(2)}\right)b_{dev}^{(1)},$$

$$b_m^{(1)} = \frac{1}{1 + c^{(2)} q_m \left(m^{(1)} - m^{(2)}\right)}, \quad b_m^{(1)} = \frac{1}{1 + c^{(2)} q_{dev} \left(\theta^{(1)} - \theta^{(2)}\right)},$$

$$b_m^{(2)} = \frac{1}{c^{(2)}}\left(1 - c^{(1)} b_m^{(1)}\right), \quad b_{dev}^{(2)} = \frac{1}{c^{(2)}}\left(1 - c^{(1)} b_{dev}^{(1)}\right).$$

Regarding step 2 of the procedure, namely, the non-linear equation which stems from the choice of the "effect stress," the classical method makes use of the average stresses $\bar{\boldsymbol{\sigma}}^{(r)} = \boldsymbol{B}^{(r)} : \bar{\boldsymbol{\sigma}}$. However, since $\boldsymbol{M}_{sct}(\boldsymbol{\sigma})$ depends on $\boldsymbol{\sigma}$ only through the von Mises stress, the only useful information in the average stress is the von Mises equivalent stress $\bar{\sigma}_{eq}^{(r)} = b_{dev}^{(r)} \bar{\sigma}_{eq}$. Therefore, the nonlinear equations to be solved in the classical secant method read

$$\theta^{(r)} = \theta_{sct}^{(r)}\left(\bar{\sigma}_{eq}^{(r)}\right) = \frac{1}{\mu_0^{(r)}} + \frac{3p^{(r)}\left(\bar{\sigma}_{eq}^{(r)}\right)}{\bar{\sigma}_{eq}^{(r)}}, \quad \bar{\sigma}_{eq}^{(r)} = b_{dev}^{(r)} \bar{\sigma}_{eq} \tag{22}$$

where the coefficients $b_{dev}^{(r)}$ depend on the $\theta^{(r)}$ (this is where the nonlinearity comes into play).

As for the secant method based on the "second-order moment," using again the fact that the secant compliance depends on the stress only through the von Mises stress, it is sufficient to compute the following quantities $\bar{\bar{\sigma}}_r = \langle \sigma_{eq}^2 \rangle_r^{1/2}$. These quantities can be calculated by means of Eq. 16. The resulting nonlinear systems of equations finally read (see Suquet [9] for more details):

$$\left.\begin{aligned}
\theta^{(r)} = \theta_{sct}^{(r)}\left(\bar{\bar{\sigma}}_{eq}^{(r)}\right) = \frac{1}{\mu_0^{(r)}} + \frac{3p^{(r)}\left(\bar{\bar{\sigma}}_{eq}^{(r)}\right)}{\bar{\bar{\sigma}}_{eq}^{(r)}}, \quad \bar{\bar{\sigma}}_{eq}^{(r)} = b_{dev}^{(1)} \bar{\sigma}_{eq}, \\
\bar{\bar{\sigma}}_{eq}^{(2)} = \left(a\bar{\sigma}_m^2 + b\bar{\sigma}_{dev}^2\right)^{1/2} \\
a = \frac{3}{c^{(2)}\theta^{(2)}}\left(\tilde{m} - c^{(1)} m^{(1)} b_m^{(1)^2} - c^{(2)} m^{(2)} b_m^{(2)^2}\right) \\
b = \frac{1}{c^{(2)}\theta^{(2)}}\left(\tilde{\theta} - c^{(1)}\theta^{(1)} b_{dev}^{(1)^2} - \frac{12}{5} c^{(1)} c^{(2)} m^{(2)} b_{dev}^{(1)^2} \left(\frac{\theta^{(1)} - \theta^{(2)}}{3\theta^{(2)} + 4m^{(2)}}\right)^2\right)
\end{aligned}\right\} \tag{23}$$

Note that the effective stress in phase 2 (matrix) is now sensitive to the overall hydrostatic pressure (which is not the case in the classical secant method).

## 10.3.3 VARIATIONAL BOUNDS

### 10.3.3.1 Effective Potentials

In this section it is further assumed that the constitutive behavior of the individual phases derives from a potential, or strain-energy function $w(\varepsilon)$, or equivalently a stress-energy function $w^*(\boldsymbol{\sigma})$, in such a way that the (infinitesimal) strain and stress fields, $\boldsymbol{\varepsilon}$ and $\boldsymbol{\sigma}$, are related by

$$\boldsymbol{\sigma} = \frac{\partial w}{\partial \boldsymbol{\varepsilon}}(\boldsymbol{\varepsilon}), \quad \text{or equivalently} \quad \boldsymbol{\varepsilon} = \frac{\partial w^*}{\partial \boldsymbol{\sigma}}(\boldsymbol{\sigma}),$$

$$\text{with } w(\boldsymbol{\varepsilon}) + w^*(\boldsymbol{\sigma}) = \boldsymbol{\sigma} : \boldsymbol{\varepsilon} \tag{24}$$

In the composite, the potentials $w$ and $w^*$ depend on the phase under consideration (and are denoted by $w^{(r)}$ and $(w^{(r)})^*$, respectively).

The solutions $\boldsymbol{u}$ and $\boldsymbol{\sigma}$ of the local problem (Eq. 2) have variational properties which are essential to deriving bounds on the effective energy of nonlinear composites. These properties, called *minimum energy principles*, also permit us to define the effective potentials of the composite. For simplicity, we shall only consider here the "minimum complementary-energy principle":

$$\tilde{w}^*(\bar{\boldsymbol{\sigma}}) = \inf_{\boldsymbol{\tau} \in \mathcal{K}(\bar{\boldsymbol{\sigma}})} \langle w^*(\boldsymbol{\tau}) \rangle, \quad \mathcal{K}(\bar{\boldsymbol{\sigma}}) = \{\boldsymbol{\tau},\ div(\boldsymbol{\tau}) = 0 \text{ in } V,\ \boldsymbol{\tau}.\boldsymbol{n} = \bar{\boldsymbol{\sigma}}.\boldsymbol{n} \ on\ \partial V\} \tag{25}$$

It can be checked [10] that the effective constitutive relations obtained by averaging the strain field solution of Eq. 2 or equivalently Eq. 25 derives from the potential $\tilde{w}^*$, which is therefore the effective complementary-energy of the composite:

$$\bar{\boldsymbol{\varepsilon}} = \frac{\partial \tilde{w}^*}{\partial \bar{\boldsymbol{\sigma}}}(\bar{\boldsymbol{\sigma}}).$$

### 10.3.3.2 Bounds

#### 10.3.3.2.1 Ponte Castañeda's General Theory

To obtain bounds on the effective potential $\tilde{w}^*$ which are sharper than the Voigt and Reuss bounds, we introduce an inhomogeneous linear comparison composite with compliance $\boldsymbol{M}(x)$ and complementary-energy

$w_0^*(x, \tau) = \frac{1}{2}\tau : M_0(x) : \tau$. Writing $w^*$ as $w^* - w_0^* + w_0^*$, one obtains

$$\left.\begin{aligned} \tilde{w}^*(\bar{\sigma}) &= \inf_{\tau \in \mathcal{K}(\bar{\sigma})} \langle w^*(\tau) \rangle = \inf_{\tau \in \mathcal{K}(\bar{\sigma})} \left( \langle (w^* - w_0^*)(\tau) \rangle + \langle w_0^*(\tau) \rangle \right) \\ &\geq \left( \inf_{\tau \in \mathcal{K}(\bar{\sigma})} \langle w_0^*(\tau) \rangle \right) - \langle v(x, M_0(x)) \rangle \\ &= \tilde{w}_0^*(M_0, \bar{\sigma}) - V(M_0) \end{aligned}\right\} \quad (26)$$

where $\tilde{w}_0^*$ is the effective complementary energy of the linear comparison solid and

$$v(x, M_0(x)) = \sup_{\tau} \left[ w_0^*(x, \tau) - w^*(x, \tau) \right], \quad V(M_0) = \langle v(x, M_0(x)) \rangle \quad (27)$$

Since Eq. 26 is valid for any choice of $M_0(x)$, one has

$$\tilde{w}^*(\bar{\sigma}) \geq \sup_{M_0(x) > 0} \left[ \tilde{w}_0^*(M_0, \bar{\sigma}) - V(M_0) \right] \quad (28)$$

The inequality (Eq. 28), due to Ponte Castañeda, gives a rigorous bound[1] on the nonlinear effective properties of the composite (through the potential $\tilde{w}^*$) in terms of two functions:

- $\tilde{w}_0^*(M_0, \bar{\sigma}) = \frac{1}{2}\bar{\sigma} : \tilde{M}_0 : \bar{\sigma}$ is the elastic energy of a *linear comparison composite* made up of phases with compliance $M_0(x)$ at point $x$; the linear comparison solid is chosen among all possible comparison composites by solving the optimization problem (Eq. 28). The difficulty lies in the precise determination of the energy $\tilde{w}_0^*$ for a linear comparison solid consisting of infinitely many different phases.
- The role of $v(x, .)$ is to measure the difference between the nonquadratic potential $w^*(x, .)$ and the quadratic energy $w_0^*(x, .)$ of the linear comparison solid. This function is difficult to compute in general, but a bound can be easily computed for the class of materials considered in following text.

The problem of bounding the effective properties of a linear composite with infinitely many phases being too difficult, we reduce it by minimizing over a smaller set of compliance fields $M_0(x)$, namely, those fields which are uniform on each subdomain $V_r$. With this smaller set, the supremum

[1] For a broad class of material behavior (discussed in Section 10.3.3.3), the inequality (Eq. 28) is in fact an equality and is strictly equivalent to the variational characterization of $\tilde{w}^*$ given in Eq. 25.

in Eq. 28 even smaller:

$$\tilde{w}^*(\bar{\boldsymbol{\sigma}}) \geq \sup_{M_0^{(r)}>0} \left( \frac{1}{2} \bar{\boldsymbol{\sigma}} : \tilde{\mathbf{M}}_0 \left( \left\{ \mathbf{M}_0^{(r)} \right\} \right)_{r=1,\ldots,N} : \bar{\boldsymbol{\sigma}} - V \left( \left\{ \mathbf{M}_0^{(r)} \right\} \right)_{r=1,\ldots,N} \right) \tag{29}$$

The linear comparison composite is now an $N$-phase composite, with compliance $M_0^{(r)}$ uniform throughout phase $r$. There is a similarity (and even more, as will be discussed in the next paragraph) with the secant methods in that a linear comparison solid is introduced in both approaches. Note, however, that here the elastic moduli in the linear composite are determined by means of an optimization procedure, whereas they were deduced from the (somehow arbitrary) choice of an effective stress in the case of the secant method. This optimization procedure leads to a rigorous bound for the potential $\tilde{w}^*$.

*Remark*: Most of the nonlinear bounds available to date are bounds on the energy of the composite. Only in specific situations do these bounds give bounds on the stress-strain relations of the composite. This is the case for power-law materials for which, due to the Euler theorem for homogeneous functions of degree $n+1$, one has $\bar{\boldsymbol{\sigma}} : \bar{\boldsymbol{\varepsilon}} = (n+1)\tilde{w}^*(\bar{\boldsymbol{\sigma}})$. Therefore, any bound on $\tilde{w}^*$ gives a bound on the overall strain $\bar{\boldsymbol{\varepsilon}}$ in the direction of the applied stress $\bar{\boldsymbol{\sigma}}$. However, no information on the other components of the strain is provided. Interestingly, a method for bounding directly the stress-strain relation has been recently proposed by Milton and Serkov [19].

#### 10.3.3.2.2 Complementary Energies Depending Only on the Quadratic Stress

We consider here a broad class of behaviors corresponding to potentials $w^*$ which depend on the stress tensor $\boldsymbol{\sigma}$ only through the quadratic stress $\mathscr{S}$, $w^*(\boldsymbol{\sigma}) = G(\mathscr{S})$, for some appropriately chosen function $G$ and where $\mathscr{S} = \frac{1}{2}\boldsymbol{\sigma} \otimes \boldsymbol{\sigma}$. $G$ is further assumed to be a *convex* function of $\mathscr{S}$. This class of materials contains in particular all materials with a complementary energy in the form $w^*(\boldsymbol{\sigma}) = \frac{1}{2k}\sigma_m^2 + \psi(\sigma_{eq})$.

For this class of materials, the function $v^{(r)}$ (corresponding to $v$ in phase $r$) can be bounded from above by

$$v^{(r)}\left(\mathbf{M}_0^{(r)}\right) = \sup_{\tau} \left[ \mathbf{M}_0^{(r)} :: \mathscr{T} - G^{(r)}(\mathscr{T}) \right] \leq \left(G^{(r)}\right)^* \left(\mathbf{M}_0^{(r)}\right), \tag{30}$$

where $\mathscr{T} = \frac{1}{2}\boldsymbol{\tau} \otimes \boldsymbol{\tau}$

$(G^{(r)})^*$ denotes the Legendre transform of the convex function $G^{(r)}$:

$$\left(G^{(r)}\right)^*(\mathbf{M}) = \sup_{\mathscr{T}} \left[\mathbf{M} :: \mathscr{T} - G^{(r)}(\mathscr{T})\right] \tag{31}$$

Note that in Eq. 31 the supremum is taken over all symmetric fourth-order $\mathscr{T}$, whereas the supremum in Eq. 30 is restricted to rank-one symmetric $\mathscr{T}$. The lower bound (Eq. 29) reduces to

$$\tilde{w}^*(\bar{\boldsymbol{\sigma}}) \geq \sup_{\mathbf{M}_0^{(r)}>0} \left(\frac{1}{2}\bar{\boldsymbol{\sigma}} : \tilde{\mathbf{M}}_0\left(\left\{\mathbf{M}_0^{(r)}\right\}\right)_{r=1,\ldots,N} : \bar{\boldsymbol{\sigma}} - \sum_{r=1}^{N} c^{(r)}\left(G^{(r)}\right)^*\left(\mathbf{M}_0^{(r)}\right)\right) \tag{32}$$

## 10.3.3.3 Connection with the Secant Method Based on Second-Order Moments

When the complementary energy of the constitutive phases depends only on the "quadratic stress" $\mathscr{S}$, the constitutive relation (Eq. 24) can alternatively be written as

$$\begin{aligned} \boldsymbol{\varepsilon} = \mathbf{M}_{sct}(\mathscr{S}) : \boldsymbol{\sigma}, \quad &\text{where } \mathbf{M}_{sct}(\mathscr{S}) = \frac{\partial G}{\partial \mathscr{S}}(\mathscr{S}), \\ \text{or equivalently } \mathscr{S} &= \frac{\partial G^*}{\partial \mathbf{M}}(\mathbf{M}_{sct}) \end{aligned} \tag{33}$$

We are now going to inspect in more detail the optimality conditions for the moduli $\mathbf{M}_0^{(r)}$ derived from the optimization problem (Eq. 32). Assuming stationarity with respect to these moduli, the optimality conditions read as

$$\frac{1}{2}\bar{\boldsymbol{\sigma}} : \frac{\partial \tilde{\mathbf{M}}_0}{\partial \mathbf{M}_0^{(r)}} : \bar{\boldsymbol{\sigma}} = c^{(r)} \frac{\partial\left(G^{(r)}\right)^*}{\partial \mathbf{M}}\left(\mathbf{M}_0^{(r)}\right)$$

But according to Eq. 16, the first term in this equality is nothing other than the average second-order moment of the stress in the linear comparison solid. Therefore:

$$\tilde{\mathscr{S}}^{(r)} = \frac{1}{2}\langle \boldsymbol{\sigma} \otimes \boldsymbol{\sigma} \rangle_r = \frac{\partial\left(G^{(r)}\right)^*}{\partial \mathbf{M}}\left(\mathbf{M}_0^{(r)}\right)$$

Making use of Eq. 33, the optimally conditions finally amount to solving the following systems of nonlinear equations:

$$\mathbf{M}_0^{(r)} = \mathbf{M}_{sct}^{(r)}\left(\tilde{\mathscr{S}}^{(r)}\right), \quad \text{with} \quad \tilde{\mathscr{S}}^{(r)} = \frac{1}{2c^{(r)}}\bar{\boldsymbol{\sigma}} : \frac{\partial \tilde{\mathbf{M}}_0}{\partial \mathbf{M}_0^{(r)}} : \bar{\boldsymbol{\sigma}}$$

This systems coincides with Eq. 17.

In conclusion, it has been shown that the optimal moduli $M_0^{(r)}$ in the variational procedure coincide with the secant moduli $M^{(r)}$ determined by the (more heuristic) secant method based on the second-order moment described in Section 10.3.2.5. In other words, the variational procedure can be interpreted as a secant method. It has, however, the definite advantage of delivering a clear rigorous bound on the effective properties of the nonlinear composite.

## ACKNOWLEDGEMENTS

This paper was written while the author was a Visiting Associate at the California Institute of Technology for the year 2000–2001. The financial support of Caltech is gratefully acknowledged. Stimulating discussions with P. Ponte Castañeda are gratefully acknowledged.

## REFERENCES

1. Willis, J. R. (1989). The structure of overall constitutive relations for a class of nonlinear composites. *IMA J. Appl. Math.* **43**: 231–242.
2. Ponte Castañeda, P. (1991). The effective mechanical properties of nonlinear isotropic composites. *J. Mech. Phys. Solids* **39**: 45–71.
3. Willis, J. (1991). On methods for bounding the overall properties of nonlinear composites. *J. Mech. Phys. Solids* **39**: 73–86.
4. Suquet, P. (1993). Overall potentials and extremal surfaces of power law or ideally plastic materials. *J. Mech. Phys. Solids* **41**: 981–1002.
5. Ponte Castañeda, P. (1992). New variational principles in plasticity and their application to composite materials. *J. Mech. Phys. Solids* **40**: 1757–1788.
6. Ponte Castañeda, P. (1996). Exact second-order estimates for the effective mechanical properties of nonlinear composite materials. *J. Mech. Phys. Solids* **44**: 827–862.
7. Masson, R., and Zaoui, A. (1999). Self-consistent estimates for the rate-dependent elastoplastic behavior of polycrystalline materials. *J. Mech. Phys. Solids* **47**: 1543–1568.
8. Suquet, P. (1995). Overall properties of nonlinear composites: A modified secant moduli theory and its link with Ponte Castañeda's nonlinear variational procedure. *C.R. Acad. Sc. Paris*, **320**, Série IIb: 563–571.
9. Suquet, P. (1997). Effective properties of nonlinear composites, in *Continuum Micromechanics*, vol. **337** of *CISM Lecture Notes*. pp. 197–264, Suquet, P. ed., New York: Springer-Verlag.
10. Ponte Castañeda, P. Ponte, and Suquet, P. (1998). Nonlinear composites, in *Advances in Applied Mechanics*, pp. 171–302, vol. **34**, van der Giessen, E., and Wu, T.Y. eds., New York: Academic Press.
11. Willis, J. R. (2000). The overall response of nonlinear composite media. *Eur. J. Mech. A/Solids* **19**: S165–S184.
12. Suquet, P. (1997). *Continuum Micromechanics*, vol. **337** of *CISM Lecture Notes*, Wien: Springer-Verlag.
13. Gilormini, P. (1996). A critical evaluation of various nonlinear extensions of the self-consistent model, in *Micromechanics of Plasticity and Damage of Multiphase Materials*, pp. 67–74, Pineau, A., and Zaoui, A., eds., Dordrecht: Kluwer Acad. Pub.

14. Masson, R., Bornert, M., Suquet, P., and Zaoui, A. (2000). An affine formulation for the prediction of the effective properties of nonlinear composites and polycrystals. *J. Mech. Phys. Solids* **48**: 1203–1227.
15. Buryachenko, V. (1993). Effective strength properties of elastic physically nonlinear composites, in *Micromechanics of Materials*, pp. 567–578, Marigo, J. J., and Rousselier, G., eds., Paris: Eyrolles.
16. Qiu, Y. P., and Weng, G. J. (1992). A theory of plasticity for porous materials and particle-reinforced composites. *J. Appl. Mech.* **59**: 261–268.
17. Hu, G. (1996). A method of plasticity for general aligned spheroidal void of fibre-reinforced composites. *Int. J. Plasticity* **12**: 439–449.
18. Kreher, W. (1990). Residual stresses and stored elastic energy of composites and polycrystals. *J. Mech. Phys. Solids* **38**: 115–128.
19. Milton, G. W., and Serkov, S. K. (2000). Bounding the current in nonlinear conducting composites. *J. Mech. Phys. Solids* **48**: 1295–1324.

SECTION 10.4

# Nonlocal Micromechanical Models

J. R. WILLIS
*Department of Mathematical Sciences, University of Bath, Bath BA2 7AY, United Kingdom*

Contents



## 10.4.1 INTRODUCTION

Virtually any composite material has a microstructure that needs to be treated as random. Many composites are obviously disordered: polycrystalline metal or glass-fibre reinforced material is an example. Even composites with highly correlated microstructures can be viewed as random. Consider, for instance, the idealised case of a specimen comprising a matrix containing a perfectly periodic array of identical inclusions. The spacing of the inclusions is a very small fraction of a typical dimension of the specimen, and it is most unlikely that the exact location of any one inclusion (in terms of which the positions of all the others are fixed) will be known. Thus it is appropriate to regard the specimen as one chosen at random from a set of macroscopically identical specimens, distinguished from one another by the exact location of one chosen inclusion within one particular periodic cell. In practice, observation



*Handbook of Materials Behavior Models*. ISBN 0-12-443341-3.

of stress or strain at some position $x$ within the specimen will result in the measurement of some average in the vicinity of $x$ — for instance, by use of a strain guage or some optical device. It is appropriate, therefore, to seek an effective constitutive relation between such local averages of stress and strain. Unfortunately, this is difficult to accomplish, and it is usual instead to seek a relation between the ensemble means of the stress and the strain. This expedient is adopted throughout virtually the whole of physics, and it will be adopted without further comment here, although for composites at least the distinction between local spatial average and ensemble average can be confronted explicitly; it could also in principle be evaluated quantitatively in some particular case, but no such study is known to the author.

## 10.4.2 FORMULATION

The composite, in general, will be taken to occupy a domain $\Omega$ and to be made of $n$ constituents, or phases, perfectly bonded across interfaces. The stress and strain in material of type $r$ $(r = 1, 2, \ldots, n)$ are related so that

$$\boldsymbol{\sigma} = \boldsymbol{L}_r \boldsymbol{e}, \quad \sigma_{ij} = (L_r)_{ijkl} e_{kl}, \tag{1}$$

the second form (in which the usual summation convention on repeated suffixes is employed) giving the explicit meaning of the first more concise form which will be used in the sequel. The elastic constant tensor $\boldsymbol{L}(x)$ of the composite then takes the form

$$\boldsymbol{L}(x) = \sum_{r=1}^{n} \boldsymbol{L}_r \chi_r(x) \tag{2}$$

where $\chi_r(x)$ is the characteristic function of the region occupied by phase $r$, taking the value 1 when $x$ lies in phase $r$ and zero otherwise. The elastic constant tensors $\boldsymbol{L}_r$ are known and correspond to particular materials, but the functions $\chi_r$ that specify the microgeometry are random. The probability $p_r(x)$ for finding material of type $r$ at $x$, and the probability $p_{rs}(x, x')$ for finding simultaneously material of type $r$ at $x$ and of type $s$ at $x'$ are

$$p_r(x) = \langle \chi_r(x) \rangle, \quad p_{rs}(x, x') = \langle \chi_r(x) \chi_s(x') \rangle, \tag{3}$$

the angled brackets signifying the ensemble average. If the composite is statistically uniform, $p_r$ becomes a constant, equal to the volume fraction of phase $r$, and $p_{rs}$ becomes a function of $(x - x')$ only. (Actually, these conditions by themselves only require the medium to be "statistically second-order stationary.")

The (vector) equation of equilibrium of the actual medium is

$$\operatorname{div} \boldsymbol{\sigma} + \boldsymbol{f} = 0, \quad \boldsymbol{x} \in \Omega \tag{4}$$

which is to be solved in conjunction with the constitutive relation

$$\boldsymbol{\sigma}(\boldsymbol{x}) = \boldsymbol{L}(\boldsymbol{x})\boldsymbol{e}(\boldsymbol{x}) \tag{5}$$

and suitable boundary conditions applied over the boundary $\partial\Omega$ of $\Omega$. Prescribing the displacement, $\boldsymbol{u} = \boldsymbol{u}^0$, say, over $\partial\Omega$ is one possibility; another is to prescribe the traction $\boldsymbol{\sigma} \cdot \boldsymbol{n} = \boldsymbol{\sigma}^0 \cdot \boldsymbol{n}$, where $\boldsymbol{n}$ is the (outward) normal to $\partial\Omega$.

Under such conditions, the stress, strain, and displacement of the composite are defined. They could, in principle, be determined for each realisation of the composite and then their ensemble means calculated. The purpose of developing an effective constitutive relation is to allow the direct determination of the ensemble means. First, averaging Eq. 4 gives

$$\operatorname{div}\langle \boldsymbol{\sigma} \rangle + \boldsymbol{f} = 0, \quad \boldsymbol{x} \in \Omega \tag{6}$$

It is assumed here that the body force $\boldsymbol{f}$ is sure, for simplicity. The more general case of configuration-dependent body force has been treated recently by Luciano and Willis [1]. The exact constitutive relation (Eq. 5) is replaced by the effective relation

$$\langle \boldsymbol{\sigma} \rangle = \boldsymbol{L}^{eff} \langle \boldsymbol{e} \rangle \tag{7}$$

where $\boldsymbol{L}^{eff}$ is a nonlocal operator. If it were known, it is a reasonable assumption that Eqs. 6 and 7 could be solved in conjunction with the ensemble averaged boundary condition, either $\langle \boldsymbol{u} \rangle = \boldsymbol{u}^0$ or $\langle \boldsymbol{\sigma} \rangle \cdot \boldsymbol{n} = \boldsymbol{\sigma}^0 \cdot \boldsymbol{n}$ over $\partial\Omega$.

There is, however, a serious limitation. It will be seen that the effective nonlocal operator can be constructed in terms of a certain Green's function for the domain $\Omega$, and this Green's function depends on the form of the boundary conditions. Thus, implicitly, at least the type of boundary condition has an influence on the form of $\boldsymbol{L}^{eff}$. This influence will only manifest itself, however, in some "boundary layer" in the vicinity of $\partial\Omega$. Outside of this layer, $\boldsymbol{L}^{eff}$ will take the same form as it would in an infinite body. This is the only case for which explicit results are known. Any attempt to use the infinite-body form right up to the boundary $\partial\Omega$ would require the replacement of the exact (mean) boundary condition by some "effective" boundary condition, if such a condition could be proved to have a meaning. This question has not even been addressed to date. In the sequel, the problem will be avoided by considering only an infinite domain, loaded by a body force $\boldsymbol{f}$ which has compact support, coupled with the requirement that all fields should tend to zero as $|x| \to \infty$. The question of boundary conditions will be briefly returned to at the end of the article.

## 10.4.3 THE EFFECTIVE RELATION

It is useful to introduce a comparison medium, with elastic constant tensor $L_0$, and to replace Eq. 5 by

$$\sigma = L_0 e + \tau \tag{8}$$

The "stress polarisation" $\tau$ should satisfy

$$\tau(x) = (L(x) - L_0)e(x) \tag{9}$$

Substituting Eq. 9 into Eq. 4 gives

$$\operatorname{div}(L_0 e) + \operatorname{div} \tau + f = 0, \quad x \in \Omega \tag{10}$$

which, together with the relevant boundary condition, implies that

$$u = u^0 - E_0^\dagger \tau \tag{11}$$

Here, $u^0 = G_0 f$ is the solution of the problem defined by Eq. 10 together with the boundary condition, $G_0$ representing the corresponding Green's function operator; $E_0^\dagger$ is the operator obtained from it by integration by parts to move the operation corresponding to the divergence on $\tau$ onto the Green's function. Differentiating Eq. 11 with respect to $x$ and symmetrising then gives the expression

$$e = e^0 - \Gamma_0 \tau \tag{12}$$

for the strain $e$. Explicitly, the operator $\Gamma_0$ has kernel

$$(\Gamma_0)_{ijkl}(x, x') = \left. \frac{\partial^2 (G_0)_{ik}(x, x')}{\partial x_j \partial x'_l} \right|_{(ij),(kl)} \tag{13}$$

where the bracketing of suffixes implies symmetrisation. Combining Eqs. 9 and 12 now gives an equation that must be satisfied by $\tau$ or, equivalently, an equation that must be satisfied by $e$.

In the case that the medium is weakly heterogeneous, so that $L(x)$ differs only slightly from $L_0$, the equation for $e$ was solved by iteration, many years ago, by Beran and McCoy [2]. Such analysis demonstrates that the operator $L^{eff}$ is unavoidably nonlocal. Here, however, concern will be for the general case of large fluctuations. For this purpose, it is useful to consider the equation for $\tau$ in the form

$$(L - L_0)^{-1} \tau + \Gamma_0 \tau = e^0 \tag{14}$$

Averaging Eq. 12 allows $e^0$ in Eq. 14 to be expressed in terms of $\langle e \rangle$, so that

$$(L - L_0)^{-1}\tau + \Gamma_0(\tau - \langle \tau \rangle) = \langle e \rangle \tag{15}$$

Formally, the solution of this equation is

$$\tau = T\langle e \rangle \tag{16}$$

and it follows by averaging Eq. 8 that

$$L^{eff} = L_0 + \langle T \rangle \tag{17}$$

The perturbation theory contained in Reference [2] provides a series representation for $T$, yielding

$$L^{eff} = \langle L \rangle - \langle (L - L_0)\Gamma_0(L - \langle L \rangle) \rangle + \cdots \tag{18}$$

The operator $\Gamma_0$ may be expressed in the form

$$\Gamma_0 = \Gamma_0^{\infty} + \Gamma_0^{im} \tag{19}$$

where $\Gamma_0^{\infty}$ is the corresponding operator for an infinite body and $\Gamma_0^{\mathrm{im}}$ is an "image" correction, which comes from the boundary conditions. The kernel of $\Gamma_0^{\infty}$ is a homogeneous function of degree $-3$ (in three-dimensional space). Except when the point $x$ of interest is close to the boundary $\partial\Omega$ (relative to the scale $l$ of the microstructure), the kernel of the "image" term remains bounded, even when $x'$ is close to $x$. In this case, $\Gamma_0$ may be replaced, asymptotically, when $l/L \ll 1$, where $L$ is a length characteristic of the domain $\Omega$, by $\Gamma_0^{\infty}$, and then the effective operator $L^{eff}$ reduces to the form appropriate to an infinite body. As remarked previously, it is only in this case that any explicit calculations have been performed.

## 10.4.4 BOUNDS

Even in the infinite-body limit, $L^{eff}$ cannot be found exactly. It is, however, possible to obtain useful information by bounding the energy in the composite. Bounds which involve no more information than the one- and two-point probabilities $p_r$ and $p_{rs}$ can be found by using the Hashin–Shtrikman formalism, as developed in Reference [3] and exploited further in Reference [1]. Considering the case of zero prescribed displacement over $\partial\Omega$ (including, as a limiting case, the problem for an infinite body), the energy in one realisation of the composite (elastic stored energy plus the energy associated with the body-force loading) is

$$\mathscr{E} = -\tfrac{1}{2}\int_{\Omega} f^T G f \, \mathrm{d}x \tag{20}$$

where $G$ represents the Green's function operator of the actual composite.

Ensemble averaging gives

$$\langle \mathscr{E} \rangle = -\tfrac{1}{2}\int_\Omega f^T G^{eff} f \, dx \tag{21}$$

Here, $G^{eff} = \langle G \rangle$ is precisely the Green's function operator associated with the effective medium, which is defined by the effective modulus operator $L^{eff}$. Application of the Hashin–Shtrikman formalism permits the development of bounds for $\langle G \rangle$, implying corresponding restrictions on the operator $L^{eff}$, as detailed in References [1] and [3], for example. The result is that a Hashin–Shtrikman approximation $G_{HS}$ to $G^{eff}$ follows by solving, in place of the exact equation (Eq. 15) for $\boldsymbol{\tau}$, the Hashin–Shtrikman equations

$$p_r(L_r - L_0)^{-1}\boldsymbol{\tau}_r + \sum_{s=1}^{n} \{\boldsymbol{\Gamma}_0(p_{rs} - p_r p_s)\}\boldsymbol{\tau}_s = p_r\langle e_{HS}\rangle \tag{22}$$

where $\boldsymbol{\tau}$ is approximated as

$$\boldsymbol{\tau}(x) = \sum_{r=1}^{n} \boldsymbol{\tau}_r(x)\chi_r(x) \tag{23}$$

the functions $\boldsymbol{\tau}_r$ being nonrandom, and $e_{HS}$ is the corresponding approximation to $e$. The notation $\{K(p_r s - p_r p_s)\}$ is used for an integral operator whose kernel is $K(x, x')[p_{rs}(x, x') - p_r(x)p_s(x')]$.

If the solution of Eq. 22 is expressed as

$$\boldsymbol{\tau}_r = \sum_{s=1}^{n} T_{rs}(p_s\langle e_{HS}\rangle) \tag{24}$$

the corresponding Hashin–Shtrikman approximation to $L^{eff}$ is

$$L_{HS} = L_0 + \sum_{r=1}^{n}\sum_{s=1}^{n} p_r T_{rs} p_s \tag{25}$$

The associated Hashin–Shtrikman approximation $G_{HS}$ to $G^{eff}$ provides the approximation

$$\mathscr{E}_{HS} = -\tfrac{1}{2}\int_\Omega f^T G_{HS} f \, dx \tag{26}$$

for $\langle \mathscr{E} \rangle$. It is an upper bound to $\langle \mathscr{E} \rangle$ if $L_0$ is chosen so that $L_r - L_0$ defines a positive (semi-)definite quadratic form for each $r$, and a lower bound if $L_r - L_0$ defines a negative (semi-)definite form for each $r$.

## 10.4.5 STATISTICALLY UNIFORM MEDIA

For a statistically uniform medium, $p_r$ is constant, and $p_{rs}$ is a function of $(x - x')$ only. Also, taking the medium as infinite in extent, $\mathbf{\Gamma}_0$ is replaced by $\mathbf{\Gamma}_0^\infty$, whose kernel is also a function of $(x - x')$. All operators become convolutions, and Eq. 22 can be reduced to algebraic form by taking Fourier transforms. Also, Plancherel's theorem allows Eq. 21 to be expressed

$$\langle \mathscr{E} \rangle = -\tfrac{1}{2} \int \tilde{f}^*(k)\tilde{G}^{eff}(k)\tilde{f}(k)\,\mathrm{d}k \tag{27}$$

where the tilde denotes the Fourier transform and * denotes complex conjugate transpose.

It follows that $(\tilde{G}^{eff} - \tilde{G}_{HS})$ defines a Hermitian form that is negative or positive semidefinite when $L_0$ is such that $(L_r - L_0)$ defines a quadratic form that is positive or negative semidefinite for each $r$.

## 10.4.6 GRADIENT APPROXIMATIONS

We write the overall constitutive relation

$$\langle \sigma \rangle(x) = (L^{eff}\langle e \rangle)(x) = \int_\Omega K(x, x')\langle e \rangle(x')\,\mathrm{d}x' \tag{28}$$

The kernel function $K$ decays to zero rapidly as $|x - x'|/l \to \infty$, where $l$ is a characteristic length associated with the microstructure. Therefore, when the loading is such that the resulting mean strain $\langle e \rangle$ varies slowly relative to the scale $l$, the result of applying the operator $L^{eff}$ to $\langle e \rangle$ can be evaluated, asymptotically, by approximating $\langle e \rangle(x')$ by the first few terms in its Taylor expansion about $x' = x$: in symbolic notation,

$$\begin{aligned}\langle e(x') \rangle \approx{}& \langle e \rangle(x) + (x' - x) \cdot \nabla \langle e \rangle(x) \\ &+ \tfrac{1}{2}[(x' - x) \otimes (x' - x)] : [\nabla \otimes \nabla]\langle e \rangle(x) + \cdots\end{aligned} \tag{29}$$

It follows that

$$\langle \sigma \rangle(x) \approx L^{hom}(x)\langle e \rangle(x) + M(x) \cdot \nabla \langle e \rangle(x) + N(x) : (\nabla \otimes \nabla)\langle e \rangle(x) + \cdots \tag{30}$$

where

$$L^{hom}(x) = \int K(x, x')\,\mathrm{d}x'$$

$$M(x) = \int K(x, x')(x' - x)\,\mathrm{d}x'$$

$$N(x) = \tfrac{1}{2}\int K(x,x')[(x'-x)\otimes(x'-x)]\,dx' \tag{31}$$

The tensor $L^{hom}$ is the tensor of effective moduli in the "homogenisation limit" $l/L \to 0$.

If the medium is statistically uniform, then $K$ depends on $(x - x')$. The tensors $L^{hom}$, $M$, and $N$ become constants; it will usually be the case that $K$ is an even function of its argument, so that $M = 0$. An alternative representation is also possible:

$$L^{hom} = \tilde{K}(0)$$

$$N = -\tfrac{1}{2}(\nabla_k \otimes \nabla_k)\tilde{K}(0) \tag{32}$$

Thus the gradient approximation represented by Eq. 30 follows from the small-$\boldsymbol{k}$ expansion of the Fourier transform of $L^{eff}$. The corresponding Hashin–Shtrikman approximations follow from parallel treatment of the Fourier transform version of Eq. 25. The constants were developed from this standpoint by Drugan and Willis [4].

## 10.4.7 VARIATIONAL FORMULATION FOR THE EFFECTIVE MEDIUM

For the infinite medium or, more generally, for the finite medium with displacements prescribed over the boundary, the solution of Eqs. 4, 5, and the boundary conditions minimises the energy functional

$$\mathcal{F}(u) = \int_\Omega \left\{\tfrac{1}{2}e : L : e - f \cdot u\right\} dx \tag{33}$$

over displacement fields $u$ that satisfy the boundary conditions. It follows that the family of solutions, defined over the entire set of realisations, minimises

$$\langle \mathcal{F}(u)\rangle = \left\langle \int_\Omega \{\tfrac{1}{2}e : Le - f \cdot u\}\, dx \right\rangle \tag{34}$$

over all random fields $u$, defined over $\Omega$ *and* the sample space, and satisfying the given boundary conditions on $\partial\Omega$. The minimum value of $\mathcal{F}$ is $\mathcal{E}$, and the minimum value of $\langle \mathcal{F} \rangle$ is $\langle \mathcal{E} \rangle$.

It follows from simple mathematics (not physics) that the solution of Eqs. 6 and 7 minimises the functional

$$\mathcal{F}^{eff}(\langle u\rangle) = \int_\Omega \{\tfrac{1}{2}\langle e\rangle : L^{eff}\langle e\rangle - f \cdot \langle u\rangle\}\, dx \tag{35}$$

over fields $\langle u\rangle$ that satisfy the given boundary conditions ($\langle u\rangle = u^0$ on $\partial\Omega$).

The minimum value of $\mathscr{F}^{eff}$ is

$$(\mathscr{F}^{eff})_{min} = -\tfrac{1}{2}\int_{\Omega} f \cdot \langle u \rangle \, dx \tag{36}$$

that is, $\langle \mathscr{E} \rangle$, precisely. There is, however, no relation between $\mathscr{F}^{eff}$ and $\langle \mathscr{F} \rangle$ more generally.

Consider now the gradient approximation,

$$\langle \boldsymbol{\sigma} \rangle = L^{hom}\langle e \rangle + N : (\nabla \otimes \nabla)\langle e \rangle \tag{37}$$

which applies under the assumptions specified in deriving Eqs. 32. If this constitutive relation is *assumed* to hold, right up to the boundary $\partial\Omega$, then the problem specified by Eqs. 6 and 37 requires another boundary condition in addition to the given displacement condition. It is natural to associate with it the functional

$$\mathscr{G}(\langle u \rangle) = \int_{\Omega} \{\tfrac{1}{2}\langle e \rangle L^{hom}\langle e \rangle - \tfrac{1}{2}(\nabla\langle e \rangle) \cdot N \cdot (\nabla\langle e \rangle) - f \cdot \langle u \rangle\} \, dx \tag{38}$$

which is stationary at the solution if the additional boundary condition on $\partial\Omega$ is taken as the "natural" one, $\boldsymbol{n} \cdot N \cdot (\nabla\langle e \rangle) = 0$. There is, however, no direct physical reason why this condition should be applied: its need is a consequence of having adopted the constitutive relation (Eq. 37) up to the boundary, where its derivation has lost validity.

There is another complication, perhaps still more severe. The functional $\mathscr{G}$ may or may not be convex, depending on the constant tensor N. Thus the stationary value need not be a minimum, and, when it is not, it is possible that the boundary value problem just proposed, consisting of Eqs. 6, 37, the given boundary condition, and the "natural" boundary condition, may not have a unique solution; it is also possible that the desired solution could be unstable. This is not in conflict with the basic physics or mathematics: the gradient approximation is *only* good when the gradients are small. This problem was highlighted before, in the context of "weakly heterogeneous" analysis, by Beran and McCoy [2].

## 10.4.8 CONCLUDING REMARKS

This article has shown how nonlocal effective constitutive response can be *deduced* from explicit consideration of the micromechanics of deformation of a composite. The relation in general involves an integral operator. In the case of a statistically uniform medium, the integral takes convolution form, and the effective response can equally well be described in terms of the Fourier transform of the integral operator. It cannot be determined exactly, but bounds on the energy can be developed and these, in turn, provide restrictions on the integral operator, which can be expressed purely algebraically in Fourier space, in the case of statistical uniformity. Figure 10.4.1 (reproduced from Reference [5]) presents a sample result. The example is two-dimensional (representing, for example, a cross section of a fibre-reinforced material under conditions of plane strain). The composite is a matrix ($r = 2$) containing square inclusions ($r = 1$), at volume fraction 0.16, arranged periodically so that each period is a square. The length of each side of the square period is normalised and taken equal to $2\pi$. Both matrix and inclusions have Poisson's ratio 0.25. Young's modulus is 16 for the inclusions and 8 for the matrix. The two-point probability $p_{11}$ (from which the other $p_{rs}$ can be derived, for a two-phase composite) is exactly periodic, its period being the square. The composite is statistically uniform because, as discussed in Section 10.4.1, the location of any one inclusion is treated as random, with uniform distribution over one period. Thus the effective properties can be developed in Fourier space. The figure gives three pairs of curves. One of each pair was obtained by taking $\boldsymbol{L}_0 = \boldsymbol{L}_1$ (to give an upper bound for the energy), and the other was obtained by taking $\boldsymbol{L}_0 = \boldsymbol{L}_2$ (to give a lower bound for the energy). The component $(\tilde{\boldsymbol{L}}_{HS})_{1111}(\boldsymbol{k})$ is displayed, plotted in the three cases $\boldsymbol{k} = (k, 0)$, $\boldsymbol{k} = (k, k)/\sqrt{2}$, $\boldsymbol{k} = (0, k)$. The associated gradient approximations follow from the quadratic approximations to the curves, in the vicinity of $\boldsymbol{k} = 0$. It can be seen that the signs of the quadratic terms are not always the same; it can also be seen that the gradient approximation is accurate only when the magnitude of $\boldsymbol{k}$ is 0.1 or smaller.

The results shown in Figure 10.4.1 are considered to be of some interest because the model is completely unambiguous and (in principle) realisable. In particular, the statistics of the medium are defined without any approximation, and the meaning of the ensemble average is completely explicit. Some results that were presented as bounds for nonlocal properties were produced several years ago by Diener, Hürrich, and Weissbarth [6]. However, their assumed two-point probability was at best an approximation, and their results did not show the complexity displayed in Figure 10.4.1.

The results apply, of course, only to the types of elastic composite for which they have been derived. They may, nevertheless, throw some light more generally on nonlocal models. Most nonlocal models rely on some underlying averaging. It is common practice (see, for instance, Reference [7]) to postulate some relation involving an integral which may, in turn, be developed under an assumption of small gradients to give a gradient approximation. Gradient approximations may also be postulated directly (for instance, Reference [8]). The micromechanical considerations presented here show that the form of the nonlocal kernel may be quite complicated, and the associated gradient approximation may result in a problem that is globally unstable, or ill-posed:

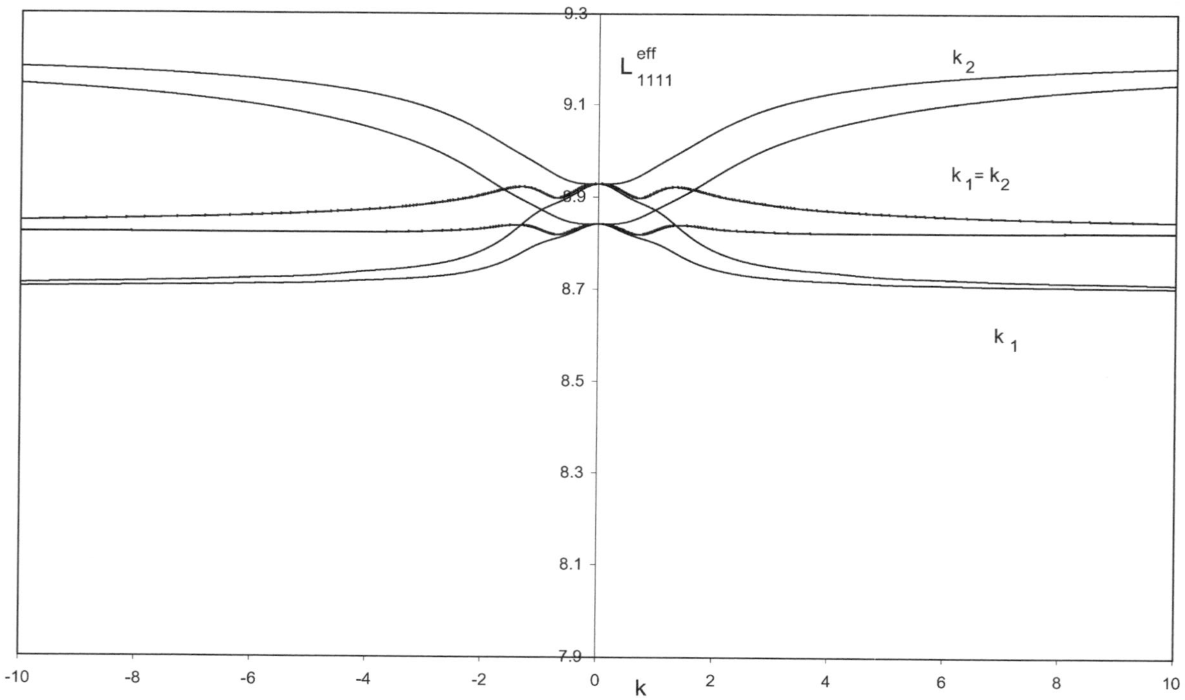


FIGURE 10.4.1 Plots of "upper-bound" and "lower-bound" estimates of ($\boldsymbol{k}$), for a particular two-dimensional, two-phase periodic composite with square periodic cell, for the cases $\boldsymbol{k} = (k, 0)$, $\boldsymbol{k} = (k, k)/\sqrt{2}$, $\boldsymbol{k} = (0, k)$.

it depends on the signs, and these do not always come out as might be desired!

In all such theories, there remains the serious question of assessing the influence of the boundary, which itself interacts nonlocally with adjacent inhomogeneities.

Although only linear elasticity has been considered here, it is, in fact, possible to develop a similar formulation for certain classes of nonlinear problems (starting from a Hashin–Shtrikman formulation as initiated by Talbot and Willis [9]; see also Reference [10]). However, for such problems, the resulting equations are nonlinear, and nonlocality can only be addressed by iteration, leading directly to a gradient approximation. The more exact form of the effective relation in the nonlinear case is at present entirely unknown.

## REFERENCES

1. Luciano, R., and Willis, J. R. (2000). Bounds on non-local effective relations for random composites loaded by configuration-dependent body force. *J. Mech. Phys. Solids* 48: 1827–1849.
2. Beran, M. J., and McCoy, J. J. (1970). The use of strain gradient theory for analysis of random media. *Int. J. Solids Struct.* 6: 1267–1275.
3. Willis, J. R. (1983). The overall response of composite materials. *J. Appl. Mech.* 50: 1202–1209.
4. Drugan, W. J., and Willis, J. R. (1996). A micromechanics-based nonlocal constitutive equation and estimates of representative volume element size for elastic composites. *J. Mech. Phys. Solids* 44: 497–524.
5. Luciano, R., and Willis, J. R. (2001). Non-local effective relations for fibre reinforced composites loaded by configuration-dependent body forces (in preparation).
6. Diener, G., Hürrich, A., and Weissbarth, J. (1984). Bounds for the non-local effective elastic properties of composites. *J. Mech. Phys. Solids* 32: 21–39.
7. Barenblatt, G. I., and Prostokrishin, V. M. (1993). A model of damage taking into account microstructural effects. *Eur. J. Appl. Math.* 4: 225–240.
8. Leroy, Y., and Molinari, A. (1993). Spatial patterns and size effects in shear zones: A hyperelastic model with higher-order gradients. *J. Mech. Phys. Solids* 41: 631–663.
9. Talbot, D. R. S., and Willis, J. R. (1985). Variational principles for inhomogeneous nonlinear media. *IMA J. Appl. Math.* 35: 39–54.
10. Talbot, D. R. S., and Willis, J. R. (1997). Bounds of third order for the overall response of nonlinear composites. *J. Mech. Phys. Solids* 45: 87–111.

SECTION **10.5**

# Transformation Field Analysis of Composite Materials

GEORGE J. DVORAK
*Rensselaer Polytechnic Institute, Troy, New York*

Contents



## 10.5.1 OVERVIEW OF THE METHOD

The transformation field analysis (TFA) method evaluates averages of local and overall strain and stress fields caused in elastic heterogeneous solids, such as composite materials, laminates, and polycrystals, by piecewise uniform distributions of transformation strains or eigenstrains; these may be caused by thermal changes, phase transformations, moisture absorption, or inelastic deformation. Also, equivalent eigenstrains applied in modeling of local cracks, voids, and other damage modes can be included in the analysis. Moreover, the method provides a system of differential equations describing evolution of the local transformation fields under varying overall loads, according to local constitutive relations prescribed in the constituents. Here we discuss applications to inelastic deformation and damage evolution in composites. A more complete description of the method can be found in References [1–4].


*Handbook of Materials Behavior Models*. ISBN 0-12-443341-3.

## 10.5.2 GOVERNING EQUATIONS

Consider a sufficiently large representative volume $V$ of a composite material consisting of two or more perfectly bonded phases. The phases may have any physically admissible elastic symmetry and microstructural geometry. Each phase $r$ in volume $V_r$ is a homogeneous elastic solid, of constant stiffness $\boldsymbol{L}_r$ and compliance $\boldsymbol{M}_r = \boldsymbol{L}_r^{-1}$ defined in a fixed overall Cartesian coordinate system $x$ in $V$. The volume $V$ may be subdivided into several local volumes $V_r$, $r$=1,2, ..., N, $\Sigma\ V_r = V$, such that each contains only one phase material, although, in a more refined evaluation of the local fields, any given phase may reside in more than one volume $V_r$. The volume $V$ is subjected either to surface displacements on $S$ of $V$ that cause uniform overall strain $\boldsymbol{\varepsilon}^0$ in $V$, or to tractions derived from a uniform overall uniform stress $\boldsymbol{\sigma}^0$ on $S$. Moreover, a uniform temperature change $\Delta\theta$ and a piecewise uniform field of transformation strains $\boldsymbol{\mu}_r \in V_r$ may be prescribed in $V$. The local constitutive relations in the phases are then written as

$$\begin{aligned} \boldsymbol{\sigma}_r(x) &= \boldsymbol{L}_r \boldsymbol{\varepsilon}_r(x) + \boldsymbol{l}_r \Delta\theta + \boldsymbol{\lambda}_r \\ \boldsymbol{\varepsilon}_r(x) &= \boldsymbol{M}_r \boldsymbol{\sigma}_r(x) + \boldsymbol{m}_r \Delta\theta + \boldsymbol{\mu}_r \end{aligned} \tag{1}$$

where $\boldsymbol{m}_r$ is a thermal strain vector of linear coefficients of thermal expansion, $\boldsymbol{l}_r = -\boldsymbol{L}_r \boldsymbol{m}_r$ is a thermal stress vector, and $\boldsymbol{\lambda}_r = -\boldsymbol{L}_r \boldsymbol{\mu}_r$ is a local transformation stress or eigenstress.

Overall constitutive relations are written in the analogous form

$$\boldsymbol{\sigma} = \boldsymbol{L}\boldsymbol{\varepsilon}^0 + \boldsymbol{l}\Delta\theta + \boldsymbol{\lambda} \quad \boldsymbol{\varepsilon} = \boldsymbol{M}\boldsymbol{\sigma}^0 + \boldsymbol{m}\Delta\theta + \boldsymbol{\mu} \tag{2}$$

where $\boldsymbol{L}$ and $\boldsymbol{M}$ are overall stiffness and compliance satisfying $\boldsymbol{LM} = \boldsymbol{I}$, $\boldsymbol{l} = -\boldsymbol{L}\boldsymbol{m}$ is the overall thermal stress vector, and $\boldsymbol{\lambda} = -\boldsymbol{L}\boldsymbol{\mu}$ is the overall transformation stress caused by the $\boldsymbol{\mu}_r$ distribution in a fully constrained volume $V$. The thermal strains and transformation strains in Eq. 1 are assumed to be distributed such that both the overall stress $\boldsymbol{\sigma}$ at $\boldsymbol{\varepsilon}^0 = \boldsymbol{0}$ and strain $\boldsymbol{\varepsilon}$ at $\boldsymbol{\sigma}^0 = \boldsymbol{0}$ are uniform in $V$. All strains are assumed to be small.

Local averages of the stress and strain fields in Eq. 1 can be related to prescribed sets of the independent load contributions $\boldsymbol{\varepsilon}^0, \Delta\theta, \boldsymbol{\mu}_r$, or $\boldsymbol{\sigma}^0, \Delta\theta, \boldsymbol{\lambda}_r$, in the form [1]

$$\boldsymbol{\varepsilon}_s = \boldsymbol{A}_s \boldsymbol{\varepsilon}^0 + \sum_{r=1}^{N} \boldsymbol{D}_{sr}(\boldsymbol{m}_r \Delta\theta + \boldsymbol{\mu}_r) \quad \boldsymbol{\sigma}_s = \boldsymbol{B}_s \boldsymbol{\sigma}^0 + \sum_{r=1}^{N} \boldsymbol{F}_{sr}(\boldsymbol{l}_r \Delta\theta + \boldsymbol{\lambda}_r) \tag{3}$$

where $r$, s=1,2 ,..., N. The $\boldsymbol{A}_s$ and $\boldsymbol{B}_s$ are the mechanical concentration factor tensors [5] and $\boldsymbol{D}_{sr}$, $\boldsymbol{F}_{sr}$ are certain eigenstrain and eigenstress concentration factor tensors [1–4] which evaluate the contribution of a uniform transformation in $V_r \in V$ to the field average in $V_s \in V$. Both are averages of the respective influence functions that define the local fields; they depend on the local and overall elastic moduli and on the shape and volume fractions of the phases, and are therefore constant within certain temperature intervals. Like the Eshelby tensor $S$, the self–induced factors $D_{ss}$ and $F_{ss}$ contribute both the residual field caused in $V_s$ by the eigenstrains and the eigenstrain themselves [6].

In the heterogeneous aggregates considered herein, the transformation concentration factors can be related to the mechanical factors. In two-phase systems $r$, $s=\alpha$, $\beta$, where each phase contains a single, uniform transformation strain or stress, there are exact connections between the respective influence functions; c.f. Reference [1], Eqs. 123–126:

$$\begin{aligned} D_{r\alpha}(x) &= [I - A_r(x)]\,(L_\alpha - L_\beta)^{-1}L_\alpha \\ D_{r\beta}(x) &= -[I - A_r(x)]\,(L_\alpha - L_\beta)^{-1}L_\beta \\ F_{r\alpha}(x) &= [I - B_r(x)]\,(M_\alpha - M_\beta)^{-1}M_\alpha \\ F_{r\beta}(x) &= -[I - B_r(x)]\,(M_\alpha - M_\beta)^{-1}M_\beta \end{aligned} \tag{4}$$

In multiphase systems, and in two-phase systems loaded by several piecewise uniform local transformations, such connections can be found in terms of estimates of the mechanical concentration factors and overall stiffness or compliance that are provided by certain averaging methods [2, 3, 7], which are briefly reviewed in Section 10.5.4.

$$\begin{aligned} D_{sr} &= (I - A_s)\,(L_s - L)^{-1}(\delta_{sr}I - c_r A_r^{\mathrm{T}})L_r \\ F_{sr} &= (I - B_s)\,(M_s - M)^{-1}(\delta_{sr}I - c_r B_r^{\mathrm{T}})L_r \end{aligned} \tag{5}$$

where $I$ is a $(6 \times 6)$ identity matrix, $c_r=V_r/V$, and $\delta_{sr}$ is the Kronecker symbol, but no summation is indicated by repeated subscripts. The averaging methods are valid for either two-phase systems of arbitrary phase geometry or for multiphase systems where all phases have the same shape and orientation [8]. In more detailed evaluations of the local fields in unit cells and other material models, the transformation concentration factors are best evaluated with the finite element method, as described in References [4, 9].

Volume averages of the local fields (Eq. 3), obtained with substitutions from Eq. 4, or with any other admissible evaluation of the respective concentration factors, are related to the overall averages by

$$\begin{aligned} \varepsilon = \sum_{r=1}^{N} c_r \varepsilon_r \quad & m\Delta\theta + \mu = \sum_{r=1}^{N} c_r B_r^{\mathrm{T}}(m_r\Delta\theta + \mu_r) \\ \sigma = \sum_{r=1}^{N} c_r \sigma_r \quad & l\Delta\theta + \lambda = \sum_{r=1}^{N} c_r A_r^{\mathrm{T}}(l_r\Delta\theta + \lambda_r) \end{aligned} \tag{6}$$

These are distinct connections that must be satisfied separately by the averages of total fields and the thermal and other transformation fields [3, 5].

Finally, by invoking the well-known expressions for overall stiffness and compliance [5],

$$L = \sum_{r=1}^{N} c_r L_r A_r \quad M = \sum_{r=1}^{N} c_r M_r B_r \tag{7}$$

evaluation of the local fields in Eq. 3 using Eq. 4 or 5 is reduced to finding the mechanical concentration factors $A_r$ $B_r$; see Section 10.5.5. Note that Eq. 5 is valid for $A_s$, $A_r$, $L$ and $B_s$, $B_r$, $M$ derived with an averaging method; however, relations Eqs. 4, 6 and 7 must be satisfied by all admissible evaluations of concentration factors or influence functions.

## 10.5.3 EVOLUTION OF INELASTIC DEFORMATION

Inelastic, thermal, and any other transformation strains are deformations remaining at a material point after an instantaneous elastic unloading from the current to zero local stress. This agrees with the additive strain and stress decompositions (Eq. 1) and implies that elastic-plastic, viscoelastic, viscoplastic, and any other inelastic strains $\boldsymbol{\varepsilon}_r^{in}(x)$ can be regarded as eigenstrains acting on an elastic solid in superposition with mechanical and thermal loads. The local constitutive equations (Eq. 1) can thus be replaced by

$$\begin{aligned} \mathrm{d}\boldsymbol{\sigma}_r(x) &= \boldsymbol{L}_r \mathrm{d}\boldsymbol{\varepsilon}_r(x) + \boldsymbol{l}_r \mathrm{d}\theta + \mathrm{d}\boldsymbol{\sigma}_r^{re}(x) \\ \mathrm{d}\boldsymbol{\varepsilon}_r(x) &= \boldsymbol{M}_r \mathrm{d}\boldsymbol{\sigma}_r(x) + \boldsymbol{m}_r \mathrm{d}\theta + \mathrm{d}\boldsymbol{\varepsilon}_r^{in}(x) \end{aligned} \tag{8}$$

where the relaxation stress $\mathrm{d}\boldsymbol{\sigma}_r^{re}(x) = -\boldsymbol{L}_r \mathrm{d}\boldsymbol{\varepsilon}_r^{in}(x)$. As long as $\boldsymbol{L}_r, \boldsymbol{l}_r, \boldsymbol{M}_r$, and $\boldsymbol{m}_r$ remain constant, Eq. 8 holds for both instantaneous and accumulated increments.

Inelastic constitutive laws typically specify when a material starts to deviate from the linear elastic response, and then relate either the local relaxation stress to the past history and current increment of the local strain, or the inelastic strain to the history and current increment of the local stress. If these fields are uniform in $V_r$, this can be written as

$$\begin{aligned} \mathrm{d}\boldsymbol{\sigma}_r &\doteq \mathscr{L}_r(\boldsymbol{\varepsilon}_r - \boldsymbol{\beta}_r)\mathrm{d}\boldsymbol{\varepsilon}_r + \ell_r \mathrm{d}\theta \\ \mathrm{d}\boldsymbol{\varepsilon}_r &\doteq \mathscr{M}_r(\boldsymbol{\sigma}_r - \boldsymbol{\alpha}_r)\mathrm{d}\boldsymbol{\sigma}_r + \mathscr{m}_r \mathrm{d}\theta \end{aligned} \tag{9}$$

indicating that $\mathscr{L}_r$ or $\mathscr{M}_r$ are, respectively, functions of the current back strain $\boldsymbol{\beta}_r$ or back stress $\boldsymbol{\alpha}_r$, which together with $\ell_r$ and $\mathscr{m}_r$ depend on past deformation history. Separating the inelastic from the total components and introducing instantaneous inelastic stiffness and compliance tensors yields

$$d\boldsymbol{\sigma}_r^{re} = \mathscr{L}_r^p d\boldsymbol{\varepsilon}_r + \ell_r^p d\theta \quad d\boldsymbol{\varepsilon}_r^{in} = \mathscr{M}_r^p d\boldsymbol{\sigma}_r + m_r^p d\theta \tag{10}$$

When substituted into the incremental forms of Eq. 3, this provides the following equations for evaluation of the local strain and stress increments:

$$\begin{aligned} d\boldsymbol{\varepsilon}_s + \sum_{r=1}^{N} \mathbf{D}_{sr}\mathbf{M}_r\mathscr{L}_r^p d\boldsymbol{\varepsilon}_r &= \mathbf{A}_s d\boldsymbol{\varepsilon}^0 + \sum_{r=1}^{N} \mathbf{D}_{sr}(\mathbf{m}_r - \mathbf{M}_r\ell_r^p)d\theta \\ d\boldsymbol{\sigma}_s + \sum_{r=1}^{N} \mathbf{F}_{sr}\mathbf{L}_r\mathscr{M}_r^p d\boldsymbol{\sigma}_r &= \mathbf{B}_s d\boldsymbol{\sigma}^0 + \sum_{r=1}^{N} \mathbf{F}_{sr}(\ell_r - \mathbf{L}_r m_r^p)d\theta \end{aligned} \tag{11}$$

In multiphase systems, and in all systems which undergo a significant departure from elastic response, these equations are best written for refined subdivisions of $V$, with all concentration factors evaluated by the finite element method and solved numerically. Of course, closed-form solutions can be easily obtained for two- or three-phase systems; these appear useful in an analysis of contained inelastic deformation [9]. In any case, it is desirable to write the result in the form

$$\begin{aligned} d\boldsymbol{\varepsilon}_r &= \mathscr{A}_r d\boldsymbol{\varepsilon} + a_r d\theta \\ d\boldsymbol{\sigma}_r &= \mathscr{B}_r d\boldsymbol{\sigma} + b_r d\theta \end{aligned} \tag{12}$$

where $\mathscr{A}_r$, $a_r$, and $\mathscr{B}_r$, $b_r$ are instantaneous mechanical an thermal strain and stress concentration factor tensors for the local volumes $V_r$. The overall response of the inelastic composite medium can be derived as described in [2, §4].

## 10.5.4 MODELING OF DAMAGE EVOLUTION

As a simple application of the TFA method to damage analysis, consider a two-phase aggregate of volume $V$ consisting of a matrix ($r = 1$), the still bonded reinforcements ($r = b$), and debonded reinforcements ($r = d$). Incremental loading is applied to the undamaged aggregate by an overall uniform stress $\boldsymbol{\sigma}^0$ and by an equivalent eigenstrain $\boldsymbol{\mu}_d$ that simulates the effect of debonding in $r = d$. The average stress $\boldsymbol{\sigma}_d$ is usually zero, but any other value may be supported in $r=d$ if obtained from an independent analysis, preferably as a function of $\boldsymbol{\sigma}^0$ or $\boldsymbol{\varepsilon}$. Also, $\boldsymbol{\mu}_1^k = \boldsymbol{\mu}_b^k = 0$. Adjusting Eq. 3 (second equation) to these requirements yields three equations for the local stresses at any given loading step $k$:

$$\begin{aligned} \boldsymbol{\sigma}_1^k &= \mathbf{B}_1\boldsymbol{\sigma}_k^0 - \mathbf{F}_{1d}^k\mathbf{L}_2\boldsymbol{\mu}_d^k \\ \boldsymbol{\sigma}_d^k &= \mathbf{B}_2\boldsymbol{\sigma}_k^0 - \mathbf{F}_{dd}^k\mathbf{L}_2\boldsymbol{\mu}_d^k \\ \boldsymbol{\sigma}_b^k &= \mathbf{B}_2\boldsymbol{\sigma}_k^0 - \mathbf{F}_{bd}^k\mathbf{L}_2\boldsymbol{\mu}_d^k \end{aligned} \tag{13}$$

This respects the fact that both $r=b$ and $r=d$ volumes have the same shape and stiffness $L_2$ and thus also the same concentration factor $B_2$. However, since two different eigenstrains, $\mu_d^k$ and $\mu_d^k$, reside in phase $L_2$, all $F_{rs}$ are distinct and evaluated using Eq. 5 (second equation). Partial debonding in $r=d$ is simulated by the equivalent eigenstrain obtained from Eq. 13 (second equation):

$$\mu_d^k = M_2(F_{dd}^{-1})_k(B_2\sigma_k^0 - \sigma_d^k) \tag{14}$$

Averages of the local stresses are then to be found as

$$\sigma_1^k = B_1\sigma_k^0 + c_d^k B_1(I - c_d^k B_2)^{-1}(B_2\sigma_k^0 - \sigma_d^k) \tag{15}$$

$$\sigma_b^k = B_2\sigma_k^0 + c_d^k B_2(I - c_d^k B_2)^{-1}(B_2\sigma_k^0 - \sigma_d^k) \tag{16}$$

where $\sigma_d^k$ is the known local stress in the debonded phase, possibly equal to zero. The concentration factors $B_1$ and $B_2$ are evaluated in the undamaged two–phase composite. The first right–hand terms are the stresses in the undamaged system, and the second terms are contributed by the eigenstrain (Eq. 14) simulating partial debonding in $r=d$. It can be verified that Eq. 6 (third equation), written here as $c_1\sigma_1^k + c_d^k\sigma_d^k + c_b^k\sigma_b^k = \sigma_k^0$, is satisfied by Eqs. 15 and 16.

The local strain averages are obtained using the stresses in Eq. 1 (second equation as)

$$\varepsilon_1^k = M_1\sigma_1^k \quad \varepsilon_d^k = M_2\sigma_d^k + \mu_d^k \quad \varepsilon_b^k = M_2\sigma_b^k \tag{17}$$

The overall strain is found from $\varepsilon = \sum c_r\varepsilon_r$, or from the overall constitutive relation

$$\varepsilon_k = M\sigma_k^0 + \mu_k \tag{18}$$

where $M$ is the overall elastic compliance of the undamaged system, and the overall eigenstrain $\mu_k$ simulates the contribution to the overall strain by the current state of debonding in $r=d$; it can be evaluated from the generalized Levin formula (second part of Eq. 6) as $\mu_k = c_d^k B_2^{\mathrm{T}}\mu_d^k$. Together with a criterion that specifies the probability of debonding as a function of stress in the bonded phase, Eqs. 15 and 16 can be converted to an incremental form and solved for damage increments $\mathrm{d}c_d^k$ under overall stress change from $\sigma_k^0$ to $\sigma_k^0 + \mathrm{d}\sigma_k^0$.

## 10.5.5 CONCENTRATION FACTOR ESTIMATES

Heterogeneous solids are often modeled as aggregates of phases of ellipsoidal shape, possibly embedded in a common matrix. Under the homogeneous boundary conditions described in Section 10.5.1, the local strain or stress averages in the phases are the first right-hand terms in Eq. 3, with estimates of the concentration factors provided by the self-consistent [5] or Mori-Tanaka [10] averaging methods. These belong to a much wider, recently identified family of methods [7] which consider each phase as a solitary ellipsoidal inhomogeneity embedded in a large volume of a certain comparison medium $L_0$ that is remotely loaded by a uniform image strain or stress field. Estimates of the mechanical strain concentration factors follow from

$$A_r = A_r^I \left[ \sum_{r=1}^{N} c_r A_r^I \right]^{-1} \qquad (A_r^I)^{-1} = I - S L_0^{-1}(L_0 - L_r) \tag{19}$$

where $S$ denotes the Eshelby tensor evaluated in $L_0$. The $L_0$ must be chosen such that the matrices $\bar{L}_0 - \bar{L}_r$ and $L_0 - L_0$ are both positive or negative semidefinite. The overall stiffness estimate then follows from Eq. 7 (first equation) and 19; if $L_0$ is used in Eq. 19, then Eq. 7 (first equation) provides the Walpole [11] form of the Hashin–Shtrikman bounds. A convenient form that approximates the self–consistent estimate of the overall stiffness in two–phase systems where $L_2$-$L_1$ is positive semi definite is $L_0 = L_1 + c_2^2(L_2 - L_1)$

## ACKNOWLEDGEMENTS

Work leading to the present results was sponsored by grants from the Army Research Office and the Office of Naval Research. Drs. Yapa D.S. Rajapakse and Mohammed Zikry served as program monitors.

## REFERENCES

1. Dvorak, G. J. (1990). On uniform fields in heterogeneous media. *Proc. Roy. Soc. (London)* **A 431**: 89–110.
2. Dvorak, G. J. (1992). Transformation field analysis of inelastic composite materials. *Proc. Roy. Soc. (London)* **A 437**: 311–327.
3. Dvorak, G. J., and Benveniste, Y. (1992). On transformation strains and uniform fields in multiphase elastic media. *Proc. Roy. Soc. (London)* **A 437**: 291–310.
4. Dvorak, G. J., Bahei–El–Din, Y. A., and Wafa, A. (1994). Implementation of the transformation field analysis for inelastic composite materials. *Computational Mechanics* 14: 201–228.
5. Hill, R. (1965). A self-consistent mechanics of composite materials. *J. Mech. Phys. Solids* **13**: 213–222.
6. Eshelby, J. D. (1957). The determination of the elastic field of an ellipsoidal inclusion and related problems. *Proc. Roy. Soc. (London)* **A 241**: 376–396.
7. Dvorak, G. J., and Srinivas, M.V. (1999). New estimates of overall properties of heterogeneous solids. *J. Mech. Phys. Solids* **47**: 899–920, 2207–2208.

8. Benveniste, Y., Dvorak, G. J., and Chen, T. (1989). On the diagonal and elastic symmetry of the approximate stiffness tensor of heterogeneous media. *J. Mech. Phys. Solids* **39**: 927–946.
9. Fish, J., Pandheeradi, M., and Shephard, M. S. (1997). Computational plasticity for composite structures based on mathematical homogenization: Theory and practice. *Comp. Meth. Appl. Mech. Engng.* **148**: 53–73.
10. Benveniste, Y. (1987). A new approach to the application of Mori-Tanaka's theory in composite materials. *Mech. Mater.* **6**: 147–157.
11. Walpole, L. J. (1966). On bounds for overall elastic moduli of inhomogeneous systems: I. *J. Mech. Phys. Solids* **14**: 151–162; II. *ibid.*, 289–301.

SECTION **10.6**

# A Damage Mesomodel of Laminate Composites

PIERRE LADEVÈZE

*LMT-Cachan, ENS de Cachan/CNRS/Université Paris 6, 61 avenue du Président Wilson, 94235 Cachan Cedex, France*

Contents



## 10.6.1 DOMAIN OF APPLICATION

A major challenge in composite design is to calculate the damage state of a composite structure subjected to complex loading at any point and at any time until final fracture. *Damage* refers to the more or less gradual developments of microcracks which lead to macrocracks and then to rupture; macrocracks are simulated as completely damaged zones. A solution for composites, especially laminate composites, is based on what we call a



*Handbook of Materials Behavior Models.* ISBN 0-12-443341-3.

damage mesomodel. It is a semidiscrete model for which the damage state is locally uniform within the mesoconstituents. For laminates, it is uniform throughout the thickness of each single layer; as a complement, continuum damage models with delay effects are introduced.

The mesomodel has been used for various continuous fiber laminates, essentially for aeronautical and spatial applications.

## 10.6.2 BASIC ASPECTS

An initial step is to define what we call a laminate mesomodel [13, 14]. At the mesoscale, characterized by the thickness of the ply, the laminate structure is described as a stacking sequence of homogeneous layers through the thickness and of interlaminar interfaces (see Figure 10.6.1). The main damage mechanisms are described as fiber breaking, matrix microcracking, and debonding of adjacent layers (see Figure 10.6.2). The single-layer model includes both damage and inelasticity. The interlaminar interface is defined as a two-dimensional mechanical model which ensures traction and displacement transfer from one ply to the next. Its mechanical behavior depends on the angle between the fibers of two adjacent layers. *A priori*, 0°/0° interfaces are not introduced. Herakovich, in his book [11], calls this theory the "mesoscale composite damage theory."

The damage mechanisms are taken into account by means of internal damage variables. A mesomodel is then defined by adding another property: a uniform damage state is prescribed throughout the thickness of the elementary ply. This point plays a major role when trying to simulate a crack with a damage model. As a complement, delayed damage models are introduced.

One limitation of the proposed mesomodel is that material fracture is described by means of only two types of macrocracks:

- delamination cracks within the interfaces;
- cracks orthogonal to the laminate's midplane, each cracked layer being completely cracked through its thickness.

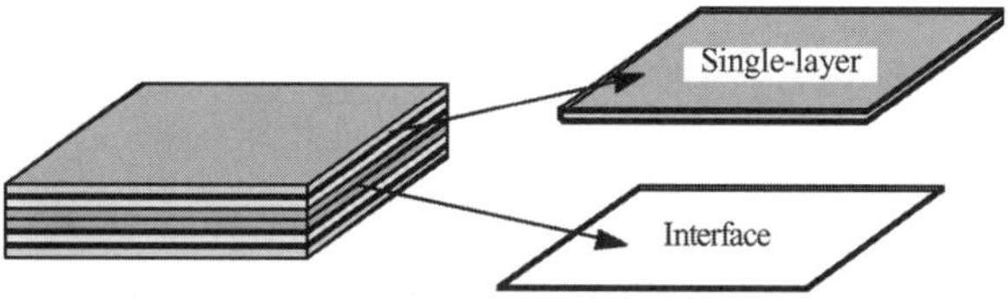


FIGURE 10.6.1 Laminate modeling

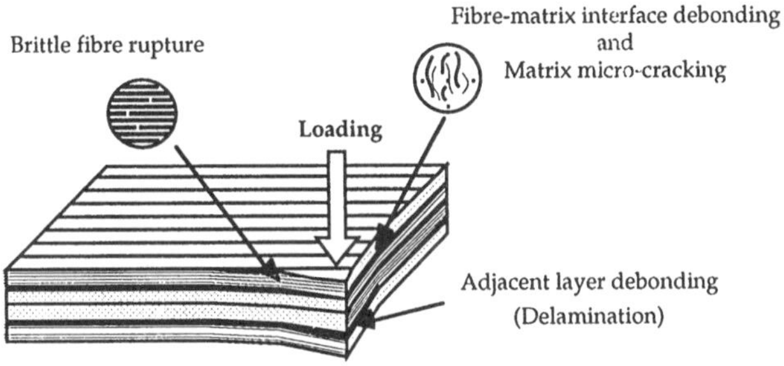


FIGURE 10.6.2 Damage and failure mechanisms.

The layers — in our sense — are assumed to be not too thick. Another limitation is that very severe dynamic loadings cannot be studied; the dynamic wavelength must be larger than the thickness of the plies.

Two models have to be identified: the single-layer model [16] and the interface model [1,6]. The appropriate tests used consist of tension, bending, and delamination. Each composite specimen, which contains several layers and interfaces, is analyzed in order to derive the material quantities intrinsic to the single layer or to the interlaminar interface.

Various comparisons with experimental results have been performed to show the possibilities and limits of our proposed computational damage mechanics approach for laminates [2, 8, 19, 20].

The single-layer model is presented here. A similar model is used for the interface.

## 10.6.3 THE SINGLE-LAYER MODEL

The single-layer model and its identification procedure are detailed in Herakovich [11]. Complements concerning the compression behavior and the temperature-dependent behavior can be found in Allix *et al.* [3, 4]. Here we present the latest version, which takes into account the layer size effects [23].

### 10.6.3.1 Damage Kinematics

The composite materials under investigation in this study (e.g., carbon-fiber/epoxy-resin) have only one reinforced direction. In the following, subscripts 1, 2, and 3 designate the fiber direction, the transverse direction inside the layer, and the normal direction, respectively. The energy of the damaged material defines the damage kinematics. Using common notations, this energy is

$$
\begin{aligned}
E_D = \frac{1}{2(1-d_F)} & \left[ \frac{\langle \sigma_{11} \rangle^2}{E_1^0} + \frac{\varnothing(\langle -\sigma_{11} \rangle)}{E_1^0} - \left( \frac{\nu_{21}^0}{E_2^0} + \frac{\nu_{12}^0}{E_1^0} \right) \sigma_{11}\sigma_{22} \right. \\
& \left. - \left( \frac{\nu_{31}^0}{E_3^0} + \frac{\nu_{13}^0}{E_1^0} \right) \sigma_{11}\sigma_{33} - \left( \frac{\nu_{32}^0}{E_3^0} + \frac{\nu_{23}^0}{E_2^0} \right) \sigma_{22}\sigma_{33} \right] \\
& + \frac{\langle -\sigma_{22} \rangle^2}{2E_2^0} + \frac{\langle -\sigma_{33} \rangle^2}{2E_3^0} + \frac{1}{2} \left[ \frac{1}{(1-d')} \left( \frac{\langle \sigma_{22} \rangle^2}{E_2^0} + \frac{\langle \sigma_{33} \rangle^2}{E_3^0} \right) \right. \\
& \left. + \frac{1}{(1-d)} \left( \frac{\sigma_{12}^2}{G_{12}^0} + \frac{\sigma_{23}^2}{G_{23}^0} + \frac{\sigma_{31}^2}{G_{31}^0} \right) \right] \qquad (1)
\end{aligned}
$$

$\varnothing$ is a material function which takes into account the nonlinear response in compression [3], and $d_F$, $d$, and $d'$ are three scalar internal variables which remain constant throughout the thickness of each single layer and which serve to describe the damage mechanisms inside. The unilateral aspect of microcracking is taken into account by splitting the energy into a "tension" energy and a "compression" energy; $\langle . \rangle$ denotes the positive part. The thermodynamic forces associated with the mechanical dissipation are

$$
Y_d \equiv \frac{\partial}{\partial d} \langle\langle E_D \rangle\rangle_{|\,:cst} = \frac{1}{2(1-d)^2} \left\langle \left\langle \frac{\sigma_{12}^2}{G_{12}^0} + \frac{\sigma_{23}^2}{G_{23}^0} + \frac{\sigma_{31}^2}{G_{31}^0} \right\rangle \right\rangle
$$

$$
Y_{d'} \equiv \frac{\partial}{\partial d'} \langle\langle E_D \rangle\rangle_{|\,:cst} = \frac{1}{2(1-d')^2} \left\langle \left\langle \frac{\langle \sigma_{22} \rangle^2}{E_2^0} + \frac{\langle \sigma_{33} \rangle^2}{E_3^0} \right\rangle \right\rangle
$$

$$
\begin{aligned}
Y_F \equiv \frac{\partial}{\partial d_F} \langle\langle E_D \rangle\rangle_{|\,:cst} = \frac{1}{2(1-d_F)^2} & \left\langle \left\langle \frac{\langle \sigma_{11} \rangle^2}{E_1^0} + \frac{\varnothing(\langle -\sigma_{11} \rangle)}{E_1^0} \right. \right. \\
& - \left( \frac{\nu_{12}^0}{E_1^0} + \frac{\nu_{21}^0}{E_2^0} \right) \sigma_{11}\sigma_{22} - \left( \frac{\nu_{13}^0}{E_1^0} + \frac{\nu_{31}^0}{E_3^0} \right) \sigma_{11}\sigma_{33} \\
& \left. \left. - \left( \frac{\nu_{32}^0}{E_3^0} + \frac{\nu_{23}^0}{E_2^0} \right) \sigma_{22}\sigma_{33} \right\rangle \right\rangle \qquad (2)
\end{aligned}
$$

Here, unlike in previous papers, $\langle\langle . \rangle\rangle$ denotes the integral value through the thickness, not the mean value.

### 10.6.3.2 Damage Evolution Law

From experimental results, it follows that the governing forces of damage evolution are

$$Y = [Y_d + bY_{d'}],\ Y' = [Y_{d'} + b' Y_d],\ Y_F \tag{3}$$

where $b$ and $b'$ are material constants which balance the influence of the transverse energy and the shear energy. For small damage rates and quasi-static loading, we get

$$\begin{aligned} d &= f_d(\underline{Y}^{1/2}) && \text{for } d \le 1 \\ d' &= f_{d'}(\underline{Y}'^{1/2}) && \text{for} d' \le 1 \\ d_F &= f_F(\underline{Y}_F^{1/2}) && \text{for } d_F \le 1 \end{aligned} \tag{4}$$

where

$$\underline{r}|_t = sup\ r|_\tau,\ \tau \le t$$

and $f_d$, $f_{d'}$, and $f_F$ are material "functions"; progressive damage evolution (generally defined by a linear function) and brittle damage evolution (defined by a threshold) are both present. $f_F$ is generally associated with a brittle damage mechanism. The model stays valid for a rather large temperature range [4]. At room temperature, a typical material function $f_d$ is given in Figure 10.6.3. For large damage rates, we have introduced a damage model with delay effects:

$$\begin{aligned} \dot{d} &= \frac{1}{\tau_c}[1 - exp(-a\langle f_d(Y^{1/2}) - d\rangle)] \quad \text{if } d<1,\ d = 1 \text{ otherwise} \\ \dot{d}' &= \frac{1}{\tau_c}[1 - exp(-a\langle f_{d'}(Y'^{1/2}) - d'\rangle)] \quad \text{if } d'<1,\ d' = 1 \text{ otherwise} \\ \dot{d}_F &= \frac{1}{\tau_c}[1 - exp(-a\langle f_F(Y_F'^{1/2}) - d_F\rangle)] \quad \text{if } d_F<1,\ d_F = 1 \text{ otherwise} \end{aligned} \tag{5}$$

The same material constants, $\tau_c$ and $a$, are taken for the three damage evolution laws. For this damage model with delay effects, the variations of the forces $Y$, $Y'$ and $Y_F$ do not lead to instantaneous variations of the damage variables $d$, $d'$ and $d_F$. There is a certain delay, defined by the characteristic time $\tau_c$. Moreover, a maximum damage rate, which is $1/\tau_c$, does exist. A first identification consists of taking half the Rayleigh wave speed combined with the critical value of the energy release rate. Let us also point out here that a clear distinction can be made between this damage model with delay effects and viscoelastic or viscoplastic models: the characteristic time introduced in the damage model with delay effects is several orders of magnitude less than

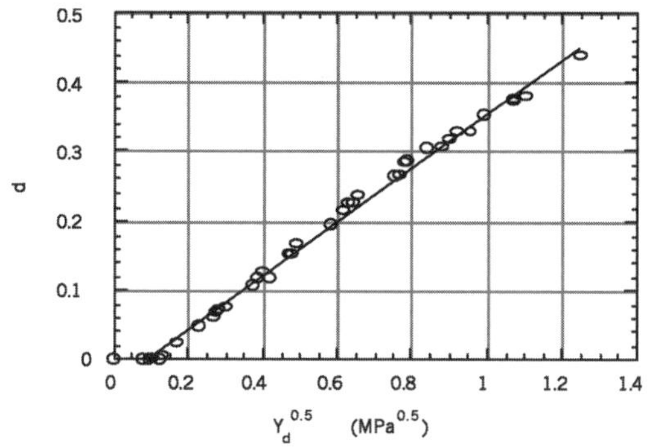


FIGURE 10.6.3 Shear damage material function $Y_d^{1/2} \rightarrow f_d(Y_d^{1/2})$ of the single-layer for the M55J/M18 material at room temperature.

in the viscous case. This characteristic time is, in fact, related to the fracture process.

#### 10.6.3.2.1 Remarks

- Two damage variables are used to describe the damage associated with matrix microcracking and fiber-matrix debonding. They seem to account for all the proposed damage kinematics, including that starting from an analysis of the microcracks. Many works have established, experimentally or theoretically, a relation between the microcrack density and our damage variable $d$, which can be very useful for the identification of a damage fatigue model.
- What we call the single layer is the assemblage of adjacent, usual elementary, plies of the same direction. The damage forces, being integral values through the thickness of the single layer, can be interpreted as energy release rates. It follows that the damage evolution law of the single layer is thickness-dependent. For single layers which are not too thick, such damage evolution laws include results coming from shear lag analyses. Consequently, the size effects — observed, for example, in tension — are produced by both the single-layer model and the interface model through a structure problem. This theory, which is very simple, works very well for most engineering laminates; however, it cannot be satisfactory for rather thick layers. A first solution is to modify the damage evolution law, using the thickness as a parameter.
- The damage variables are active for $[0^\circ, 90^\circ]_n$ laminates even if the apparent modulus does not change. The model predicts this hidden damage [15].

- For fatigue loadings, we introduce:

$$\begin{aligned} d &= d_S + d_F \\ d' &= d'_S + d'_F \end{aligned} \tag{6}$$

where $d_S$ and $d'_S$ are the quasi-static part of the damage defined by Eq. 4 or 5. $d_F$ and $d'_F$ denote the fatigue part characterized by the following fatigue evolution laws:

$$\begin{aligned} \frac{\partial d_F}{\partial N} &= a(d,\ [Yd + bY_{d'}]) \\ \frac{\partial d'_F}{\partial N} &= a'(d', [Y_{d'}]) \end{aligned} \tag{7}$$

where $a$, $a'$ are two material functions and [.] denotes the maximum value over the cycle.

## 10.6.3.3 Coupling between Damage and Plasticity (or Viscoplasticity)

The microcracks, i.e., the damage, lead to sliding with friction, and thus to inelastic strains. The effective stress and inelastic strain are defined by

$$\begin{aligned} &\tilde{\sigma}_{11} \equiv \sigma_{11}\ \tilde{\sigma}_{22} \equiv -\langle -\sigma_{22}\rangle + \frac{\langle \sigma_{22}\rangle}{(1-d')}\ \tilde{\sigma}_{33} \equiv -\langle -\sigma_{33}\rangle + \frac{\langle \sigma_{33}\rangle}{(1-d')} \\ &\tilde{\sigma}_{12} \equiv \frac{\sigma_{12}}{1-d} \quad \tilde{\sigma}_{23} \equiv \frac{\sigma_{23}}{1-d} \quad \tilde{\sigma}_{31} \equiv \frac{\langle \sigma_{31}\rangle}{1-d'} \\ &\dot{\tilde{\varepsilon}}_{11p} \equiv \dot{\varepsilon}_{11p} \quad \dot{\tilde{\varepsilon}}_{22p} \equiv \langle \dot{\varepsilon}_{22p}\rangle(1-d') - \langle -\dot{\varepsilon}_{22p}\rangle \quad \dot{\tilde{\varepsilon}}_{33p} \equiv \langle \dot{\varepsilon}_{33p}\rangle(1-d') - \langle -\dot{\varepsilon}_{33p}\rangle \\ &\dot{\tilde{\varepsilon}}_{12p} \equiv \dot{\varepsilon}_{12p}(1-d) \quad \dot{\tilde{\varepsilon}}_{23p} \equiv \dot{\varepsilon}_{23p}(1-d) \quad \dot{\tilde{\varepsilon}}_{31p} \equiv \dot{\varepsilon}_{31p}(1-d) \end{aligned} \tag{8}$$

$\varepsilon_{ijp}$ for $ij \in \{1, 2, 3\}$ denotes the usual inelastic strain. The idea is to apply classical plasticity or viscoplasticity models to effective quantities. A very simple plasticity model is defined by the following elastic domain:

$$f(\tilde{\sigma}, R) = [\tilde{\sigma}_{12}^2 + \tilde{\sigma}_{23}^2 + \tilde{\sigma}_{31}^2 + a^2(\tilde{\sigma}_{22}^2 + \tilde{\sigma}_{33}^2)]^{1/2} - R - R_0 \tag{9}$$

Hardening is assumed to be isotropic, which means that the threshold $R$ is a function of the cumulated strain $p$; $p \rightarrow R(p)$ is a material function, $p$ being defined by

$$p = \int_0^t dt[\dot{\tilde{\varepsilon}}_{12p}^2 + \dot{\tilde{\varepsilon}}_{23p}^2 + \dot{\tilde{\varepsilon}}_{31p}^2 + \frac{1}{a^2}(\dot{\tilde{\varepsilon}}_{22p}^2 + \dot{\tilde{\varepsilon}}_{33p}^2)]^{1/2} \tag{10}$$

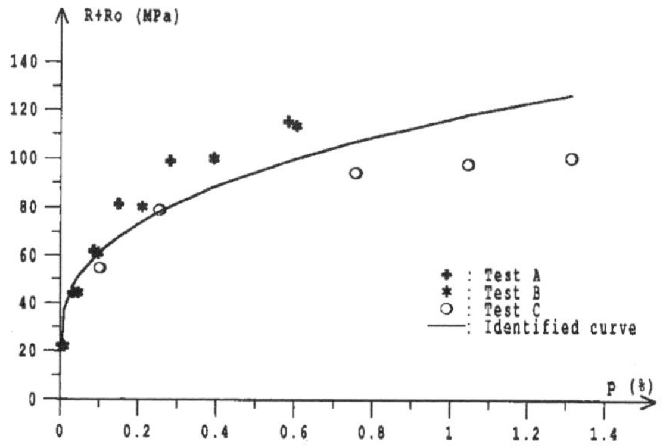


FIGURE 10.6.4 Hardening curve at room temperature for IM6/914.

$a$ is a material coupling constant. The yield conditions are

$$\begin{aligned} &\bullet\quad \dot{p}f = 0,\ \dot{p} \geq 0,\ f \leq 0 \\ &\bullet\quad \dot{\tilde{\varepsilon}}_{ijp} = \frac{1}{2}\dot{p}\,\frac{\tilde{\sigma}_{ij}}{R + R_0} \quad \text{for } i \neq ji, j \in \{1,2,3\} \\ &\bullet\quad \dot{\tilde{\varepsilon}}_{ijp} = \dot{p}\,\frac{a^2\tilde{\sigma}_{ij}}{R + R_0} \quad \text{for } i \in \{2,3\} \end{aligned} \tag{11}$$

An example of such a hardening curve is given for the T300-914 material in Figure 10.6.4.

## 10.6.4 IDENTIFICATION OF THE MATERIAL PARAMETERS

The single-layer model and the interface model have been identified for various materials. Aside from the elastic constants, the model depends on:

- three coupling coefficients $b$, $b'$, $a^2$;
- the damage "functions" $f_d$, $f_{d'}$, $f_F$ describing progressive and brittle evolutions;
- the hardening function $p \rightarrow R(p)$;
- the function $\varnothing$ defined practically by one parameter characterizing the compressive stiffness loss in the fiber direction (see Reference [1]).

The identification is developed here for low-stiffness matrixes. It is based on three canonic tests: $[0°, 90°]_{2s}$, $[+45°,-45°]_{2s}$, $[+67.5°,-67.5°]_{2s}$. The measured experimental quantities are:

- the tension $F$, which is related to the macrostress $\sigma_L^*$ by $\sigma_L^* = F/S$, where $S$ is a specimen's section;
- the laminate's axial strain $\varepsilon_L^*$ and the transverse one $\varepsilon_T^*$.

Consequently, the inelastic strains and moduli variations are determined.

#### 10.6.4.1 TENSILE TEST ON $[0°, 90°]_{2S}$ LAMINATE

This test defines $f_F$, i.e., in many cases, the fiber's limit tensile strain.

#### 10.6.4.2 TENSILE TEST ON $[+45°, -45°]_{2S}$ LAMINATE

The following relations allow us to reconstitute the ply's shear behavior:

$$\begin{aligned} \sigma_{12} &= \frac{\sigma_L^*}{2} \\ \varepsilon_{12} &= (\varepsilon_L^* - \varepsilon_T^*)/2 = \varepsilon_{12e}^* + \varepsilon_{12p}^* \\ \sigma_{11} &= \sigma_L^* \end{aligned} \tag{12}$$

For many materials, one has

$$\begin{aligned} \varepsilon_{12} &\geq (\varepsilon_{11},\ \varepsilon_{22}) \\ \sigma_L^* &\geq \sigma_{22} \end{aligned}$$

Consequently, the transverse stress and strain do not affect the behavior. The damage function $f_d$ is defined using

$$\begin{aligned} Y &= \sqrt{2G_{12}^0}\ \varepsilon_{12e}^* \\ (1-d) &= \sigma_L^*/2G_{12}^0\ \varepsilon_{12e}^* \end{aligned} \tag{13}$$

The hardening function $p \to R(p)$ is identified from

$$\begin{aligned} R + R_0 &= G_{12}^0 \varepsilon_{12e}^* \\ p &= \int_0^{\varepsilon_{12e}^*} 2(1-d)d\varepsilon_p \end{aligned} \tag{14}$$

#### 10.6.4.3 TENSILE TEST ON A $[+67.5°,-67.5°]_{2S}$ LAMINATE

$b'$ can be taken to zero for many materials. Then, the stresses and strains in the upper layer are

$$\begin{aligned} &\sigma_{11} = s\sigma_L^* \quad \sigma_{22} = s'\sigma_L^* \quad \sigma_{12} = s''\sigma_L^* \quad \theta = 67.5° \\ &\varepsilon_{11} \approx 0\ \varepsilon_{22} = cos^2\theta\varepsilon_L^* + sin^2\theta\varepsilon_T^*\ \varepsilon_{12} = cos\theta\, sin\theta\,(\varepsilon_L^* - \varepsilon_T^*) \end{aligned} \tag{15}$$

where s, s′, s″ are coefficients depending on $\theta$ and the single layer's characteristics. This test allows one to identify the two coupling coefficients $b$ and $a^2$ and the material function $f_{d'}$. A complete numerical simulation of the model is needed here.

### 10.6.5 HOW DOES ONE USE THE MODEL?

The complete damage mesomodel of laminate composites can be introduced in a finite element code. Some commercial codes propose it.

However, the mesomodel of the single layer is sufficient when delamination does not occur. A simplified approach which has been used extensively is to introduce it in a finite element postprocessor in order to predict the intensities of the different damage mechanisms inside the different layers. The data is a finite element solution calculated under the assumption of elastic behavior.

### 10.6.6 EXTENSIONS

A first extension, in which microcracks can occur orthogonally to fiber directions, was introduced for ceramics composites [10]. Models for three-dimensional and four-dimensional carbon–carbon composite materials are given in Reference [9, 18, 21]. More refined damage models are necessary for most ceramics composites [19, 20]. They follow the anisotropic damage theory, which includes microcracks opening and closure effects introduced in Reference [12].

The extension to impact problems has been studied in Allix *et al.* [5, 6].

## REFERENCES


1. Allix, O., and Ladevèze, P. (1992). Interlaminar interface modelling for the prediction of laminate delamination. *Composite Structures* **22**; 235–242.
2. Allix, O. (1992). Damage analysis of delamination around a hole, in *New advances in Computational Structural Mechanics*, pp. 411–421, Ladevèze, P., and Zienkiewicz, O.C., eds, Elsevier Science Publishers B.V.
3. Allix, O., Ladevèze P., and Vitecoq, E. (1994). Modelling and identification of the mechanical behaviour of composite laminates in compression. *Composite Science and Technology* **51**; 35–42.
4. Allix, O., Bahlouli, N., Cluzel, C., and Perret, L. (1996). Modelling and identification of temperature-dependent mechanical behaviour of the elementary ply in carbon/epoxy laminates. *Composite Science and Technology* **56**; 883–888.

5. Allix, O., and Deü, J. F. (1997). Delay-damage modeling for fracture prediction of laminated composites under dynamic loading. *Engineering Transactions* **45**; 29–46.
6. Allix, O., Guedra-Degeorges, D., Guinard, S., and Vinet, A. (1999). 3D analysis applied to low-energy impacts on composite laminates. *Proceedings ICCM12*, pp. 282–283, Masard, T., and Vautrin, A., eds.
7. Allix, O., Leveque, D., and Perret, L. (1998). Interlaminar interface model identification and forecast of delamination in composite laminates. *Composite Science and Technology* **56**; 671–678.
8. Daudeville, L., and Ladevèze, P. (1993). A damage mechanics tool for laminate delamination. *Journal of Composite Structures* **25**; 547–555.
9. Dumont, J. P., Ladevèze, P., Poss, M., and Remond, Y. (1987). Damage mechanics for 3D composites. *Int. J. Composite Structures,* 119–141.
10. Gasser, A., Ladevèze, P., and Peres, P. (1998). Damage modelling for a laminated ceramic composite. *Materials Science and Engineering A* **250**(2); 249–255.
11. Herakovich, C. T. (1998). *Mechanics of Fibrous Composites*, J. Wiley.
12. Ladevèze, P. (1983). On an anisotropic damage theory; *Report no. 34-LMT-Cachan (in French); Failure Criteria of Structured Media,* Boehler, J. P. ed., Balkema (1996), 355–364.
13. Ladevèze, P. (1986). Sur la mécanique de l'endommagement des composites, in *Comptes-Rendus des JNC5,* pp. 667–683, Bathias, C., and Menkès, D. eds., Paris: Pluralis Publications.
14. Ladevèze, P. (1989). About a damage mechanics approach, in *Mechanics and Mechanisms of Damage in Composite and Multimaterials.* pp. 119–142, Baptiste, D. ed., MEP.
15. Ladevèze, P. (1992). A damage computational method for composite structures. *J. Computer and Structure* **44**(1/2); 79–87.
16. Ladevèze, P., and Le Dantec, E. (1992). Damage modeling of the elementary ply for laminated composites. *Composite Science and Technology* **43**(3); 257–267.
17. Ladevèze, P. (1992). Towards a fracture theory, in *Proceedings of the Third International Conference on Computational Plasticity,* pp. 1369–1400, Owen, D. R. J., and Hinton, E. eds., Cambridge: Pineridge Press.
18. Ladevèze, P., Allix, O., and Cluzel, C. (1993). Damage modelling at the macro- and meso-scales for 3D composites, in *Damage Composite Materials,* pp. 195–215, Voyiadjis G. ed., Elsevier.
19. Ladevèze, P. (1995). A damage computational approach for composites: Basic aspects and micromechanical relations. *Computational Mechanics* **8**; 142–150.
20. Ladevèze, P. (1995). Modeling and simulation of the mechanical behavior of CMCs, in *High-Temperature Ceramic–Matrix Composites.* pp. 53–63, Evans, A. G., and Naslain, R. eds. (Cereamic Transaction).
21. Ladevèze, P., Allix, O., Gornet, L., Leveque, D., and Perret, L. (1998). A computational damage mechanics approach for laminates: Identification and comparison with experimental results, in *Damage Mechanics in Engineering Materials*, pp. 481–500, Voyiadjis, G. Z., Wu, J. W., and Chaboche, J. L. eds., Amsterdam: Elsevier.
22. Ladevèze, P., Aubard, X., Cluzel, C., and Guitard, L. (1998). Damage and fracture modeling of 4D CC composites, in *Damage Mechanics in Engineering Materials*, pp. 351–367, Voyiadjis, G., Wu, J., Chaboche, J. L., eds., Elsevier.
23. Ladevèze, P. (2000). Modelling and computation until final fracture of laminate composites, in *Recent Developments in Durability Analysis of Composite Systems*, pp. 39–47, Cardon, A. H., *et al.* eds., Balkema.

SECTION **10.7**

# Behavior of Ceramic–Matrix Composites under Thermomechanical Cyclic Loading Conditions

FREDERICK A. LECKIE[1], ALAIN BURR[2] and FRANÇOIS HILD[3]

[1] *Department of Mechanical and Environmental Engineering, University of California, Santa Barbara, California*

[2] *Laboratoire de Physico-Chimie Structurale et Macromoléculaire, UMR 7615, ESPCI, 10 rue Vauquelin, 75231 Paris Cedex 05, France*

[3] *Université Paris 6, LMT-Cachan, 61 avenue du Président Wilson, 94235 Cachan Cedex, France*

Contents



## 10.7.1 OVERVIEW

A constitutive law is proposed for ceramic–matrix composites (CMCs) which models matrix cracking, interface sliding and wear, fiber breakage, and fiber pullout and creep (of the matrix). These different mechanisms induce loss of stiffness, inelastic strains, creep strains, hysteresis loops, and crack closure. The features are analyzed within the framework of

*Handbook of Materials Behavior Models.* ISBN 0-12-443341-3.

continuum damage mechanics (CDM) by the introduction of physical internal variables identified previously in material science investigations. The intention, then, of the present study is to develop a continuum description of the damage processes which is mechanism-based and which may be used to describe the behavior of CMCs under the conditions of multiaxial stress occurring in practice. Since crack spacing at saturation is small in most CMCs, CDM is an appropriate means of describing degradation, since changes in elastic moduli measured on a macroscopic level provide a simpler and more robust means of measuring damage than does microscopic measurement of crack density, which requires the average of many readings before reliable values are established.

By combining CDM with micromechanical studies which are mechanism-based, constitutive equations are developed which lend themselves to the finite element procedures commonly used in practice. The CDM formulation applied to reinforced composites is written within the framework of the continuum thermodynamics. The first step in establishing such a model is to identify the internal variables which define the state of the material. The second is to determine the expression of the state potential in terms of the state variables, and the third to define the evolution laws of the internal variables.

Composites consisting of a ceramic matrix reinforced by continuous ceramic fibers are candidates for application in components which operate at temperatures in excess of those which are normal for metallic structures. In spite of the fact that the constituents of the CMC are both brittle, it has been demonstrated that following matrix cracking, sliding occurs at the fiber–matrix interface which causes inelastic deformations. The presence of matrix cracks and inelastic deformations may impart to the material the ability to redistribute stresses. The ability to redistribute stress is an important property since design studies indicate that working stresses for CMC components are sufficiently high for matrix cracking to be unavoidable in regions of stress concentration occurring at the junctions and penetrations which are a feature of engineering components.

The model is integrated into a finite element system (ABAQUS) and is used to estimate the behavior of a representative structures under monotonic [1], cyclic [2], and creep [3] loading conditions. Design and lifting procedures are able to deal with thermal loading, creep, and cyclic loading just by extending the original formulation to isochronous analysis [4]. It is observed that some CMCs have the ability to redistribute stress so that, as for plasticity, the presence of initial stress concentrations does not compromise the performance of the component, because it is reduced during the loading.

## 10.7.2 BACKGROUND: STATE POTENTIAL OF A [0, 90] FIBER-REINFORCED COMPOSITE

This part deals with the derivation of a constitutive law for a [0,90] (laminated or woven) composite submitted to multiaxial loads in plane stress conditions. The method is based upon the construction of the properties of the composite from the properties of the constituents and the stacking sequence of the layers.

The initial behavior of the matrix is assumed to be isotropic. The presence of cracks leads the behavior to become anisotropic. The assumption is made that cracking occurs normal to the $y$-direction (e.g., maximum principal strain direction) in the matrix. Under the hypothesis of a monotonic loading condition, only one damage variable is needed to model matrix cracking, and that is denoted by $D_{my}$. The study of a cracked system normal to one direction shows that the Young's modulus along that direction and the shear modulus are altered and that the expression of the elastic energy density of the matrix is

$$\psi_m = \frac{1}{2}\frac{E_m\left[\varepsilon_{mxx}^2 + 2\nu_m(1-D_{my})\varepsilon_{mxx}\varepsilon_{myy} + (1-D_{my})\varepsilon_{myy}^2\right]}{1-\nu_m^2(1-D_{my})} + 2\tilde{G}_m(D_{my})\varepsilon_{mxy}^2 \tag{1}$$

with

$$\tilde{G}_m(D_{my}) = \frac{G_m}{1+\left(\dfrac{D_{my}}{1-D_{my}}\right)\dfrac{1}{2(1+\nu_m)}} \tag{2}$$

where $E_m$, $\nu_m$, $G_m$ are the initial elastic properties of the matrix. The components of the strain tensor of the matrix $\underline{\underline{\varepsilon}}_m$ expressed in the $x{-}y$ frame are denoted by $\varepsilon_{mxx}$, $\varepsilon_{myy}$, and $\varepsilon_{mxy}$.

The fibers are aligned along the 1-direction. The fiber breaks are assumed to be perpendicular to the fiber direction and are described by a damage parameter $D_{f1}$. Therefore, the elastic energy density of the fibers is given by

$$\psi_f = \frac{1}{2}\left[E_f(1-D_{f1})\varepsilon_{f11}^2 + E_f\varepsilon_{f22}^2\right] + 2\tilde{G}_f(D_{f1})\varepsilon_{f12}^2 \tag{3}$$

with

$$\tilde{G}_f(D_{f1}) = \frac{G_f}{1+\left(\dfrac{D_{f1}}{1-D_{f1}}\right)\dfrac{1}{2(1+\nu_f)}} \tag{4}$$

where $E_f$, $\nu_f$, $G_f$ are the initial elastic properties of the fiber. The components of the strain tensor of the fiber $\underline{\underline{\varepsilon}}_f$ expressed in the $x-y$ frame are denoted by $\varepsilon_{f11}$, $\varepsilon_{f22}$, and $\varepsilon_{f12}$.

A layer consists of fibers aligned along the 1-direction embedded in the matrix. To determine the behavior of this layer, microinterface compatibility conditions are written in terms of strains $\underline{\underline{\varepsilon}}^L$ and stresses $\underline{\underline{\sigma}}^L$ on the layer level. It is more convenient to write the conditions in the 1-2 frame. The following equations are derived by using Voigt's approximation in the fiber direction and Reuss's approximation for the transverse properties:

$$\varepsilon_{m11} = \varepsilon_{f11} = \varepsilon_{11}^L \quad f_m\, \varepsilon_{m22} + f_f\, \varepsilon_{f22} = \varepsilon_{22}^L \quad f_m \varepsilon_{m12} + f_f\, \varepsilon_{f12} = \varepsilon_{12}^L \tag{5}$$

$$f_m\, \sigma_{m11} + f_f\, \sigma_{f11} = \sigma_{11}^L \quad \sigma_{m22} = \sigma_{f22} = \sigma_{22}^L \quad \sigma_{m12} = \sigma_{f12} = \sigma_{12}^L \tag{6}$$

where $f_f$, $f_m$ denote the volume fraction of the fiber and the matrix, respectively, and $\sigma_{ij}^L$ and $\varepsilon_{ij}^L$ are the components of the stress and strain tensors $\underline{\underline{\sigma}}^L$ and $\underline{\underline{\varepsilon}}^L$ in the 1-2 frame. The solution of the previous system yields

$$\underline{\underline{\sigma}}^L = \underline{\underline{\underline{\underline{E}}}}^L(D_{my}, D_{f1}) : \underline{\underline{\varepsilon}}^L \tag{7}$$

where $\underline{\underline{\underline{\underline{E}}}}^L(D_{my}, D_{f1})$ is the stiffness tensor of a layer which is dependent upon all damage variables defined at the constituent level. The elastic energy density associated with matrix cracking and fiber breakage at the layer level is expressed as

$$\psi^L = \frac{1}{2} \underline{\underline{\varepsilon}}^L : \underline{\underline{\underline{\underline{E}}}}^L(D_{my}, D_{f1}) : \underline{\underline{\varepsilon}}^L \tag{8}$$

The elastic behavior of a [0,90] composite system is determined by applying classical laminate theory

$$\underline{\underline{\varepsilon}} = \underline{\underline{\varepsilon}}^{00} = \underline{\underline{\varepsilon}}^{90} \text{ and } \underline{\underline{\sigma}} = f^{00}\, \underline{\underline{\sigma}}^{00} + f^{90}\, \underline{\underline{\sigma}}^{90} \tag{9}$$

where $f^{00}$ and $f^{90}$ denote the volume fraction of the 0° and 90° layers, and $\underline{\underline{\varepsilon}}^{00}$, $\underline{\underline{\sigma}}^{00}$ and $\underline{\underline{\varepsilon}}^{90}$, $\underline{\underline{\sigma}}^{90}$ the strain and stress tensors in the 0° and 90° layers. The overall behavior of the composite is defined as

$$\underline{\underline{\sigma}} = \underline{\underline{\underline{\underline{E}}}}(D_{my}^{00}, D_{my}^{90}, D_{f1}^{00}, D_{f1}^{00}) : \underline{\underline{\varepsilon}} \tag{10}$$

with

$$\underline{\underline{\underline{\underline{E}}}}(D_{my}^{00}, D_{my}^{90}, D_{f1}^{00}, D_{f1}^{90}) = f^{00}\, \underline{\underline{\underline{\underline{E}}}}(D_{my}^{00}, D_{f1}^{00}) + f^{90}\, \underline{\underline{\underline{\underline{E}}}}(D_{my}^{90}, D_{f1}^{90}) \tag{11}$$

where $\underline{\underline{\underline{\underline{E}}}}(D_{my}^{00}, D_{my}^{90}, D_{f1}^{00}, D_{f1}^{90})$ is the fourth order elastic tensor of the composite, which is a function of all damage variables on the constituent level for all layers. The elastic energy density associated with matrix cracking and fiber breakage can be written on the composite level as

$$\psi^D = f^{00}\psi^{00} + f^{90}\psi^{90} \tag{12}$$

where $\psi^{00}$ and $\psi^{90}$ are the elastic energy densities of the 0° and 90° layers.

Inelastic strains are essentially due the interface sliding between the fiber and the matrix. From a micromechanical point of view, sliding can take place as soon as a crack is bridged by fibers. The analysis of these sliding systems leads to the following expression of the stored energy density [5]:

$$\psi^S = \frac{1}{2}E\left(\frac{\varepsilon_{i11}^2}{d_{11}} + \frac{\varepsilon_{i22}^2}{d_{22}}\right) + \frac{1}{2}G\left(\frac{\varepsilon_{i12}^2}{d_{12}}\right) \tag{13}$$

with

$$E = \frac{4}{3}\frac{f^{00}E^{00}f^{90}E^{90}}{f^{00}E^{00} + f^{90}E^{90}} \quad \text{and} \quad G = \frac{f^{00}G^{00}f^{90}G^{90}}{f^{00}G^{00} + f^{90}G^{90}} \tag{14}$$

where $E^{00}$ is the Young's modulus of the 0° layer in the fiber direction 00, $E^{90}$ is the Young's modulus of the 90° layer in the fiber direction 90, $G^{00}$ is the shear modulus of the 0° layer, $G^{90}$ is the shear modulus of the 90° layer, $d_{11}$, $d_{22}$, and $d_{12}$ are damage quantities related to sliding, and $\varepsilon_{i11}$, $\varepsilon_{i22}$, and $\varepsilon_{i12}$ are the inelastic strains.

## 10.7.3 DESCRIPTION OF THE MODEL

The following expression of the free energy density for [0,90] CMCs can be obtained:

$$\psi = \frac{1}{2}(\underline{\underline{\varepsilon}} - \underline{\underline{\varepsilon_i}} - \underline{\underline{\varepsilon_c}})\underline{\underline{\underline{\underline{E}}}}(D_{mx}^{00}, D_{mx}^{90}, D_{f1}^{00}, D_{f1}^{90}):(\underline{\underline{\varepsilon}} - \underline{\underline{\varepsilon_i}} - \underline{\underline{\varepsilon_c}}) + \psi^S \tag{15}$$

where $\underline{\underline{\varepsilon_c}}$ is the creep strain tensor. From this expression, the associated forces to each internal variable are given by partial differentiation:

$$\underline{\underline{\sigma}} = \frac{\partial\psi}{\partial\underline{\underline{\varepsilon}}}, \quad Y = -\frac{\partial\psi}{\partial D}, \quad y = -\frac{\partial\psi}{\partial d}, \quad \underline{\underline{X}} = -\frac{\partial\psi}{\partial\underline{\underline{\varepsilon_i}}}, \quad \underline{\underline{S}} = -\frac{\partial\psi}{\partial\underline{\underline{\varepsilon_c}}} \tag{16}$$

where $D = \{D_{mx}^{00}; D_{mx}^{90}; D_{f1}^{00}; D_{f1}^{90}\}$ and $Y = \{Y_{mx}^{00}; Y_{mx}^{90}; Y_{f1}^{00}; Y_{f1}^{90}\}$, $d = \{d_{11}; d_{22}; d_{12}\}$ and $y = \{y_{11}; y_{22}; y_{12}\}$.

A priori, it is necessary to know thirteen evolution laws associated with the thirteen internal variables. But only the six following evolution laws are necessary:

- One law related to matrix cracking $D_m = D_m(Y_m)$, because the variables $D_{my}^{00}$ and $D_{my}^{90}$ correspond to only one single mechanism, and therefore have the same evolution law

$$D_m = D_{sat}\left(1 - \exp\left[-\left(\frac{Y_m}{Y_{m0}}\right)^{m_m}\right]\right) \tag{17}$$

• One law related to fiber breakage $D_f = D_f(Y_f)$, since $D_{f1}^{00}$ and $D_{f1}^{90}$ have the same evolution law, because fiber breaks are perpendicular to the fiber direction:

$$D_f = 1 - \exp\left[-\left(\frac{Y_f}{Y_{f0}}\right)^{(m_f+1)/2}\right]; \tag{18}$$

• Two laws related to the inelastic strains, since debonding and sliding mechanisms in the 0° and 90° layers are identical, and thus $\varepsilon_{i11}$ and $\varepsilon_{i22}$ have the same evolution law. The second one concerns $\varepsilon_{i12}$. In a unidimensional analysis [5] it has been observed that the back stress $\underline{\underline{X}}$ is only a function of the applied stress $\underline{\underline{\sigma}}$. If the relation is extended to a multiaxial state, then the inelastic strain tensor $\underline{\underline{\varepsilon_i}}$ is a function of the stress tensor $\underline{\underline{\sigma}}$:

$$\varepsilon_{i\alpha\alpha} = \varepsilon_{it}\left(\frac{\langle\sigma_{\alpha\alpha} - \sigma_{th}\rangle}{\sigma_{th}}\right)^{n_t} \quad \text{and} \quad \varepsilon_{i12} = \varepsilon_{is}\left(\frac{\langle\tau_{eq} - \tau_{th}\rangle}{\tau_{th}}\right)^{n_s} \operatorname{sign}(\sigma_{12}) \tag{19}$$

with an equivalent shear stress dependent upon the hydrostatic pressure

$$\tau_{eq} = \sqrt{\left|\sigma_{12}\left[\sigma_{12} + \frac{3}{2}(\sigma_{11} + \sigma_{22})\right]\right|} \tag{20}$$

where $\alpha\alpha = 11, 22$ and $\langle . \rangle$ are the Macauley brackets. When numerous cycles are performed, the evolution of the inelastic strains as a function of the number of cycles is written as

$$F[N] = \frac{\varepsilon_{i\alpha\beta}}{\varepsilon_{i\alpha\beta}(N)} - 1 = \gamma\left(1 - \exp\left[-\left(\frac{\log[N+1]}{a}\right)^{n}\right]\right) \tag{21}$$

where $\varepsilon_{i\alpha\beta}(N)$ is the inelastic strain after $N$ cycles and $\varepsilon_{i\alpha\beta}$ the inelastic strain for the first loading ($N = 0$).

• Two laws related to in-plane debonding $d = d(y)$, since $d_{11}$ and $d_{12}$ are assumed to have the same evolution law. These last evolution laws are not needed to derive the behavior of the composite. Neither the state laws nor the evolution laws are explicit functions of the damage variables $d$, which are needed to express the stored energy.

• Similarly, two laws related to creep are needed. For the normal strains, the evolution law is assumed to be independent of the temperature and is written as

$$\varepsilon_{c\alpha\alpha}[t] = \frac{2\sigma_{\alpha\alpha}}{f_f E_f}\left(1 - \exp\left[-\left(\frac{t}{t_{i0}}\right)^{n_t}\right]\right) \tag{22}$$

and for shear strains, the evolution law is explicitly dependent on the temperature:

$$\varepsilon_{c12}[t] = \varepsilon_{120} \exp\left[b\left(\frac{T}{T_0} - 1\right)\right]\left(\frac{\sigma_{12}}{\tau_0}\right)^{n_\sigma}\left(\frac{t}{t_{120}}\right)^{n_S}. \tag{23}$$

## 10.7.4 IDENTIFICATION OF THE MATERIAL PARAMETERS

The identification procedure is carried out by using the results of pure tension tests along two orientations: one in the direction of a set of fibers (0°) and one at ±45°. Each test involves a series of loading and unloading sequences.

The first step is to determine the elastic properties, which are usually given by the manufacturer. The ones of the matrix may be degraded by processing (e.g., porosity or microcavities). The initial elastic properties of the matrix are determined from measurement of the initial Young's moduli $E^{00}$ and $E^{45}$, respectively, for a 0° and ±45° tension test.

The next step is to identify the parameters of the relevant evolution laws. The fitting procedure can be summarized in the following way:

- The variation of $D_m$ with $Y_m$ obtained from the experimental data of the unloading slopes for different maximum stress levels of a tension test at ±45°.
- The evolution law of the inelastic tensile strain, $\varepsilon_{i11}$, is obtained from the value of the unloading strain at the maximum stress of each unloading loop in the tension test performed in the 0° direction (i.e., direction of the fibers).
- The evolution law of $D_f$ with $Y_f$, modeling fiber breakage, results in the fitting of the unloading slope for the maximum stress levels of a tension test at 0°.
- The evolution law of the inelastic shear strain, $\varepsilon_{i12}$, is obtained from the value of the unloading strain at the maximum stress of each unloading loop in a tension test at ±45°.
- The cyclic properties can be identified with either a tensile or a 4-point bend test. By measuring the evolution of the inelastic strain $\varepsilon_{i\alpha\beta}$ with the number of cycles, the parameters of the function $F$ can be identified.
- To identify creep properties, one needs more tests because creep is sensitive to stress levels and temperature range. Four different tests are used: one tensile test at 0° and three tensile tests at ±45°.

The different steps are summarized in Table 10.7.1.

TABLE 10.7.1. Experimental Tests Needed for the Identification of the Model

| Number of experiments | Type of test | | Evolution law | |
|---|---|---|---|---|
| 1 | Tension tests | @ 0° | $\varepsilon_{i\alpha\alpha}$ | $D_f$ |
| 1 | with unloading | @ ±45° | $D_m$ | $\varepsilon_{i12}$ |
| 1 | Fatigue \| Tensile / 4-point bend \| test | @ 0° | $F[N]$ | |
| 1 | Creep test | @ 0° | $\varepsilon_{c\alpha\alpha}$ | |
| 3 | Creep test | @ ±45° for 3 sets \| $(\sigma_1, T_1)$ $(\sigma_2, T_1)$ $(\sigma_3, T_3)$ | $\varepsilon_{c12}$ | |

## 10.7.5 HOW TO USE THE MODEL

This model is fully implemented and maintained in the commercial finite element code ABAQUS via a UMAT routine. Integrated laws are easy to implement and reduce the computation cost, especially for fatigue and creep loading conditions, by using the isochronous analysis [4].

Several applications of the model can be found for SiC/SiC composites [1–4]. All these examples have a full description, in terms of the distribution of tensile and shear stresses, inelastic strains, and matrix damage, of tension tests on plates with a hole and bend test on a plate. Some of the computations are compared to experimental measurements.

To simplify the identification procedure, an assistant was written by using the same routine developed for the finite element code. This quasi-automatic identification closely follows the different steps summarized in Table 10.7.1.

## 10.7.6 TABLE OF PARAMETERS

The identification procedure is applied to various [0,90] laminated or woven CMCs. Some examples are given in Table 10.7.2.

It can be noted that the material parameters modeling elasticity, matrix cracking, interface sliding, fiber-breakage, and fiber pullout of SiC/SiC composites are valid for the whole temperature range (i.e., from the room temperature to 1200°C). Furthermore, the chemical degradation of the fiber

TABLE 10.7.2 Material Parameters for Various CMCs

| | | SiC/SiC | | C/C | | $Al_2O_3/Al_2O_3$ | |
|---|---|---|---|---|---|---|---|
| Material parameter | | Value | Unit | Value | Unit | Value | Unit |
| Matrix elastic modulus | $E_m$ | 157 | GPa | 9.5 | GPa | 40 | GPa |
| Matrix Poisson's ratio | $\nu_m$ | 0.25 | | 0.1 | | 0.25 | |
| Matrix volume fraction | $f_m$ | 0.55 | | 0.56 | | 0.60 | |
| Weibull modulus | $m_m$ | 1.6 | | 1.15 | | NA | |
| Normalizing energy | $Y_{m0}$ | 0.60 | MPa | 1.21 | MPa | NA | MPa |
| Saturation parameter | $D_{sat}$ | 0.99 | | 0.99 | | 0.01 | |
| Fiber elastic modulus | $E_f$ | 200 | GPa | 280 | GPa | 380 | GPa |
| Fiber Poisson's ratio | $\nu_f$ | 0.25 | | 0.1 | | 0.2 | |
| Fiber volume fraction | $f_f$ | 0.45 | | 0.44 | | 0.40 | |
| Weibull modulus | $m_f$ | 4.5 | | NA | | NA | |
| Normalizing energy | $Y_{f0}$ | 42.5 | MPa | large | MPa | large | MPa |
| Tensile inelastic strain constant | $\varepsilon_{int}$ | $4.99\ 10^{-5}$ | | NA | | $2.68\ 10^{-5}$ | |
| Threshold tensile stress | $\sigma_{th}$ | 135 | MPa | NA | MPa | 66.4 | MPa |
| Exponent | $n_{ii}$ | 1.0 | | NA | | 1.74 | |
| Shear inelastic strain constant | $\varepsilon_{ins}$ | $5.0\ 10^{-5}$ | | $4.52\ 10^{-2}$ | | $2.61\ 10^{-6}$ | |
| Threshold shear stress | $\tau_{th}$ | 59.1 | MPa | 41.9 | MPa | 22.5 | MPa |
| Exponent | $n_{12}$ | 2.0 | | 3.45 | | 25.3 | |
| Fatigue amplitude | $\gamma$ | 0.984 | | NA | | NA | |
| Normalizing value | $a$ | 6.96 | | NA | | NA | |
| Exponent | $n$ | 1.56 | | NA | | NA | |
| Time power law exponent | $n_t$ | 0.451 | | NA | | NA | |
| Normalizing time | $t_{i0}$ | 70.2 | h | NA | h | NA | h |
| Shear creep strain constant | $\varepsilon_{120}$ | $1.35\ 10^{-3}$ | | NA | | NA | |
| Temperature coefficient | $b$ | 28.1 | | NA | | NA | |
| Stress power law exponent | $n_\sigma$ | 4.2 | | NA | | NA | |
| Time power law exponent | $n_s$ | 0.4 | | NA | | NA | |
| Normalizing temperature | $T_0$ | 1204 | °C | NA | °C | NA | °C |
| Normalizing stress | $\tau_0$ | 34.5 | MPa | NA | MPa | NA | MPa |
| Normalizing time | $t_{120}$ | 100 | h | NA | h | NA | h |

coating is not accounted for by the present model. For the two other materials, only room temperature data are reported in Table 10.7.2.

## REFERENCES

1. Burr, A., Hild, F., and Leckie, F. A. (1997). Continuum description of damage in ceramic-matrix composites. *Eur. J. Mech. A/Solids* **16**: 53–78.

2. Burr, A., Hild, F., and Leckie, F. A. (1998). The mechanical behaviour under cyclic loading of ceramic-matrix composites. *Mater. Sci. Eng.* **A250**: 256–263.
3. Burr, A., Hild, F., and Leckie, F. A. (1998). Behaviour of ceramic-matrix composites under thermomechanical cyclic loading conditions. *Comp. Sci. Tech.* **58**: 779–783.
4. Burr, A., Hild, F., and Leckie, F. A. (1998). Isochronous analysis applied to the behavior of ceramic-matrix composites, *in* Proceedings of JNC11, pp. 1343–1349, Vol. 3, Baptiste, D., and Lamon, J., eds., Paris: AMAC.
5. Burr, A., Hild, F., and Leckie, F. A. (1995). Micro-mechanics and continuum damage mechanics. *Arch. Appl. Mech.* **65**: 437–456.

SECTION **10.8**

# Limit and Shakedown Analysis of Periodic Heterogeneous Media

GIULIO MAIER, VALTER CARVELLI, and ALBERTO TALIERCIO

*Department of Structural Engineering, Technical University (Politecnico) of Milan, Piazza Leonardo Da Vinci 32, 20133 Milano, Italy*

Contents



## 10.8.1 INTRODUCTION: BASIC CONCEPTS AND APPLICABILITY DOMAIN

The engineering motivations for what follows may be clarified first by referring to two typical examples of ductile heterogeneous media: (i) perforated steel plates frequently employed in power plants; (ii) structural components made of metal-matrix fiber-reinforced composites (MMCs). In both these representative cases, overall inelastic analyses must be carried out in terms of average (or "macroscopic") stresses $\underline{\underline{\Sigma}} = \langle \underline{\underline{\sigma}}(\underline{x}) \rangle$ and strains $\underline{\underline{E}} = \langle \underline{\underline{\varepsilon}}(\underline{x}) \rangle$. Therefore, at first a homogenization procedure is required, i.e., the

*Handbook of Materials Behavior Models*. ISBN 0-12-443341-3.

characterization of a homogeneous "equivalent material" through a microstructural analysis based on the local constitutive models for each constituent (or "phase"), say, $\underline{\underline{\sigma}}(\underline{x}, t) = f(\underline{\underline{\varepsilon}}(\underline{x}, \tau),\ 0 \leq \tau \leq t)$, and on the geometry of the texture at the "microscale"; see, e.g., Reference [1]. From here onwards, $\langle \cdot \rangle$ denotes averaging over a suitable representative volume; underlined symbols denote vectors, doubly underlined ones second-order tensors, and dots time derivatives.

Consider engineering situations in which deformations up to collapse can be reasonably expected to be "small" (this implies linear kinematic compatibility and rules out the influence of deformations on equilibrium) and the material behavior can be realistically interpreted as elastic-plastic and stable in Drucker's sense (hence: convex yield surfaces; associated flow rules; no softening), either perfectly plastic or hardening with saturation. In such situations, if it is experimentally, or otherwise, ascertained that the dominant dissipative phenomenon is plastic yielding (no debonding at interfaces between phases; no damage or microfractures), then, in view of overall structural analyses, the main feature to be assessed for the equivalent homogeneous material is its "strength," i.e., the plastic failure locus (or "yield surface") in the space of the average stresses $\underline{\underline{\Sigma}}$ and, hence, the "macroscopic strength domain," $G^{hom}$, defined by that locus.

If the structure is subjected to thermal and/or mechanical external actions ("loads") fluctuating in time ("variable repeated," in particular cyclic), it is of interest to assess in the $\underline{\underline{\Sigma}}$ space also "shakedown domains" $G_{SD}^{hom}$ with respect to sets of assigned "loading domains" $\Omega$ of variable repeated $\underline{\underline{\Sigma}}(t) \in \Omega$. Specifically, the question to answer is as follows: given one $\Omega$ ("basic" loading domain), what is the amplification factor, or "shakedown limit," $s_{SD}$, such that the cumulative (integral over space and time) dissipated energy $D$ is unbounded (i.e., $\lim_{t\to\infty} D(t) = \infty$) for factors $\mu > s_{SD}$, and it is not so if $\mu \leq s_{SD}$, namely, if yielding processes eventually vanish (i.e., the system "shakes down"). Unboundedness of $D$ characterizes three distinct "ultimate limit states": (a) incremental collapse (otherwise called "ratcheting"); (b) alternating plasticity (or "plastic shakedown"), with no divergence of deformed configuration in time; and (c) "plastic collapse" in the special case where the load domain shrinks to a point, say, $\underline{\underline{\Sigma}}_0$ (in this case, $s_{SD}$ coincides with the yield limit $s_0$, which amplifies $\underline{\underline{\Sigma}}_0$ leading to a point $\underline{\underline{\Sigma}}_0^c$ on the yield surface of the homogenized material).

The following circumstances are worth noting: (i) $G^{hom}$ contains every $G_{SD}^{hom}$ and represents an essential feature of the homogenized material model, whereas the shakedown domains $G_{SD}^{hom}$ depend also on the set of basic loading domains considered; (ii) shakedown (SD) theories and methodologies apt to compute $s_{SD}$ (and hence any $G_{SD}^{hom}$) cover as special cases those meant to provide $s_0$ and, hence, $G^{hom}$; (iii) the present conception of strength is in principle nonconservative (though practically quite realistic for prevailing ductile behaviors), since it does not allow for undesirable mechanical events (such as excessive deformations or cracks) that might occur for $\mu \leq s_{SD}$; (iv) "direct" methods for shakedown analysis (SDA) and limit analysis (LA), with respect to evolutive (step-by-step) inelastic analysis techniques are consistent with frequent lack of information on the loading history (only fluctuations intervals are often known); also, they are generally advantageous in terms of computational effort and therefore more suitable to multiple parametric studies in design processes.

The fundamentals of direct (nonevolutive) methods for LA and SDA of heterogeneous media will be presented in the subsequent sections according to the following criteria and further restrictions: (i) periodicity assumptions and consequent reference to a representative volume (RV) with periodicity boundary conditions, typical of homogenization theory; (ii) von Mises perfectly elastic-plastic local material model, defined by a single parameter (e.g., tensile strength), in addition to the elastic moduli.

The present subject is rooted in classical plasticity and homogenization theories and, hence, the relevant literature is enormous. The few items concisely presented herein are selected by application-oriented criteria, and only a few closely related references are cited for details. On direct methods, a state-of-the-art and a fairly comprehensive up-to-date pertinent bibliography can be found in Reference [2].

## 10.8.2 LIMIT ANALYSIS BY THE STATIC APPROACH

The classical static (or "lower bound," or "safe") limit theorem can be stated as follows: a solid will not collapse if and only if there is a ("statically admissible") stress field which everywhere fulfills equilibrium and local strength limitations (or "yield conditions"). Its validity is guaranteed, as mentioned in Section 10.8.1, by linear kinematics and perfect plasticity, stable in the sense of Drucker's postulate. By applying this statement, the strength domain $G^{hom}$ in the space of average stresses $\underline{\underline{\Sigma}}$ for a periodic heterogeneous medium whose RV occupies the volume $V$ with boundary $\partial V$ can be defined as follows (see, e.g., Reference [1]):

$$\begin{aligned} G^{hom} = \{\underline{\underline{\Sigma}} | \underline{\underline{\Sigma}} = \langle \underline{\underline{\sigma}}(\underline{x}) \rangle; \quad & \underline{\underline{\sigma}}(\underline{x}) \in G(\underline{x}) \forall \underline{x} \in V; \\ \text{div } \underline{\underline{\sigma}}(\underline{x}) = \underline{0} \text{ in } V; \quad & \underline{\underline{\sigma}}(\underline{x}) \cdot \underline{n}(\underline{x}) \text{ antiperiodic on } \partial V\} \end{aligned} \tag{1}$$

where $G(\underline{x})$ denotes the (convex) strength domain of the heterogeneous material at any point $\underline{x}$ of the RV, and $\underline{n}$ is the outward unit normal to $\partial V$. A possible discontinuity surface for the microscopic stress field can be also accounted for in Eq. 1 by interpreting div $\underline{\underline{\sigma}}$ in the sense of the theory of distributions.

A domain $G_o$ contained in $G^{hom}$ (and, hence, conservative, i.e., representing lower bounds to the strength of the homogenized material) can be computed by selecting over the RV special statically admissible microscopic stress fields that comply with the periodicity boundary conditions on $\partial V$. For unidirectional fiber-reinforced composites, a simple domain inside $G^{hom}$ is [3, 4]:

$$G_L = \left\{ \underline{\underline{\Sigma}} = \underline{\underline{\sigma}}_m + \sigma \underline{e}_x \otimes \underline{e}_x; \quad \underline{\underline{\sigma}}_m \in G_m; \quad -\hat{\sigma}^- \leq \sigma \leq \hat{\sigma}^+ \right\}, \quad \text{where } \hat{\sigma}^{\pm} = \nu_f \left( \bar{\sigma}_f^{\pm} - \bar{\sigma}_m^{\pm} \right) \tag{2}$$

having set $G_m$ as the strength domain of the matrix material; $\bar{\sigma}_f^{\pm}$, $\bar{\sigma}_m^{\pm}$ as the uniaxial strength of fibers and matrix in tension (+) and compression (−); $\nu_f$ as the fiber volume fraction; $\sigma$ as a scalar stress variable; and $\otimes$ as the dyadic product. Under several (also multiaxial) stress conditions, Eq. 2 gives fairly accurate estimates of the macroscopic strength of the material [4, 5].

In particular, if both phases comply with von Mises strength criterion (which is the case, e.g., for MMCs), Eq. 2 analytically provides a lower bound, say $\Sigma_L$, to the macroscopic uniaxial strength of the composite, $\Sigma^{hom}$, for any orientation $\theta$ of the applied stress $\Sigma$ to the fibers (the superscripts $\pm$ are omitted, the material behavior being symmetric in tension and compression). These bounds, visualized by the dashed line in Figure 10.8.1b for the MMC case of Figure 10.8.1a, reads [3]:

$$\Sigma^{hom}(\theta) \geq \Sigma_L(\theta) = \begin{cases} \hat{\sigma}\left(1 - \frac{3}{2}\sin^2\theta\right) + \sqrt{\bar{\sigma}_m^2 - 3\hat{\sigma}^2 \sin^2\theta \left(1 - \frac{3}{4}\sin^2\theta\right)} \\ \text{if } 0 \leq \theta \leq \theta^* \\ \dfrac{\bar{\sigma}_m}{\sin\theta \sqrt{3\left(1 - \frac{3}{4}\sin^2\theta\right)}} \quad \text{if } \theta^* \leq \theta \leq \frac{\pi}{2} \end{cases} \tag{3}$$

The orientation $\theta^*$ can be computed simply by equating the two expressions for $\Sigma_L$.

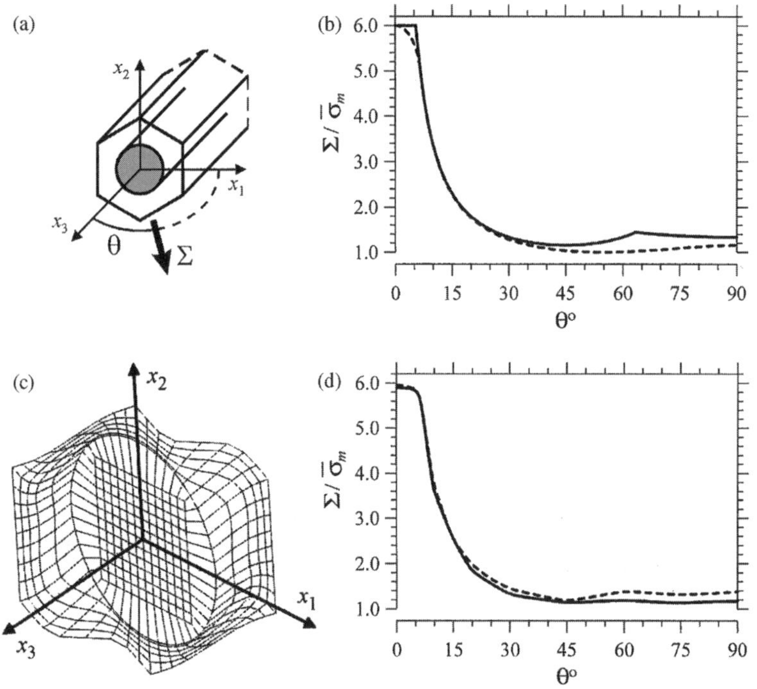


FIGURE 10.8.1 Limit and shakedown analysis of a metal-matrix composite. a. Representative volume. b. Bounds on the macroscopic uniaxial strength by the kinematic approach, Eq. 6 (solid line) and static approach, Eq. 3 (dashed line) [4]. c. Finite element mesh and incremental collapse mechanism for $\theta = 30^\circ$. d. Uniaxial plastic collapse (dashed line) [11] and shakedown limit (solid line) vs. direction $\theta$.

## 10.8.3 LIMIT ANALYSIS BY THE KINEMATIC APPROACH

The kinematic (or "upper bound," or "unsafe") limit theorem, subject to the same validity restrictions as the static one, can be expressed in the following terms: a solid will collapse if and only if there exists a mechanism (i.e., strain and displacement rates obeying geometric compatibility) such that the work done by the external forces exceeds the total dissipated energy (the local density of which is uniquely defined by the plastic strains).

On the basis of this statement, the strength domain $G^{\text{hom}}$ can be formally defined as follows:

$$G^{hom} = \left\{ \underline{\underline{\Sigma}} | \underline{\underline{\Sigma}} : \dot{\underline{\underline{E}}} \leq \pi^{hom}\left(\dot{\underline{\underline{E}}}\right), \ \forall \dot{\underline{\underline{E}}} \right\} \tag{4}$$

$\pi^{hom}$ being the dissipated power over the RV associated to $\dot{\underline{\underline{E}}}$, namely:

$$\pi^{hom}\left(\dot{\underline{\underline{E}}}\right) = \inf_{\dot{\underline{v}}(\underline{x})} \frac{1}{|V|} \int_V \pi\left(\dot{\underline{\underline{\varepsilon}}}(\underline{x})\right) dV, \text{ subject to:}$$

$$\dot{\underline{\underline{\varepsilon}}}(\underline{x}) = \frac{1}{2}\left(\nabla\dot{\underline{u}} + \nabla^t\dot{\underline{u}}\right);$$

$$\dot{\underline{u}}(\underline{x}) = \dot{\underline{\underline{\Sigma}}} \cdot \underline{x} + \dot{\underline{v}}(\underline{x}); \ \dot{\underline{v}} \text{ periodic over } V \tag{5}$$

where $|V|$ is a volumetric measure of the RV; $\dot{\underline{u}}$ denotes the microscopic velocity field; $\dot{\underline{\underline{\varepsilon}}}$ is the associated strain rate field; and $\pi$ is the specific dissipated (plastic) power (i.e., mathematically, the support function of the local strength domain $G$ at point $\underline{x} \in V$). Here again, discontinuity surfaces for the microscopic velocity field can be accounted for in Eq. 5 by interpreting $\nabla\dot{\underline{u}}$ in the sense of the theory of distributions.

Domains that bound $G^{hom}$ from outside can be obtained by considering simple "failure mechanisms" for the RV, either characterized by uniform strain rates or by "failure planes" that split the RV element into blocks rigidly moving with respect to each other [4]. For instance, upper bounds $\Sigma_U$ to the macroscopic uniaxial strength of MMCs can be computed by means of mechanisms characterized by slip planes cutting the matrix and intercepting the midpoints of a couple of opposite or adjacent sides of the RV (see Reference [4] for further details). These bounds depend on the fiber volume fraction $v_f$ and the spatial arrangement of the reinforcement. For fiber evenly spaced according to a regular hexagonal pattern, provided that the percentage of reinforcement is not too high ($v_f < \pi\sqrt{3}/8$), an upper bound to $\Sigma^{hom}$ is

$$\Sigma^{hom}(\theta) \leq \Sigma_U(\theta) = \min \left\{ \hat{\sigma} + \bar{\sigma}_m; \ \frac{2\bar{\sigma}_m}{\sqrt{3}\sin 2\theta}; \ \frac{2\bar{\sigma}_m}{\sin\theta\sqrt{3\left(1 - \frac{1}{4}\sin^2\theta\right)}} \right\} \tag{6}$$

It is worth noting that the two bounds, $\Sigma_L$, Eq. 3, and $\Sigma_U$, Eq. 6, coincide at $\theta = 0$; that is, the uniaxial strength of the composite along the fibers can be exactly predicted and turns out to be

$$\Sigma^{hom}(0) = \hat{\sigma} + \bar{\sigma}_m \equiv (1 - v_f)\bar{\sigma}_m + v_f\bar{\sigma}_f \tag{7}$$

This is a well-known formula which is widely employed in applications and is usually called the "rule of mixtures." The bounds presented in this and the preceding section can be extended to materials with different strengths in tension and compression [4]. Also, debonding at the fiber–matrix interface can be accounted for, provided it does not imply a softening behavior [5].

## 10.8.4 A GENERAL FINITE ELEMENT LINEAR PROGRAMMING PROCEDURE

The applicability of the convenient closed-form bounds was seen to be limited (they are particularly accurate for uniaxial stresses and moderate $v_f$). A numerical method for LA of composites can be developed by a static approach through the following sequence of operation phases (for basic concepts of the method, see, e.g., References [6, 7]; for specific details, see References [8]). (a) Perform a piecewise linear approximation (PWL) of the local material model. (b) Generate a finite element model of the RV. (c) Consider as loads an average stress state $\underline{\underline{\Sigma}}_0$ amplified by a load factor $\mu$. (d) Maximize $\mu$ under statical admissibility constraints on the discretized stresses, namely, subject to: linear equations expressing the average link between $\mu\underline{\underline{\Sigma}}_0$ and the stress field; equilibrium and periodicity on $\partial V$; and linear inequality which enforces the PWL yield conditions. (e) Interpret the optimal (max) value as collapse multiplier (or safety factor) $s^s$ with respect to $\underline{\underline{\Sigma}}_0$ (i.e., $s\underline{\underline{\Sigma}}_0$ as a point belonging to the boundary of $G^{hom}$) and the optimal vector as one of the possible stress states at collapse.

If a kinematic approach is adopted, the last two phases become: ($d'$) Minimize the total dissipated power (a linear function of plastic multiplier rates) under kinematic admissibility constraints on the modeled velocity fields and non-negative plastic multiplier rates, namely: linear equation which expresses splitting of velocity $\dot{\underline{u}}$ into an addend linear in space, related to $\dot{\underline{\underline{E}}}$, and a periodic addend $\dot{\underline{v}}$; periodicity on $\partial V$ and geometric compatibility of $\dot{\underline{v}}$; normalization of the "external power" $\underline{\underline{\Sigma}}_0 : \dot{\underline{\underline{E}}}$. ($e'$) Interpret the optimal (min) value as safety factor $s^k$ and the optimal vector as one of the possible collapse mechanisms. This sketchy outline can be supplemented by the following remarks.

(1) At the finite element interfaces discontinuities can be modeled for stresses under equilibrium constraints on tractions, for strains under neither detachment nor compenetration conditions on displacements (see Reference [8]).

(2) If the local PWL strength loci in (a) are inscribed in the original ones, and if the field modeling (b) and the constraint in (d) are such that static admissibility is complied with everywhere, then $s^s$ is a lower bound on the exact, "safety factor" $s$. Similarly, if phase (a) generates a circumscribed PWL approximation and ($d'$) satisfies kinematic admissibility exactly, then $s^k = s_U \geq s$. However, often in practice a tight bracketing of $s$ is computationally cumbersome to achieve; then secant approximation (a) and only approximate enforcement in ($d$), or ($d'$), are preferable, leading to $s^s \approx s$, or $s^k \approx s$, with discrepancy unknown in sign but reducible by refinements in phases (a) and (b) at the price of increased computing cost (which for a linear programming (LP) problem in "normal form" is roughly proportional to $m\, n^3$ if $m$ and $n$ are the numbers of variables and constraints, respectively).

(3) The previously outlined LA procedures are centered on LP with the following peculiar features: (i) LP software (based on the classical Simplex method and its variants) is widely available and often adopted to large-scale computing; (ii) any LP problem uniquely defines its "dual" LP problem, which can be interpreted as a kinematic formulation if the primal is a static one, and vice versa. Modern LP solvers provide the solution to both, but process the more economical one.

(4) The PWL phase (a) leading to LP implies a drastic increase in the number of variables (an increase which can be reduced by a suitable trial-and-error procedure). If (a) is omitted, the phases ($d$) and ($d'$) lead to (convex) nonlinear programming (NLP) problems, by far more laborious to solve than the LP problem. The trade-off between LP and NLP formulations of LA is not resolved in general, much depending in practice on the availability of ad hoc software.

## 10.8.5 SHAKEDOWN ANALYSIS

For periodic heterogeneous elastic-plastic solids subjected to an assigned loading domain $\Omega$ of (uniform) variable-repeated average stresses $\underline{\underline{\Sigma}}(t) \in \Omega$, the following theorems can be proved (see, e.g., References [7, 9, 10]).

(a) Shakedown will occur if, and only if, there is a field of self-stress $\underline{\underline{\sigma}}^s$ such that, at any time and everywhere in the RV, the sum of it and of the (local) elastic stresses $\underline{\underline{\sigma}}^e$ due to $\underline{\underline{\Sigma}}$ complies with the (local) yield conditions and the periodicity on the boundary of the RV.

(b) Shakedown will occur if, and only if, for all "admissible plastic strain cycles" (APSC) the cumulative dissipated energy is larger than, or equal to, the ("external") work done by the elastic stresses $\underline{\underline{\sigma}}^e$ due to the concomitant external actions. By APSC is meant a history over a time interval $T$ of a fictitious imposed plastic strain field, the time integral of which over $T$ is compatible with displacements complying with the periodicity conditions as in Eq. 5.

(c) If $\Omega$ is a polyhedrical region (possibly as a PWL approximation of the input data), for SD analysis purposes it can be replaced by the set of its, say $m$, vertices: this reduces $\underline{\underline{\sigma}}^e(t)$ to a sequence $\underline{\underline{\sigma}}_j^e$ and any APSC to a corresponding sequence of finite increments $\underline{\underline{\varepsilon}}_j^p$, $j = 1, \ldots, m$.

The kinematic method outlined in following text turns out to be computationally rather effective, in the writers' experience. The formulation based on (b) and (c), on traditional finite element discretization (displacement modeling), and on Mises elastoplasticity reads:

$$\tilde{s} = \min_{\underline{\underline{\varepsilon}}_{ij}^p, \underline{\underline{E}}^p, \underline{V}} \sum_i \sum_j c_i J_i \pi(\underline{\underline{\varepsilon}}_{ij}^p), \text{ subject to :} \tag{8}$$

$$tr\underline{\underline{\varepsilon}}_{ij}^p = 0, \; \forall i, \, j; \quad \sum_j \underline{\underline{\varepsilon}}_{ij}^p = \underline{\underline{B}}_i \underline{V} + \underline{\underline{E}}^p, \; \forall i; \quad \sum_i \sum_j c_i J_i \underline{\underline{\sigma}}_{ij}^e : \underline{\underline{\varepsilon}}_{ij}^p = 1 \tag{9}$$

Here indices $i$ and $j$ run over the sets of Gauss points ($c_i$ and $J_i$ being Gauss weights and Jacobians) and vertices of $\Omega$, respectively; Eq. 9 enforces plastic incompressibility, compatibility at the APSC end ($\underline{V}$ being the nodal displacement vector allowing for periodicity, $\underline{\underline{B}}_i$ the compatibility matrix for point $i$), and normalization of the total "external work" along the APSC, respectively.

Eqs. 8 and 9 represent a convex NLP problem with nonsmooth objective function (since the Mises dissipated power $\pi$ is nonsmooth). The solution algorithm, described and satisfactorily employed in References [11, 12], is centered on the enforcement of Eq. 9a (with empirical suitable choice of the relevant penalty factor $\alpha$) and on an iterative procedure for solving the (nonlinear) Kuhn-Tucker optimality conditions.

Remarks: (1) The optimal value $\tilde{s}$ is an approximation of $s_{SD}$, not a bound. (2) If $\Omega$ shrinks to a point, SDA reduces to LA and elasticity could be proved to become immaterial. (3) If a PWL approximation is adopted for the local material models, SDA becomes an LP problem, as a generalization of the LA formulations of Section 10.8.4. (4) History-dependent post-SD quantities (e.g., relative residual displacement of two points of the RV) can be bounded from above by bounding techniques, which, in PWL formulations (possibly accounting for hardening), can be reduced to a quadratic programming followed by an LP problem (see, e.g., Reference [7]).

## 10.8.6 EXAMPLES

The first illustrative example concerns a ductile MMC characterized by the RV shown in Figure 10.8.1a and by fiber vs. matrix ratios 1.86 for volumes and 8.7 for yield stresses (both are Mises materials). The homogenized uniaxial plastic collapse limits and SD limits (for fluctuations between 0 and $\Sigma$) versus the angle $\theta$ from the fiber axis have been computed by the bounding formulae of Sections 10.8.2 and 10.8.3 and by the finite element procedure summarized in Section 10.8.5 (432 finite element with bilinear interpolation, $\alpha = 10^6$). These results are comparatively visualized in Figure 10.8.1b, d.

The second and third example concern LA and SDA of perforated metal plates, regarded as two-dimensional plane-stress systems and depicted in Figure 10.8.2a and Figure 10.8.3a as for RV and finite element mesh. Representative results are visualized in Figure 10.8.2b and Figure 10.8.3b, and compared in the latter to experimental data with excellent agreement.

## 10.8.7 CONCLUSIONS

What precedes is intended to provide introductory and orientative information on the title subject. In closing, the reader's attention is drawn to the following remarks.

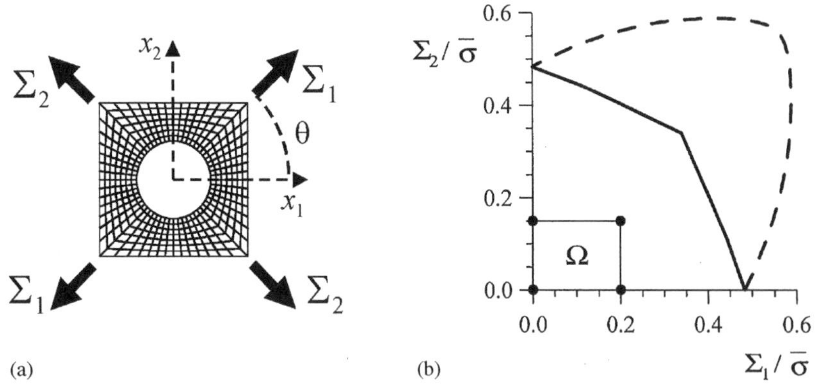


**FIGURE 10.8.2** Perforated plate with circular holes (volumetric ratio 0.2). a. Representative volume, mesh of 392 finite element and biaxial average stress state. b. For $\theta = 45°$, loci of plastic collapse (dashed line) and of shakedown limit (solid line) for rectangular load domains $\Omega$ ($m = 4$).

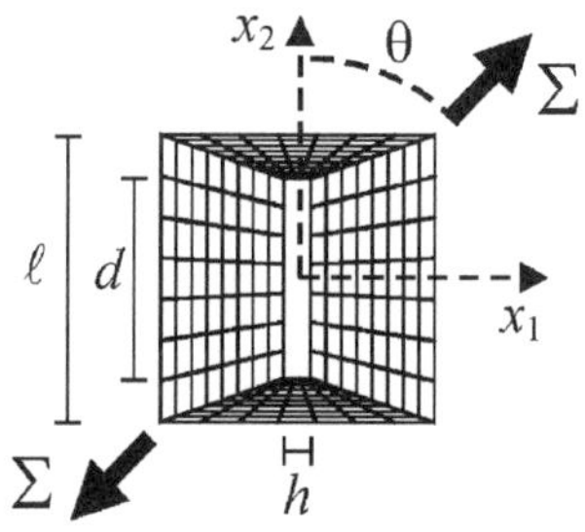


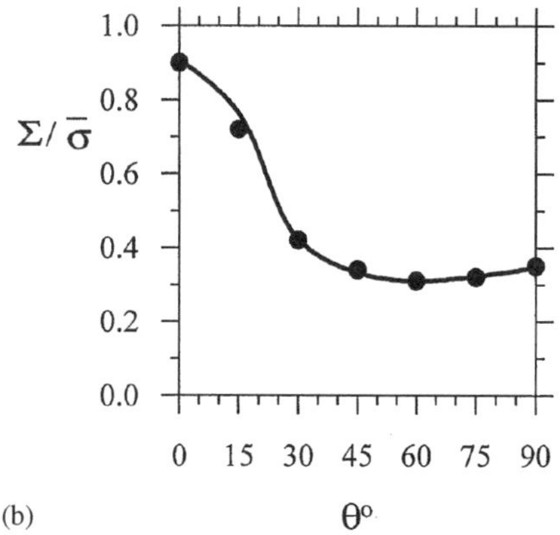


FIGURE 10.8.3 Perforated plate with rectangular holes ($h = 0.1\ell$; $d = 0.7\ell$). a. Representative volume, mesh of 224 finite element and uniaxial average stress state. b. Plastic collapse limit vs. direction $\theta$, present method (solid line) and experimental results by Litewka *et al.* [13] (dots).

(a) For understanding and implementation of the computer-oriented methods outlined in Sections 10.8.4 and 10.8.5, details can be found in References [8, 11, 12]. Moreover, a variety of alternative techniques is available nowadays in the literature: most of them are referenced in the review paper [2]; specifically on composites, important contributions (not cited here for space limitations) are due to G. Dvorak, F. Leckie, A. Ponter, F. Rammerstorfer, D. Weichert, and others.

(b) Besides plastic collapse and lack of shakedown, susceptible to be analyzed by the direct (nonevolutive) methods dealt with herein, there are other ultimate or unserviceability limit states to be studied by generally more laborious time-stepping (evolutive) methods (expounded elsewhere in this book) for the integrity assessments, based on their texture, of heterogeneous materials.

## REFERENCES


1. Suquet, P. (1985). Elements of homogenization for inelastic solid mechanics, in *Homogenization Techniques for Composite Media*, CISM Lecture Notes in Physics pp. 193–278, vol. 272, Sancheg-Palencia, E., and Zarei, A., eds., Wien: Springer.
2. Maier, G., Carvelli, V., and Cocchetti, G. (2000). On direct methods for shakedown and limit analysis. *Eur. J. Mech. A/Solids* **19**: S79–S100.
3. de Buhan, P., and Taliercio, A. (1991). A homogenization approach to the yield strength of composite materials. *Eur. J. Mech. A/Solids* **10**: 129–154.

4. Taliercio, A. (1992). Lower and upper bounds to the macroscopic strength domain of a fiber-reinforced composite material. *Int. J. Plasticity* **8**: 741–762.
5. Taliercio, A., and Sagramoso, P. (1995). Uniaxial strength of polymeric-matrix fibrous composites predicted through a homogenization approach. *Int. J. Solids Struct.* **32**: 2095–2123.
6. Maier, G. (1970). A matrix structural theory of piecewise-linear plasticity with interacting yield planes. *Meccanica* **5**: 55–66.
7. Cohn, M. Z., and Maier, G. (1979). *Engineering Plasticity by Mathematical Programming*, New York: Pergamon Press.
8. Francescato, P., and Pastor, J. (1997). Lower and upper numerical bounds to the off-axis strength of unidirectional fiber-reinforced composites by limit analysis methods. *Eur. J. Mech. A/Solids* **16**: 213–234.
9. Maier, G. (1969). Shakedown theory in perfect elastoplasticity with associated and nonassociated flow-laws: A finite element, linear programming approach. *Meccanica* **4**: 250–260.
10. König, J. A. (1987). *Shakedown of Elastic-Plastic Structures*, Elsevier.
11. Carvelli, V., Maier, G., and Taliercio, A. (2000). Kinematic limit analysis of periodic heterogeneous media. *Computer Modelling in Engineering and Science* **1**(2): 15–26.
12. Carvelli, V., Maier, G., and Taliercio, A. (1999). Shakedown analysis of periodic heterogeneous materials by a kinematic approach. *Mechanical Engineering (Strojnicky Časopis)* **50**: 229–240.
13. Litewka, A., Sawczuk, and A., Stanislawka, J. (1984). Simulation of oriented continuous damaged evolution. *Jour. de Méch. Théor. et Appl.* **3**: 675–688.

SECTION **10.9**

# Flow-Induced Anisotropy in Short-Fiber Composites

ARNAUD POITOU and FRÉDÉRIC MESLIN

*LMT-Cachan, ENS de Cachan, Université Paris 6, 61 avenue du Président Wilson, 94235 Cachan Cedex, France*

Contents



## 10.9.1 INTRODUCTION

Short-fiber composites are made of a thermoplastic matrix into which shopped fibers are immersed (fiber length around 0.1 mm, diameter around 0.01 mm, volumetric concentration around 15 to 30%). Composite parts are manufactured with molding processes similar to those used in the polymer industry (injection molding, extrusion). During the process, the filled polymer flows at a molten state and becomes anisotropic because of its deformation. The problem is coupled because the flow induces a rotation

*Handbook of Materials Behavior Models*. ISBN 0-12-443341-3.

of the particles and thus an anisotropy, but the orientation of the particles modifies the material's behavior and thus indirectly affects the anisotropy. The aim of this article is to give a brief overview of this problem and to outline contributions from different scientific communities (solid mechanics modeling and low Reynolds number hydrodynamics). The application of a differential model to this problem is original in this context.

## 10.9.2 CONSTITUTIVE RELATION

Two main steps permit one to derive macroscopic constitutive equations for anisotropic suspensions: (i) a now standard homogenization procedure which leads to volume average quantities if the orientation of the particles is perfectly defined, and (ii) a statistic description of the orientation.

### 10.9.2.1 Volume Average

Basic equations are obtained with a volume average procedure [1]. Let $\Omega$ be a representative volume of our macroscopic scale, which contains many particles located in $\Omega_i$. Let $\underline{\underline{\tau}}$, $\underline{u}$, and $\underline{\underline{d}}$ denote, respectively, the microscopic stress tensor, velocity vector, and strain rate tensor, and let $\underline{n}$ be a normal vector to the particle boundary (Figure 10.9.1).

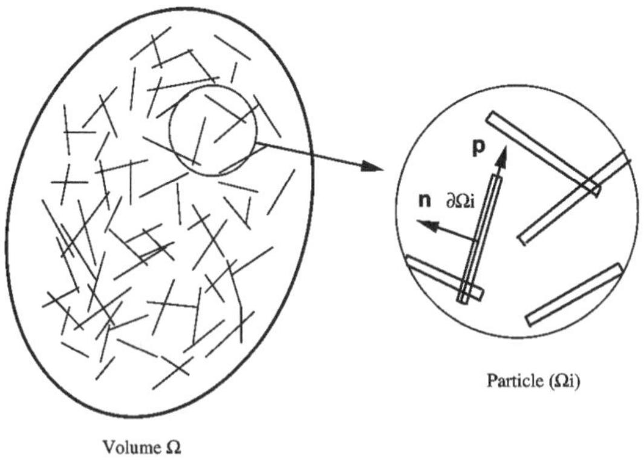


FIGURE 10.9.1 Representative volume at a macroscopic scale.

For a given fiber orientation $\underline{p}$ in the suspension, for a Newtonian ambient fluid (viscosity $\eta$), and if inertia terms can be neglected, the corresponding macroscopic variables $\underline{\underline{T}}$, $\underline{v}$, and $\underline{\underline{D}}$ are related by

$$\underline{\underline{T}} = -P\underline{\underline{Id}} + 2\eta\underline{\underline{D}}(v) + \underline{\underline{\Sigma}}_p \tag{1}$$

$$\underline{\underline{\Sigma}}_p = \frac{1}{\Omega}\sum_i \oint_{\partial\Omega_i} \underline{\underline{\tau}}\underline{n} \otimes x ds \tag{2}$$

In this expression, $P$ is the volume averaged pressure and $\underline{\underline{\Sigma}}_p$ describes the contribution of the particle to the averaged extra stress tensor.

### 10.9.2.2 Statistical Average

The fiber orientation $\underline{p}$ is given in an average sense by its probability distribution $\psi(\underline{p})$ so that the macroscopic statistical averaged stress tensor (i.e., the stress tensor used in engineering computations) writes:

$$\underline{\underline{\Sigma}} = \int \underline{\underline{T}}(\underline{p})\psi(\underline{p})dp \tag{3}$$

For most engineering applications, the exact calculation at each point of $\psi(p)$ is neither necessary nor possible. This orientation information is correctly approximated with use of the second-order and fourth-order orientation tensors:

$$\underline{\underline{a}} = \int \underline{p} \otimes \underline{p}\psi(\underline{p})dp; \quad \underline{\underline{\underline{\underline{a_4}}}} = \int \underline{p} \otimes \underline{p} \otimes \underline{p} \otimes \underline{p}\psi(\underline{p})dp \tag{4}$$

The flow around a single particle in a Newtonian fluid gives a first approximation for the evolution of one particle orientation. From Jeffery's calculations [15], for example, if $\lambda$ denotes the shape factor of the particle, and if $\underline{\underline{\Omega}}$ and $\underline{\underline{D}}$ are respectively the volume averaged vorticity and strain rate tensors:

$$\frac{d\underline{p}}{dt} = \underline{\underline{\Omega}}(V) \cdot \underline{p} + r\left\{\underline{\underline{D}}(V) \cdot \underline{p} - Tr[\underline{\underline{D}}(V) \cdot \underline{p} \cdot \underline{p}]\underline{p}\right\} \text{ with } r = \frac{\lambda^2 - 1}{\lambda^2 + 1} \tag{5}$$

The evolution equation of the probability distribution is given by

$$\frac{d\psi(\underline{p})}{dt} + \frac{\partial}{\partial \underline{p}}\left\{\psi(\underline{p})\frac{d\underline{p}}{dt}\right\} = \frac{\partial}{\partial \underline{p}}\left\{D\frac{\partial\psi(\underline{p})}{\partial \underline{p}}\right\} \tag{6}$$

$D$ is a rotation diffusion coefficient, vanishing at zero shear rates, which intends to take into account the hydrodynamic interactions between fibers.

The evolution equation for the second-order orientation tensor then writes:

$$\frac{\delta_r \underline{\underline{a}}}{\delta t} = \frac{d\underline{\underline{a}}}{dt} - \left(\underline{\underline{\Omega}} \cdot \underline{\underline{a}} - \underline{\underline{a}} \cdot \underline{\underline{\Omega}}\right) - r\left(\underline{\underline{D}} \cdot \underline{\underline{a}} + \underline{\underline{a}} \cdot \underline{\underline{D}}\right) = -2r\underline{\underline{\underline{\underline{a_4}}}} : D \tag{7}$$

This equation does not permit one to determine the orientation tensor if the velocity field is given, because it involves the fourth-order orientation tensor. The evolution equation for the fourth-order tensor would similarly involve the sixth-order tensor, etc. So, in order to close this equation, it is necessary to introduce a phenomenological relation between the second- and the fourth-order orientations (closure approximation), which can take different forms [2, 3]. For example, the simplest one is quadratic and writes:

$$\underline{\underline{\underline{\underline{a}}}}_4 = \underline{\underline{a}} \otimes \underline{\underline{a}} \tag{8}$$

## 10.9.3 EXPLICIT CALCULATIONS FOR $\underline{\underline{\Sigma}}_p$

The calculation of $\underline{\underline{\Sigma}}_p$ as a function of the macroscopic velocity field $\underline{v}$ requires one to solve a microscopic problem which can take different forms according to the level of approximation. Two main kinds of situations have been considered. Within the first one (slender bodies theory), the stress distribution at the particle boundary is reduced to a multipolar distribution. The Batchelor's model [4] is deduced, which has been modified by Shaqfeh and Fredrickson [13]. All of these models can be written in the following form:

$$\underline{\underline{\Sigma}}_p = 2\mu Tr[\underline{\underline{D}}(\underline{v})(\underline{p} \otimes \underline{p})]\underline{p} \otimes \underline{p} \tag{9}$$

In this expression, $\underline{p}$ is a unit vector aligned along the principal axis of the fiber, and $\mu$ is calculated as a function of the particle concentration and of its shape factor, which is assumed to be very large. The exact expression depends on the level of approximations. The interaction between fibers is accounted for in a certain sense but for long or very long fibers only. A second class of models deals with ellipsoidal or spheroidal particles. This case has been extensively studied for linear elastic materials [5, 6]. However, it has not been much studied for suspensions, probably because Eshelby's works [7] are not popular in fluid mechanics.

### 10.9.3.1 Suspension of Spheres

The case of a suspension of non-Brownian spheres is meaningful because the suspension remains Newtonian (viscosity $\eta_{eq}$). In this case, if $\varphi$ denotes the volumetric rate of spheres, and if there is no hydrodynamic interaction, the dilute approximation leads to Einstein's formula:

$$\underline{\underline{\Sigma}}_p = 5\eta\varphi\underline{\underline{D}}(\underline{v}) \Rightarrow \eta_{eq} = \eta(1 + 2.5\varphi) \text{ (dilute model)} \tag{10}$$

To account for hydrodynamic interactions, a very natural way consists in calculating $\underline{\underline{\Sigma}}_p$ in assuming that, instead of being immersed in the ambient fluid of viscosity $\eta$, the consequences of hydrodynamic interactions are summarized by placing the sphere in an ambient fluid of viscosity $\eta_{eq}$ (self-consistent model).

$$\underline{\underline{\Sigma}}_p = 5\eta_{eq}\varphi\underline{\underline{D}}(\underline{v}) \Rightarrow \eta_{eq} = \eta/(1 - 2.5\varphi) \text{ (self-consistent model)} \tag{11}$$

For small but noninfinitesimal volumetric rates, the self-consistent model allows one to derive (with a very different approach than the original one) the Batchelor's approximation:

$$\eta_{eq} = \eta(1 + 2.5\varphi + 6.25\varphi^2) \tag{12}$$

However, the self-consistent model suffers from strong limitations because the equivalent viscosity increases to infinity when $\varphi = 40\%$. For this reason, another approach is needed to ameliorate Eq. 11. Following Christensen [8], we assume that the spheres are added progressively, step by step. At each step, the suspension is macroscopically a viscous fluid of viscosity $\eta_{eq}(\varphi)$. We then add a quantity $d\varphi$ of spheres. This quantity is small enough that Einstein's formula for dilute suspension remains valid. The only difficulty consists in noting that the reference ambient fluid is the suspension of volumetric rate $\varphi$, so that the volumetric rate induced by $d\varphi$ is $d\varphi/(1-\varphi)$:

$$d\underline{\underline{\Sigma}}_p = 5\eta_{eq}\frac{d\varphi}{1-\varphi}\underline{\underline{D}}(\underline{v}) \Rightarrow \eta_{eq} = \eta/(1-\varphi)^{2.5} \text{(differential model)} \tag{13}$$

This differential model leads to the same kind of expression as the Krieger and Dougherty [9] one. It does not hold anymore when direct contact between particles is to be considered because the global behavior of the suspension is then neither linear nor homogeneous. In the following, we extend Eqs. 11 and 13 to ellipsoids. Two difficulties must then be overcome. The first is that a suspension of nonspherical particles is non-Newtonian. The second is that the calculation is more technical.

## 10.9.3.2 Flow around an Ellipsoidal Particle Immersed in an Anisotropic Suspension

### 10.9.3.2.1 Algebraic Preliminary Result

*Differential approach*: Particles are added progressively, so that each time one has to deal with a dilute suspension of spheroids. However, the problem is here specific, because the suspending fluid is a suspension, which is non-Newtonian. The system to solve is similar to Eq. 13:

$$\left\{\begin{aligned} d\alpha_1 &= \frac{\eta d\varphi}{(1-\varphi)S_{1212}} \\ d\alpha_2 &= \frac{2\eta d\varphi}{(1-\varphi)(S_{3333}-S_{3311})} \\ d\alpha_3 &= \frac{\eta d\varphi}{(1-\varphi)S_{1313}} \end{aligned}\right. \tag{18}$$

with initial conditions: $\alpha_i = 2\eta$ for $\varphi = 0$.

## 10.9.3.3 Results and Discussions

The numerical results [10] have been obtained by computations. In order to compare these results with other existing theories, the stress tensor will be written in a similar form as the one introduced by Tucker [11]:

$$\underline{\underline{T}} = -P\underline{\underline{Id}} + 2\eta\left\{\underline{\underline{d}} + N_s\left\lfloor(\underline{p}\otimes\underline{p})\underline{\underline{d}} + \underline{\underline{d}}(\underline{p}\otimes\underline{p})\right\rfloor + Tr\left\lfloor(\underline{p}\otimes\underline{p})\underline{\underline{d}}\right\rfloor(\underline{p}\otimes\underline{p})\right\} \tag{19}$$

with these notations:

$$\left\{\begin{aligned} N_s &= \frac{\alpha_3-\alpha_1}{\alpha_1} \\ 2\eta &= \alpha_1 \\ N_p &= \frac{3(\alpha_2\alpha_1)}{2\alpha_1} - \frac{2(\alpha_3\alpha_1)}{\alpha_1} \end{aligned}\right. \tag{20}$$

$\eta_1$ contains all the isotropic contributions to the viscosity (from both the solvent and the particles), while anisotropic contributions of the particles are represented by $N_p$ and $N_s$. However, as in Batchelor's dilute model, we find for both differential and self-consistent schemes that $N_p$ is always greater than $N_s$, even for particles of small aspect ratio. So, the anisotropic contributions by the particles are essentially described by $N_p$. The product $\eta_1$ $N_p$ is equal to the parameter $\mu$ of Shaqfeh's model, and we compare spheroidal model to slender bodies model. Figure 10.9.2 evidence the differences between the dilute

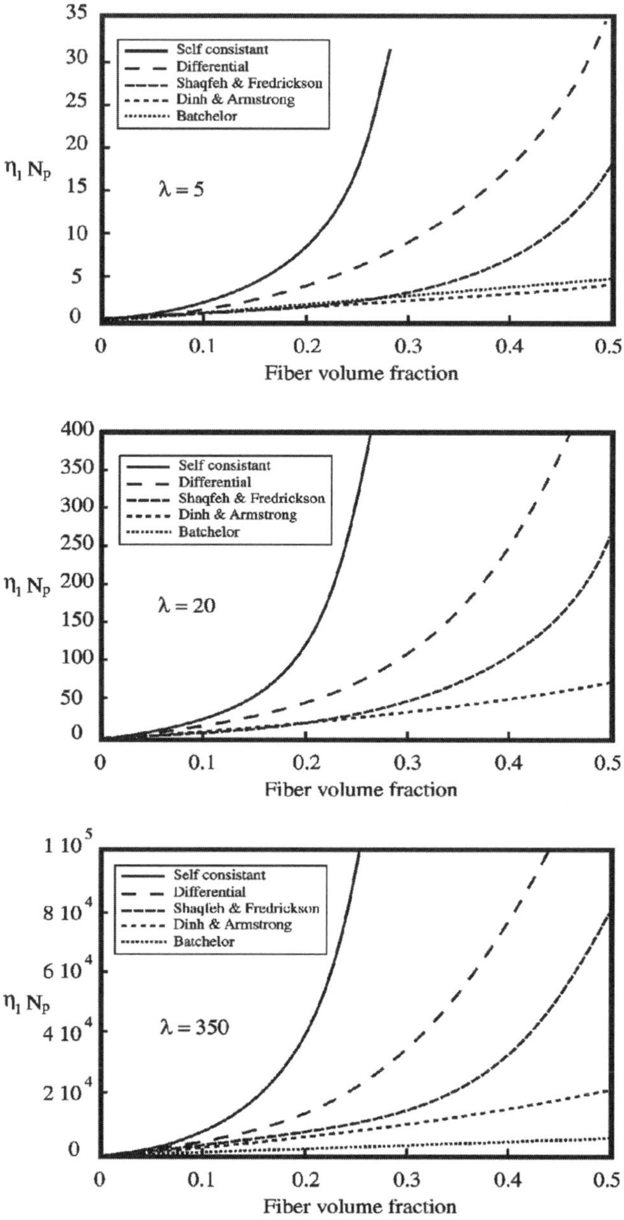


FIGURE 10.9.2 $\eta_1 N_p$ as a function of fiber volume fraction.

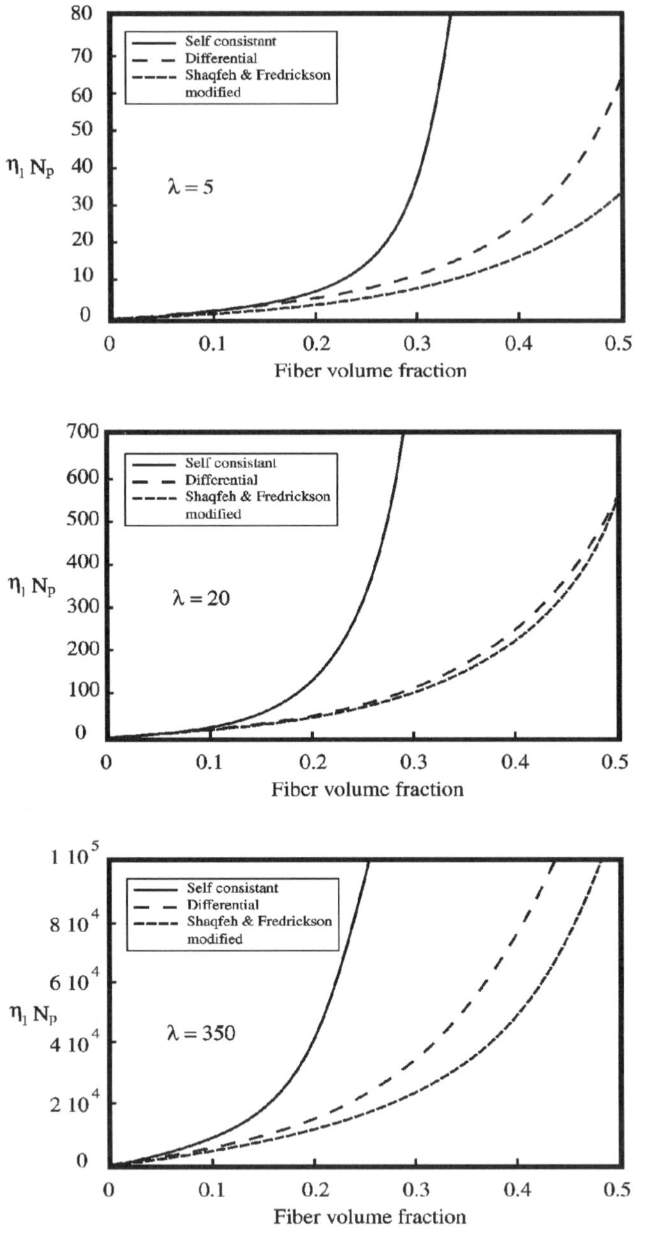


FIGURE 10.9.3 Modified $\eta_1 N_p$ as a function of fiber volume fraction.

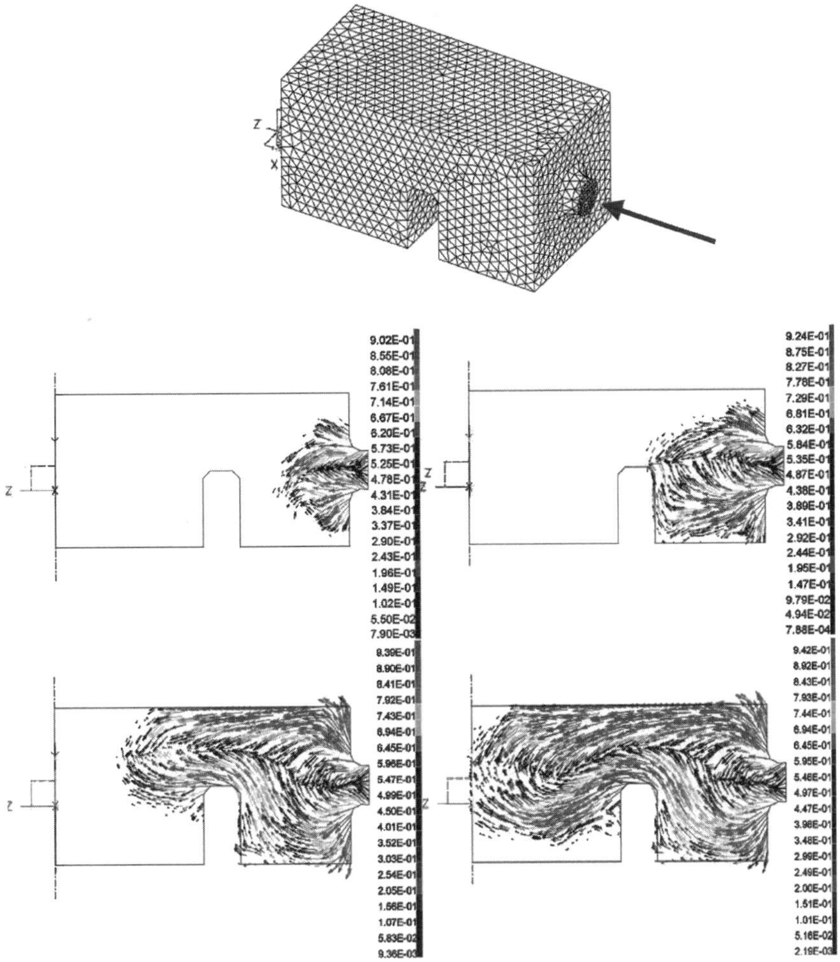


FIGURE 10.9.4 Injection molding simulation.

spheroidal approximation (Batchelor spheroid), the Dinh and Armstrong's model [14], and Shaqfeh and Fredrickson's slender bodies model and our self-consistent and differential model, for three shape factors of the particles. It can be seen that even for a shape factor $\lambda$ of 350, neither Shaqfeh and Fredrickson's nor Dinh and Armstrong's model is correct because, for very low values of $\varphi$, the curves are not tangent to Batchelor's dilute approximation. This point has been already mentioned by Ranganathan and Advani [12], who proposed a correction which accounts for the finite aspect ratio of the particles. Figure 10.9.3 compares the same models (modified for slender bodies models) for various fiber shape factors. It is to be noted that, as for hard sphere models [8], the self-consistent approach overestimates the rheological parameters (this self-consistent model evidences also a vertical asymptote for a volumic rate less than 1). The differential model lies very near the modified Shaqfeh and Fredrickson's model for the large aspect ratio (in the semidilute range). Thus, in this sense the differential model generalizes Shaqfeh and Fredrickson's model in accounting for an averaged interaction between the finite aspect ratio's fibers. In particular, we can apply the differential model for prolate spheroids.

## 10.9.4 EXAMPLE

Figure 10.9.4 illustrates a three-dimensional modeling for which an injection molding process has been simulated. Details of the computation algorithm can be found in Reference [16]. In this representation, the arrow depicts the orientation tensor $\underline{\underline{a}}$. The orientation of the arrow shows the direction of the eigenvector associated with the largest eigenvalue, and its color shows the magnitude of this eigenvalue. In other words, the arrow shows the preferential direction for the fiber orientation and the color shows the intensity of the alignment. The composite is assumed to enter the mold in an isotropic state (each eigenvalue is then 1/3). The shearing alignment is seen at the entrance, particularly near the walls. On the other hand, near the flow front a disalignment is clearly evidenced.

## 10.9.5 CONCLUSION

The flow of short-fiber composites induces an anisotropy which can be controlled by simulations. The scientific field concerned with this question lies at the interface between fluid and solids mechanics. The control of anisotropy is to be correlated with prediction of rupture for the part at a solid state.

## REFERENCES

1. Batchelor, G. K. (1970). The stress system in suspension of force free particules. *J. Fluid. Mech.* **41**: 545.
2. Advani, S. G. and Tucker, C. L. (1990). Closure approximations for three-dimensional structure tensors. *J. Rheol.* **34**: 367.
3. Dupret

4. Batchelor (1970). Slender bodies theory for particles of arbitrary cross section in Stokes flows. *J. Fluid. Mech.* **44**: 419.
5. Gilormini, P. and Vernusse, P. (1992). Tenseur d'Eshelby et problme d'inclusion dans le cas isotrope transverse incompressible. *C. R. Acad. Sci. Paris* **314**: 257.
6. Mura, T. (1982). *Micromechanics of Defects in Solids*. Martinus Nijhoff.
7. Eshelby, J. D. (1957). The determination of the elastic field of an ellipsoidal inclusion, and related problems. *Proc. Roy. Soc. London* **A241**: 376.
8. Christensen, R. M. (1990). A critical evaluation for a class of micro-mecanics models. *J. Mech. Phys. Solids* **38**: 379.
9. Krieger, I. M. and Dougherty, T. J. (1959). A mechanism for non-Newtonian flow in suspension of rigid spheres. *Trans. Soc. Rheol.* **3**: 137–152.
10. Meslin, F. (1997). Propriétés rhéologiques des composites fibres courtes  l'tat fondu. PhD thesis LMT–ENS de Cachan, France.
11. Tucker, C. L. (1991). Flow regime for fiber suspensions in narrow gaps. *J. Non-Newt. Fluid. Mech.* **39**: 239.
12. Ranganathan, S. and Advani, S. G. (1993). A simultaneous solution for flow and fiber orientation in axisymetric diverging flow. *J. Non-Newt. Fluid. Mech.* **47**: 107.
13. Shaqfeh and Fredrickson, G. (1990). The hydrodynamic stress in a suspension of rods. *Phys. Fluids* **A2**: 7.
14. Dinh, S. M. and Armstrong, R. C. (1984). A rheological equation of state for semiconcentrated fiber suspensions. *J. Rheol.* **28**: 207.
15. Jeffery, G. B. (1922). The motion of ellipsoidal particles immersed in viscous fluid. *Proc. R. Soc.* **A102**: 161.
16. Chinesta, F. (1999). Modélisation numérique en mise en forme des polyméres et céramiques: différents problóms de transport. PhD (HDR), LMT ENS de Cachan.

SECTION **10.10**

# Elastic Properties of Bone Tissue

STEPHEN C. COWIN

*New York Center for Biomedical Engineering, School of Engineering, The City College, New York*

Contents



## 10.10.1 VALIDITY

Bone is a natural composite material. It may be modeled as a linearly anisotropic elastic material for all but the rarest cases within the physiological range of loading. Large strains occur only with trauma to the bone and irreversible damage to the microstructure. There are measurable viscoelastic effects in bone, but they are generally not significant in a stress analysis of a whole bone.

A method of data analysis for a set of anisotropic elastic constant measurements of a material with variable composition has been applied to a database for cancellous bone. For these materials the identification of the type of elastic symmetry is complicated by the variable composition



*Handbook of Materials Behavior Models*. ISBN 0-12-443341-3.

of the material. A method of data analysis, described in following text, permits the identification of the type of elastic symmetry to be accomplished independent of the examination of the variable composition. This method of analysis may be applied to any set of elastic constant measurements, but it is illustrated here by application to an extraordinary database of cancellous bone elastic constants. The solid volume fraction or bulk density is the compositional variable for the elastic constants in this case. The final result is the solid-volume-fraction-dependent orthotropic Hooke's law for cancellous bone.

## 10.10.2 CORTICAL AND CANCELLOUS BONE

Bone is a remarkable and extremely complex tissue. It has at least two major functions. First, it is the material from which the skeleton is made, which provides mechanical support and protection for the organism. The bone tissue of the skeleton is continually changing to adapt its form and structure to this task. From the birth to the death of the organism it modifies its structure by laying down new bone tissue and resorbing old. A second function of bone is to store the minerals, particularly $Ca^{++}$, needed to maintain a mineral homeostasis in the body by regulating the concentrations of the key blood electrolytes, including $Ca^{++}$, $H^{+}$, and $HPO_4^{-}$.

Bone differs from the other tissues because of its greater stiffness and strength. These properties stem from its being a composite material formed by the deposition of a mineral, apatite or hydroxyapatite, in a frame of collagen. The stiffness of bone is the key to understanding why it differs so greatly from other tissues.

At the macroscopic level there are two major forms of bone tissue, called *compact* or *cortical* bone and *cancellous* or *trabecular* bone. Cortical or compact bone is a dense material with a specific gravity of almost two in humans and a little over two in cattle. It forms most of the outer shell of a whole bone, a shell of variable thickness. Cancellous bone generally exists only within the confines of the cortical bone coverings. Cancellous bone is also called trabecular bone because it is composed of short struts of bone material called trabeculae (from the Latin for "little beam"). The connected trabeculae give cancellous bone a spongy appearance, and it is often called spongy bone.

The mechanical properties of bone, other than the elasticity, have been documented. These include water content, viscoelastic properties, mechanical strength, permeability, mechano-electrokinetic effects, damage mechanics, etc. One will find data of this type collected in Cowin [4,6].

## 10.10.3 THE ELASTIC MODEL FOR CORTICAL BONE

The elastic model for cortical bone tissue is anisotropic linear elasticity. Tables 10.10.1 and 10.10.2 below list the components of the orthotropic elastic compliance coefficients for cortical bone. The first table is for human bone, and second is for bovine and canine bone. The elastic coefficients in the anisotropic Hooke's law [8,13,16] are the three Young's moduli $E_1$, $E_2$, and $E_3$ in the three perpendicular coordinate directions, the six Poisson's ratios denoted by subscripted $\nu$'s (only three of the six Poisson's ratios are independent since $\nu_{12}/E_1 = \nu_{21}/E_2$, $\nu_{13}/E_1 = \nu_{31}/E_3$, and $\nu_{23}/E_2 = \nu_{32}/E_3$), and the three shear moduli $G_{23}$, $G_{13}$, and $G_{12}$. In this notation the 3 direction is coincident with the long axis of the bone; the 1 and 2 directions are radial and circumferential, respectively. In the row indicating material symmetry, ORTH stands for orthotropy and TI stands for transverse isotropy. In the row indicating the testing method, U stands for ultrasound and M stands for standard machine testing. The entry NM stands for not measured.

**TABLE 10.10.1 Technical Constants for Cortical Human Bone**

| Group | Reilly and Burstein [15] | Yoon and Katz [21] | Knets and Malmeisters [11] | Ashman *et al.* [2] |
|---|---|---|---|---|
| Bone | Femur | Femur | Tibia | Femur |
| Symmetry | TI | TI | ORTH | ORTH |
| Method | M | U | M | U |
| $E_1$(GPa) | 11.5 | 18.8 | 6.91 | 12.0 |
| $E_2$(GPa) | 11.5 | 18.8 | 8.51 | 13.4 |
| $E_3$(GPa) | 17.0 | 27.4 | 18.4 | 20.0 |
| $G_{12}$(GPa) | 3.6 | 7.17 | 2.41 | 4.53 |
| $G_{13}$(GPa) | 3.3 | 8.71 | 3.56 | 5.61 |
| $G_{23}$(GPa) | 3.3 | 8.71 | 4.91 | 6.23 |
| $\nu_{12}$ | 0.58 | 0.312 | 0.49 | 0.376 |
| $\nu_{13}$ | 0.31 | 0.193 | 0.12 | 0.222 |
| $\nu_{23}$ | 0.31 | 0.193 | 0.14 | 0.235 |
| $\nu_{21}$ | 0.58 | 0.312 | 0.62 | 0.422 |
| $\nu_{31}$ | 0.46 | 0.281 | 0.32 | 0.371 |
| $\nu_{32}$ | 0.46 | 0.281 | 0.31 | 0.350 |

TABLE 10.10.2 Technical Constants for Cortical Bovine and Canine Bone

| Group | Lang [12] | Reilly and Burstein [15] | Ambardar and Ferris [1] | Van Buskirk *et al.* [17] | Burris [3] | Burris [3] | Ashman *et al.* [2] |
|---|---|---|---|---|---|---|---|
| Bone | Bovine phalanx | Bovine femur | Bovine femur | Bovine femur | Bovine femur | Bovine femur | Canine femur |
| Symmetry | TI | TI | TI | ORTH | TI | TI + | TI + |
| Method | U | M | U | U | M | U | U |
| $E_1$ GPa) | 11.3 | 10.2 | 6.97 | 11.6 | 13.3 | 10.79 | 12.8 |
| $E_2$(GPa) | 11.3 | 10.2 | 6.97 | 14.6 | 13.3 | 12.24 | 15.6 |
| $E_3$(GPa) | 22.0 | 22.6 | 20.9 | 21.9 | 18.1 | 18.90 | 20.1 |
| $G_{12}$(GPa) | 3.8 | NM* | 2.2 | 5.29 | 3.5 | 3.38 | 4.68 |
| $G_{13}$(GPa) | 5.4 | 3.6 | 6.9 | 6.29 | 5.0 | 4.47 | 5.68 |
| $G_{23}$(GPa) | 5.4 | 3.6 | 6.9 | 6.99 | 5.0 | 5.96 | 6.67 |
| $\nu_{12}$ | 0.484 | 0.51 | 0.55 | 0.302 | NM | 0.45 | 0.282 |
| $\nu_{13}$ | 0.203 | NM | 0.15 | 0.109 | NM | 0.24 | 0.289 |
| $\nu_{23}$ | 0.203 | NM | 0.15 | 0.205 | NM | 0.22 | 0.265 |
| $\nu_{21}$ | 0.484 | 0.51 | 0.55 | 0.380 | NM | 0.51 | 0.366 |
| $\nu_{31}$ | 0.396 | 0.36 | 0.44 | 0.206 | 0.22 | 0.42 | 0.454 |
| $\nu_{32}$ | 0.396 | 0.36 | 0.44 | 0.307 | 0.22 | 0.33 | 0.341 |

*Not measured.

## 10.10.4 THE SOURCE OF THE ELASTIC CONSTANT DATA FOR CANCELLOUS BONE

The elastic constant results for cancellous bone presented here are based upon an analysis of a database consisting of 141 human cancellous bone specimens. This database, reported by van Rietbergen *et al.* [18, 19] and Kabel *et al.* [9], is superior to previous databases because the authors provide the entire set of anisotropic elastic constants without an *a priori* assumption of a particular material symmetry and without an assumption of the direction in which the maximum Young's modulus occurs. This database is unique in many different ways, the most important of which is the large number of specimens and its method of construction, but particularly because it is not based entirely on measurements of real specimens. The database of elastic constants of 141 human cancellous bone specimens employed here was constructed by imaging real specimens and then computationally determining their elastic constants. This cyberspace method of construction is thought to be more accurate than the conventional mechanical testing procedures for evaluating the elastic constants of human cancellous bone. The determination of the elastic constants of cancellous bone by conventional mechanical test procedures is very difficult. The basic problem is that, because of the size of the human body, it is difficult to obtain specimens of cancellous bone that are more than 5-mm cubes. The logical way to test small cubes such as these is by compression testing. However, compression testing is highly inaccurate for cancellous bone because of (1) the frictional end effects of the platens, (2) the near impossibility of identifying, *a priori*, the grain directions in a bone specimen and thus of cutting a specimen in the grain directions, (3) the stiffening effect of the platens on the bone near the platens, and (4) the unpredictable inhomogeneity of the specimen.

The construction of the database of elastic constants of 141 human cancellous bone specimens employed here is a relatively inexpensive method of determining the full set of anisotropic elastic constants for a small specimen of cancellous bone by a combination of imaging the specimen [9,10,14] and subsequent evaluation of the effective elastic constants using computational techniques based on the finite element by van Rietbergen *et al.* [18,19]. Once the image of the specimen was in the computer and a finite element mesh was generated, a sequence of loadings [18,19] was applied to the specimen and the responses were determined. The sequence of loadings was sufficient in number to determine all 21 elastic constants. Thus no material symmetry assumptions were made in the determination of the constants. Quantitative stereological programs were used to determine the solid volume fraction $\phi$ of each specimen. These are the data employed in the determination of the elastic constants for cancellous bone recorded in following text.

In this method, the actual matrix material of the trabeculae comprising the bone specimen is assumed to have an axial Young's modulus $E_t$. The value of $E_t$ may be fixed from a knowledge of the axial Young's modulus for the tissue, or from the shear modulus about some axis, or by measuring the tissue modulus $E_t$ itself. For purposes of numerical calculation, $E_t$ was taken to be 1 Gpa [9,10,14,18,19]. However, since these are linear finite element (*FE*) models, the FE results can be scaled for any other modulus by multiplying the results with the new value of $E_t$ (in GPa). The tissue modulus $E_t$ thus is a scale factor that magnifies or reduces all the elastic constants. The cancellous bone elastic constant results are presented here as multiples of $E_t$.

## 10.10.5 THE ANALYSIS OF THE ELASTIC CONSTANT FOR CANCELLOUS BONE

The elasticity of cancellous bone is complicated to analyze, report, and record because of the highly inhomogeneous porous nature of this bone type. The elastic constants depend upon the volume fraction of solid matrix material present (one minus the porosity). In the case of porous isotropic materials, for example, it is customary to regress the Young's modulus against the solid volume fraction and obtain expressions for the Young's modulus $E$ as a function of the solid volume fraction $\phi$; for example, $E = (\text{constant})\ \phi^n$. For cancellous bone a method must be available for construction similar to representations for an inhomogeneous anisotropic porous solid. Such a method is described in the next paragraph.

Many materials are anisotropic and inhomogeneous because of the varying composition of their constituents. The identification of the type of elastic symmetry is complicated by the variable composition of the material, which makes the analysis of the elastic constant measurement data difficult. A solution to this problem in which identification of the type of elastic symmetry and analysis of the variable composition are separated, and then analyzed independently, was described in Cowin and Yang [5] and applied in Yang *et al.* [20] to bone softwood, and hardwood. The method consists of averaging eigenbases, that is to say, the bases composed of the orthogonal sets of eigenvectors of different measurements of the elasticity tensor, in order to construct an average eigenbasis for the entire data set. This is possible because the eigenbases, composed of eigenvectors, are independent of composition whereas the eigenvalues are not. The eigenvalues of all the anisotropic elastic coefficient matrices can then be transformed to the average eigenvector basis and regressed against their compositional parameters. This method treats the individual measurement as a measurement of a tensor instead of as a collection of individual elastic constant or matrix element measurements, recognizing that the measurements by different investigators will reflect the systematic invariant tensorial properties of a material, like eigenvectors and eigenvalues. This method for averaging different measurements of the anisotropic elastic constants for a specific material has advantages over the traditional method of averaging the individual matrix components of the elasticity or compliance matrices. Averaging invariants removes the effect of the reference coordinate system in the measurements, while the traditional method of averaging the components may induce errors because of the various reference coordinate systems and may distort the nature of the symmetry. This averaging process explicitly retains the orthonormality of the eigenvector basis.

The results of Cowin and Yang [5] provided a means of extending the empirical method of representing the Young's modulus $E$ as a function of the solid volume fraction $\phi$ (for example, $E = (\text{constant})\ \phi^n$) to all the elastic constants of an inhomogeneous anisotropic material. In Cowin and Yang [5] this method was applied to feldspar, and it was discovered that the eigenvectors, but not the eigenvalues, were relatively independent of material composition. That result was extended by Yang *et al.* [20] to three natural, porous materials: cancellous bone, hardwood, and softwood. These works have established this method of analysis as a valid approach to the construction of anisotropic stress-strain relations for other compositionally dependent materials.

This new method of analysis also identifies the type of elastic symmetry possessed by the material. No *a priori* assumption as to the type of elastic symmetry is made. The type of symmetry is identified from the character of the eigenvectors that are calculated. For the cancellous bone considered here, the analysis shows that human cancellous bone has orthotropic elastic symmetry at the 95% confidence level.

## 10.10.6 THE ELASTIC CONSTANT DATA FOR CANCELLOUS BONE

Applying the method of analysis just described to the 141-specimen database discussed in the section before last, explicit representations of the solid volume fraction $\phi$ dependent orthotropic elastic constants of human cancellous bone were obtained. These explicit representations are $E_1 = 1240\ E_t\phi^{1.80}$, $E_2 = 885\ E_t\phi^{1.89}$, $E_3 = 529\ E_t\phi^{1.92}$, $2G_{23} = 533\ E_t\phi^{2.04}$, $2G_{13} =$

633 $E_t\phi^{1.97}$, $2G_{12}=973\ E_t\phi^{1.98}$; $\nu_{23}=0.256\ \phi^{-0.09}$, $\nu_{32}=0.153\ \phi^{-0.05}$, $\nu_{13}=0.316\ \phi^{-0.19}$, $\nu_{31}=0.135\ \phi^{-0.07}$, $\nu_{12}=0.176\ \phi^{-0.25}$ and $\nu_{21}=0.125\ \phi^{-0.16}$. $E_t$ has the dimension of stress, and the other numbers multiplying all $\phi$'s raised to a power are dimensionless. The squared correlation coefficients ($R^2$) for the orthotropic elastic coefficients are as follows: for $1/E_1$, $R^2=0.934$; for $1/E_2$, $R^2=0.917$; for $1/E_3$, $R^2=0.879$; for $1/(2G_{23})$, $R^2=0.870$; for $1/(2G_{13})$, $R^2=0.887$; for $1/(2G_{12})$, $R^2=0.876$; for $-\nu_{12}/E_1$, $R^2=0.740$; for $-\nu_{13}/E_1$, $R^2=0.841$; and for $\nu_{21}/E_2$, $R^2=0.666$.

## REFERENCES


1. Ambardar, J. D., and Ferris, C. D. (1976). A simple technique for measuring certain elastic moduli in bone. *Biomed. Sci. Instrum.* **12**: 23.
2. Ashman, R. B., Cowin, S. C., Van Buskirk, W. C., and Rice, J. C. (1984). A continuous wave technique for the measurement of the elastic properties of bone. *J. Biomechanics* **17**: 349.
3. Burris, C. L. (1983). A Correlation of Quasistatic and Ultrasonic Measurements of the Elastic Properties of Cortical Bone. Ph.D. dissertation, Tulane University, New Orleans, Louisiana.
4. Cowin, S. C., ed. (1989). *Bone Mechanics,* Boca Raton, FL: CRC Press.
5. Cowin, S. C., and Yang, G. (1997). Averaging anisotropic elastic constants data. *J. Elasticity* **46**: 151–180.
6. Cowin, S. C., ed. (2001). *Bone Mechanics Handbook,* Boca Raton, FL: CRC Press.
7. Cowin, S. C., Van Buskirk, W. C., and Ashman, R. B. (1987). The properties of bone, in *Handbook of Bioengineering*, Skalak, R., and Chien, S., editors-in-chief, New York: McGraw–Hill.
8. Hearmon, R. F. S. (1961). *An Introduction to Applied Anisotropic Elasticity,* Oxford: Oxford University Press.
9. Kabel, J., van Rietbergen, B., Odgaard, A., and Huiskes, R. (1999). Constitutive relationships of fabric, density and elastic properties in cancellous bone architecture. *Bone* **25**: 481–486.
10. Kabel, J., van Rietbergen, B., Dalstra, M., Odgaard, A., and Huiskes, R. (1999). The role of an effective isotropic tissue modulus in the elastic properties of cancellous bone. *J. Biomechanics* **32**: 673–680.
11. Knets, I., and Malmeisters, A. (1977). Deformability and strength of human compact bone tissue, in *Mechanics of Biological Solids: Proc. Euromech Colloquium 68*, p. 133, Brankov, G., ed., Sofia: Bulgarian Academy of Sciences.
12. Lang, S. B. (1970). Ultrasonic method for measuring elastic coefficients of bone and results on fresh and dried bovine bones. *IEEE Trans. Biomed. Eng.* **17**: 101.
13. Lekhnitskii, S. G. (1963). *Theory of Elasticity of an Anisotropic Elastic Body,* San Francisco: Holden Day.
14. Odgaard, A., Kabel, J., van Rietbergen, B., and Huiskes, R. (1997). Fabric and elastic principal directions of cancellous bone are closely related. *J. Biomechanics* **30**: 487–495.
15. Reilly, D. T., and Burstein, A. H. (1975). The elastic and ultimate properties of compact bone tissue. *J. Biomechanics*, **8**: 393.
16. Saada, A. S. (1974). *Elasticity Theory and Applications*, Pergamon.
17. van Buskirk, W. C., Cowin, S. C., and Ward, R. N. (1981). Ultrasonic measurement of orthotropic elastic constants of bovine femoral bone. *J. Biomechanical Eng.* **103**: 67.

18. van Rietbergen, B., Odgaard, A., Kabel, J., and Huiskes, R. (1996). Direct mechanical assessment of elastic symmetries and properties of trabecular bone architecture. *J. Biomechanics* 29: 1653–1657.
19. van Rietbergen, B., Odgaard, A., Kabel, J., and Huiskes, R. (1998). Relationships between bone morphology and bone elastic properties can be accurately quantified using high-resolution computer reconstructions. *J. Orthop. Res.* 16: 23–28.
20. Yang, G., Kabel, J., van Rietbergen, B., Odgaard, A., Huiskes, R., and Cowin, S. C. (1999). The anisotropic Hooke's law for cancellous bone and wood. *J. Elasticity*, 53: 125–146.
21. Yoon, H. S., and Katz, J. L. (1976). Ultrasonic wave propagation in human cortical bone. 11. Measurements of elastic properties and micro-hardness. *J. Biomechanics*, 9: 459.

SECTION **10.11**

# Biomechanics of Soft Tissue

GERHARD A. HOLZAPFEL
*Institute for Structural Analysis, Computational Biomechanics, Graz University of Technology, 8010 Graz, Austria*

Contents



## 10.11.1 VALIDITY

An efficient constitutive formulation approximates all types of soft tissues with a reasonable accuracy over a large strain range. We request a simple constitutive equation with only a few material parameters involved that allow for an "explanation" of the material response of tissues in terms of their structure. In addition, we request that the constitutive formulation is fully three-dimensional and consistent with both mechanical and mathematical

*Handbook of Materials Behavior Models.* ISBN 0-12-443341-3.

requirements, applicable for arbitrary geometries, and suitable for use within the context of finite element methods in order to solve complex initial boundary-value problems.

The presented general model is a fully three-dimensional material description of soft tissues for which nonlinear continuum mechanics is used as the fundamental basis [10, 18]. It has the special feature that it is based partly on histological information (i.e., the microscopic structure of organs and tissues). The general model describes the highly nonlinear and anisotropic behavior of soft tissues as composites reinforced by two families of collagen fibers. The constitutive framework is based on the theory of the mechanics of fiber-reinforced composites [26] and is suitable to describe a wide variety of physical phenomena of soft tissues. The performance and the physical mechanism of the model are presented in Reference [11]. As a representative example, the general model for soft tissues is specified to predict the mechanical response of *healthy* and *young* arteries under physiological loading conditions [12]. The model neglects active components, i.e., contracting elements with biochemical energy supply which are controlled by biological mechanisms, and is concerned with the description of the *passive* state of arteries.

The models are suitable for predicting the anisotropic *elastic* response of soft tissues in the large strain domain. A suitable constitutive and numerical model that is general enough to describe the finite *viscoelastic* domain is documented in Reference [11]. The presented models do not consider acute and long-term changes in geometry and/or the mechanical response of tissues due to, for example, drugs, aging, and disease. When soft tissues are subjected to loads that are beyond their physiological range, the load-carrying fibers of the tissue slip relative to each other. In clinical procedures tissues may undergo irreversible (plastic) deformations [12] which are of medical importance. Constitutive equations for describing plastic deformations of, for example, arteries are proposed in References [8, 27].

## 10.11.2 BACKGROUND ON THE STRUCTURE OF SOFT TISSUES: COLLAGEN AND ELASTIN

*What do we mean by soft tissues?* A primary group of tissue which binds, supports, and protects our human body and structures such as organs is *soft connective tissue*. In contrary to other tissues, it is a wide-ranging biological material in which the cells are separated by extracellular material. Connective tissues may be distinguished from hard (mineralized) tissues such as bones for their high flexibility and their soft mechanical properties. This article discusses connective tissue from the point of view of material science, biomechanics, and structural engineering (for more details, see, for example, Reference [6], Chapter 7).

Examples of soft tissues are *tendons*, *ligaments*, *blood vessels*, *skins*, or *articular cartilages*, among many others. Tendons are muscle-to-bone linkages that stabilize the bony skeleton (or produce motion), and ligaments are bone-to-bone linkages that restrict relative motion. Blood vessels are prominent organs composed of soft tissues which have to distend in response to pulse waves. The skin is the largest single organ (16% of the human adult weight). It supports internal organs and protects our body. Articular cartilages form the surface of body joints (which is a layer of connective tissue with a thickness of 1–5 mm), distribute loads across joints, and minimize contact stresses and friction.

Soft connective tissues of our body are complex, fiber-reinforced, composite structures. Their mechanical behavior is strongly influenced by the concentration and structural arrangement of constituents such as *collagen* and *elastin*, the hydrated matrix of *proteoglycans*, and the topographical site and respective function in the organism.

#### 10.11.2.1 Collagen

Collagen is a protein which is a major constituent of the extracellular matrix of connective tissue. It is the main load-carrying element in a wide variety of soft tissues and is very important to human physiology (for example, the collagen content of the [human] Achilles tendon is about 20 times that of elastin).

Collagen is a macromolecule with a length of about 280 nm. Collagen molecules are linked to each other by covalent bonds that build collagen fibrils. Depending on the primary function and the requirement of strength of the tissue, the diameter of collagen fibrils varies (the order of magnitude is 1.5 nm [17]). In the structure of tendons and ligaments, for example, collagen appears as parallel-oriented fibers [1], whereas many other tissues have an intricate disordered network of collagen fibers embedded in a gelatinous matrix of proteoglycans.

More than 12 types of collagen have been identified [17]. The most common collagen is type I, which can be isolated from any tissue. It is the major constituent in blood vessels. The rodlike shape of the collagen molecule comes from three polypeptide chains which are composed in a right-handed triple-helical conformation. Most of the collagen molecule consists of three amino acids: glycine (33%), which enhances the stability of the molecule, proline (15%), and hydroxyproline (15%) [23].

The intramolecular crosslinks of collagen gives the connective tissues the strength which varies with age, pathology, etc. (for a correlation between the collagen content in the tissue, % dry weight, and its ultimate tensile strength, see Table 10.11.1. The function and integrity of organs are maintained by the *tension* in collagen fibers. They shrink upon heating breakdown of the crystalline structure (at 65°C, for example, mammalian collagen shrinks to about one-third of its initial length, [6], p. 263).

### 10.11.2.2 Elastin

Elastin, like collagen, is a protein which is a major constituent of the extracellular matrix of connective tissue. It is present as thin strands in soft tissues such as skin, lung, ligamenta flava of the spine, and ligamentum nuchae (the elastin content of the latter is about five times that of collagen).

The long flexible elastin molecules build up a three-dimensional (rubber-like) network, which may be stretched to about 2.5 times the initial length of the unloaded configuration. In contrast to collagen fibers, this network does not exhibit a pronounced hierarchical organization. As for collagen, 33% of the total amino acids of elastin consists of glycine. However, the proline and hydroxyproline contents are much lower than in collagen molecules.

The mechanical behavior of elastin may be explained within the concept of entropic elasticity. As for rubber, the random molecular conformations, and hence the entropy, change with deformation. Elasticity arises through entropic straightening of the chains, i.e., a decrease of entropy, or an increase of internal energy (see, for example, References [9]). Elastin is essentially a linearly elastic material (tested for the ligamentum nuchae of cattle). It displays very small relaxation effects (they are larger for collagen).

**TABLE 10.11.1 Mechanical Properties [6, 15, 25] and Associated Biochemical Data [30] of Some Representative Organs Mainly Consisting of Soft Connective Tissues**

| Material | Ultimate tensile strength (MPa) | Ultimate tensile strain (%) | Collagen (% dry weight) | Elastin (% dry weight) |
|---|---|---|---|---|
| Tendon | 50–100 | 10–15 | 75–85 | <3 |
| Ligament | 50–100 | 10–15 | 70–80 | 10–15 |
| Aorta | 0.3–0.8 | 50–100 | 25–35 | 40–50 |
| Skin | 1–20 | 30–70 | 60–80 | 5–10 |
| Articular cartilage | 9–40 | 60–120 | 40–70 | — |

## 10.11.3 GENERAL MECHANICAL CHARACTERISTICS OF SOFT TISSUES

Before describing a model for soft tissues, it is beneficial and instructive to gain some insight into their general mechanical characteristics. Soft tissues behave anisotropically because of their fibers, which tend to have preferred directions. In a microscopic sense they are nonhomogeneous materials because of their composition. The tensile response of soft tissue is nonlinear stiffening, and tensile strength depends on the strain rate. In contrast to hard tissues, soft tissues may undergo large deformations. Some soft tissues show viscoelastic behavior (relaxation and/or creep), which has been associated with the shear interaction of collagen with the matrix of proteoglycans [16] (the matrix provides a viscous lubrication between collagen fibrils).

In a simplified way we explain here the tensile stress-strain behavior for skin, an organ consisting mainly of connective tissues, which is representative of the mechanical behavior of many (collagenous) soft connective tissues. For the connective tissue parts of the skin the three-dimensional network of fibers appears to have preferred directions parallel to the surface. However, in order to prevent out-of-plane shearing, some fiber orientations also have components out of plane.

Figure 10.11.1 shows a schematic diagram of a typical J-shaped (tensile) stress-strain curve for skin. This form, representative for many soft tissues,

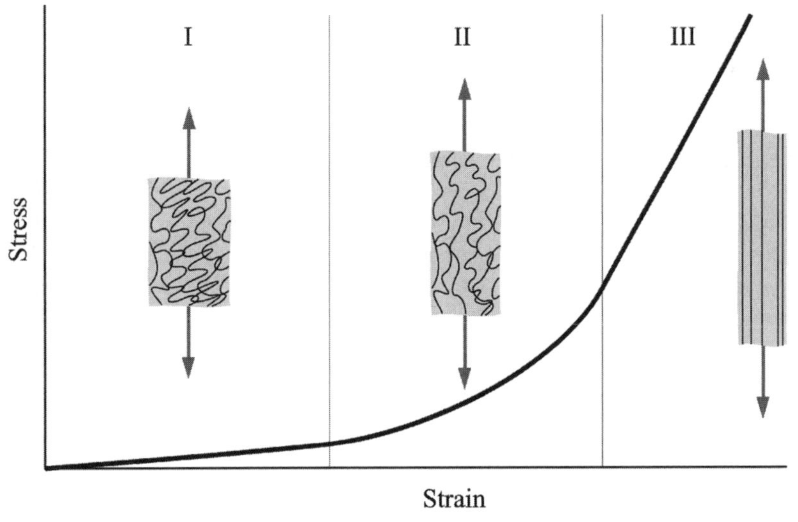


**FIGURE 10.11.1** Schematic diagram of a typical (tensile) stress-strain curve for skin showing the associated collagen fiber morphology.

differs significantly from stress-strain curves of hard tissues or from other types of (engineering) materials. In addition, Figure 10.11.1 shows how the collagen fibers straighten with increasing stress.

The deformation behavior for skin may be studied in three phases: I, II, and III.

*Phase I.* In the absence of load the collagen fibers, which are woven into a rhombic-shaped pattern, are in relaxed conditions and appear wavy and crimped. Unstretched skin behaves approximately isotropically. Initially low stress is required to achieve large deformations of the individual collagen fibers without requiring stretch of the fibers. In phase I the tissue behaves like a very soft (isotropic) rubber sheet, and the elastin fibers (which keep the skin smooth) are mainly responsible for the stretching mechanism. The stress-strain relation is approximately linear, and the elastic modulus of skin in phase I is low (0.1–2.0 MPa).

*Phase II.* In phase II, as the load is increased, the collagen fibers tend to line up with the load direction and bear loads. The crimped collagen fibers gradually elongate and interact with the hydrated matrix. With deformation the crimp angle in collagen fibrils leads to a sequential uncrimping of fibrils. Note that the skin is normally under tension *in vivo*.

*Phase III.* In phase III, at high tensile stresses, the crimp patterns disappear and the collagen fibers become straighter. They are primarily aligned with one another in the direction in which the load is applied. The straightened collagen fibers resist the load strongly, and the tissue becomes stiff at higher stresses. The stress-strain relation becomes linear again. Beyond the third phase the ultimate tensile strength is reached and fibers begin to break.

The mechanical properties of soft tissues depend strongly on the topography, risk factors, age, species, physical and chemical environmental factors such as temperature, osmotic pressure, and pH, and the strain rate. The material properties are strongly related to the quality and completeness of experimental data, which come from *in vivo* or *in vitro* tests having the aim of mimicking real loading conditions. Therefore, to present specific values for the ultimate tensile strength and strain of a specific tissue is a difficult task. Nevertheless, Table 10.11.1 attempts to present ranges of values of mechanical properties and collagen and elastin contents (% dry weight) in some representative organs mainly consisting of soft connective tissues.

## 10.11.4 DESCRIPTION OF THE MODEL

At any referential position $\mathbf{X}$ of the tissue we postulate the existence of a Helmholtz free-energy function $\Psi$. We assume the *decoupled* form

$$\Psi = U(\mathbf{X}; J) + \bar{\Psi}(\mathbf{X}; \bar{\mathbf{C}}, \mathbf{A}_1, \mathbf{A}_2) \qquad (1)$$

where $U$ is a purely *volumetric* (dilatational) contribution and $\bar{\Psi}$ is a purely *isochoric* (volume-preserving) contribution to the free energy $\Psi$. Here $\bar{\mathbf{C}} = \bar{\mathbf{F}}^T\bar{\mathbf{F}}$ denotes the modified right Cauchy-Green tensor and $\bar{\mathbf{F}} = J^{-1/3}\mathbf{F}$ is the unimodular (distortional) part of the deformation gradient $\mathbf{F}$, with $J = \det \mathbf{F} > 0$ denoting the local volume ratio. In addition, in Eq. 1, $\{\mathbf{A}_1, \mathbf{A}_2\}$ is a set of two (second-order) tensors which characterize the anisotropic properties of the tissue at any $\mathbf{X}$. The structure tensors $\mathbf{A}_1$ and $\mathbf{A}_2$ are defined as the tensor products $\mathbf{a}_{0\alpha} \otimes \mathbf{a}_{0\alpha}$, where $\mathbf{a}_{0\alpha}$, $\alpha = 1, 2$, are two unit vectors characterizing the orientations of the families of collagen fibers in the (undeformed) reference configuration of the tissue (see Figure 10.11.2).

Since most types of soft tissues are regarded as incompressible (for example, arteries do not change their volume within the physiological range of deformation [2]), we now focus attention on the description of their isochoric deformation behavior characterized by the energy function $\bar{\Psi}$. We suggest the simple additive split

$$\bar{\Psi} = \bar{\Psi}_{\text{iso}}(\mathbf{X}; \bar{I}_1) + \bar{\Psi}_{\text{aniso}}(\mathbf{X}; \bar{I}_1^*, \bar{I}_2^*) \tag{2}$$

of $\bar{\Psi}$ into a part $\bar{\Psi}_{\text{iso}}$ associated with *iso*tropic deformations and a part $\bar{\Psi}_{\text{aniso}}$ associated with *aniso*tropic deformations. This is sufficiently general to capture the salient mechanical feature of soft tissue elasticity as described in Section 10.11.3 (a more general constitutive framework is presented in References [8, 11, 12]). In Eq. 2 we used $\bar{I}_1 = \bar{\mathbf{C}} : \mathbf{I}$ for the first invariant of tensor $\bar{\mathbf{C}}$($\mathbf{I}$ is the second-order unit tensor), and the definitions

$$\bar{I}_1^*(\bar{\mathbf{C}}, \mathbf{a}_{01}) = \bar{\mathbf{C}} : \mathbf{A}_1, \quad \bar{I}_2^*(\bar{\mathbf{C}}, \mathbf{a}_{02}) = \bar{\mathbf{C}} : \mathbf{A}_2 \tag{3}$$

of the invariants, which are stretch measures for the two families of collagen fibers (see, for example, References [10, 26]). The invariants $\bar{I}_1^*$ and $\bar{I}_2^*$ are squares of the stretches in the directions of $\mathbf{a}_{01}$ and $\mathbf{a}_{02}$, respectively. Isotropy is described through the invariant $\bar{I}_1$ and anisotropy through $\bar{I}_1^*$ and $\bar{I}_2^*$.

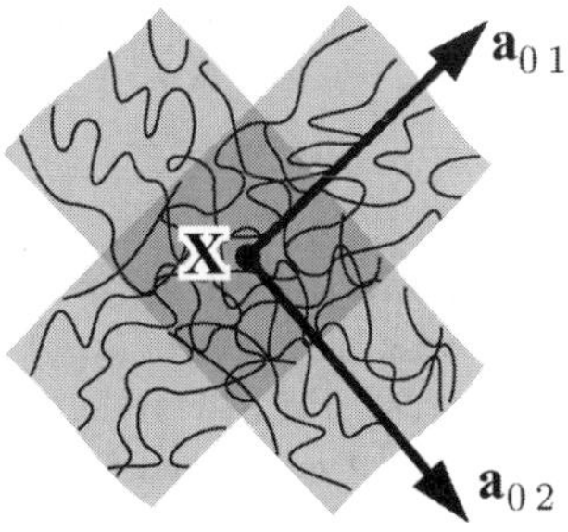


FIGURE 10.11.2 Arrangement of collagen fibers in the reference configuration characterized by two unit vectors $\mathbf{a}_{01}$, $\mathbf{a}_{02}$ at position $\mathbf{X}$.

Since the (wavy) collagenous structure of tissues is not active at low stresses (it does not store strain energy), we associate $\bar{\Psi}_{\text{iso}}$ with the mechanical response of the noncollagenous matrix of the material (which is less stiff than its elastin fiber constituent). To determine the noncollagenous matrix response, we propose to use the isotropic neo-Hookean model according to

$$\bar{\Psi}_{\text{iso}} = \frac{c}{2}(\bar{I}_1 - 3), \tag{4}$$

where $c > 0$ is a stresslike material parameter. However, to model the (isotropic) noncollagenous matrix material any Ogden-type elastic material may be applied [18].

According to morphological findings at highly loaded tissues, the families of collagen fibers become straighter and the resistance to stretch is almost entirely due to collagen fibers (the tissue becomes stiff). Hence, the strain energy stored in the collagen fibers is taken to be governed by the polyconvex (anisotropic) function

$$\bar{\Psi}_{\text{aniso}} = \frac{k_1}{2k_2}\left\{\exp\left[k_2(\bar{I}_1^* - 1)^2\right] - 1\right\} + \frac{k_3}{2k_4}\left\{\exp\left[k_4(\bar{I}_2^* - 1)^2\right] - 1\right\} \tag{5}$$

where $k_1 > 0, k_3 > 0$ are stresslike material parameters and $k_2 > 0, k_4 > 0$ are dimensionless parameters. According to Eqs. 2, 4, and 5, the collagen fibers do not influence the mechanical response of the tissue in the low stress domain. Because of the crimp structure of collagen fibers, we assume that they do not support compressive stresses, which implies that they are inactive in compression. Hence the relevant part of the anisotropic function (Eq. 5) is omitted for this case. If, for example, $\bar{I}_1^* > 1$ and $\bar{I}_2^* > 1$, then the soft tissue responds similarly to a rubberlike (purely isotropic) material described by the energy function (Eq. 4). However, in extension, that is when $\bar{I}_1^* > 1$ or $\bar{I}_2^* > 1$, and the collagen fibers are active and energy is stored in the fibers.

Equation 1 enables the Cauchy stress tensor, denoted $\boldsymbol{\sigma}$, to be derived in the *decoupled* form

$$\boldsymbol{\sigma} = \boldsymbol{\sigma}_{\text{vol}} + \bar{\boldsymbol{\sigma}} \quad \text{with} \quad \boldsymbol{\sigma}_{\text{vol}} = p\mathbf{I}, \quad \bar{\boldsymbol{\sigma}} = 2J^{-1}\text{dev}\left(\bar{\mathbf{F}}\frac{\partial\bar{\Psi}}{\partial\bar{\mathbf{C}}}\bar{\mathbf{F}}^{\text{T}}\right) \tag{6}$$

with the *volumetric* contribution $\boldsymbol{\sigma}_{\text{vol}}$ and the *isochoric* contribution $\bar{\boldsymbol{\sigma}}$ to the Cauchy stresses. In the stress relation (Eq. 6), $p = dU/dJ$ denotes the hydrostatic pressure and dev($\bullet$) furnishes the deviatoric operator in the Eulerian description. The operator is defined as $\text{dev}(\bullet) = (\bullet) - \frac{1}{3}[(\bullet) : \mathbf{I}]\mathbf{I}$, so that $\text{dev}(\bullet) : \mathbf{I} = 0$.

Using the additive split (Eq. 2) and particularizations (Eqs. 4 and 5), we get with the third part of Eq. 6 an explicit constitutive expression for the isochoric behavior of soft connective tissues in the Eulerian description, i.e.,

$$\bar{\boldsymbol{\sigma}} = c \operatorname{dev} \bar{\mathbf{b}} + \sum_{\alpha=1}^{2} 2\bar{\Psi}_\alpha \operatorname{dev} (\mathbf{a}_\alpha \otimes \mathbf{a}_\alpha) \tag{7}$$

where $\bar{\mathbf{b}} = \bar{\mathbf{F}}\bar{\mathbf{F}}^T$ denotes the modified left Cauchy-Green tensor,

$$\bar{\Psi}_1 = \frac{\partial \bar{\Psi}_{\text{aniso}}}{\partial \bar{I}_1^*} = k_1 (\bar{I}_1^* - 1) \exp\left[k_2 (\bar{I}_1^* - 1)^2\right] \tag{8}$$

$$\bar{\Psi}_2 = \frac{\partial \bar{\Psi}_{\text{aniso}}}{\partial \bar{I}_2^*} = k_3 (\bar{I}_2^* - 1) \exp\left[k_4 (\bar{I}_2^* - 1)^2\right] \tag{9}$$

are (scalar) response functions, and $\mathbf{a}_\alpha = \bar{\mathbf{F}}\mathbf{a}_{0\alpha}$, $\alpha = 1, 2$, are the Eulerian counterparts of the unit vectors $\mathbf{a}_{0\alpha}$.

The specific form of the proposed constitutive equation (Eq. 7) requires the *five* material parameters $c$, $k_1$, $k_2$, $k_3$, $k_4$ whose interpretations can be partly based on the underlying histological structure, i.e., matrix and collagen of the tissue. Note that in Eq. 7, orthotropic ($k_1 = k_3$, $k_2 = k_4$), transversely isotropic ($k_1 = 0$ or $k_3 = 0$), and isotropic hyperelastic descriptions ($k_1 = k_3 = 0$) at finite strains are included as special cases.

## 10.11.5 REPRESENTATIVE EXAMPLE: A MODEL FOR THE ARTERY

In this section we describe a model for the *passive* state of the *healthy* and *young* artery (no pathological changes in the intima, which is the innermost arterial layer frequently affected by atherosclerosis) suitable for predicting three-dimensional distributions of stresses and strains under physiological loading conditions with reasonable accuracy. It is a specification of the constitutive framework for soft tissues stated in the previous section. For an adequate model of arteries incorporating the active state (contraction of smooth muscles), see Reference [22]. For a detailed study of the mechanics of arterial walls, see the extensive review in Reference [13].

Experimental tests show that the elastic properties of the media (middle layer of the artery) and adventitia (outermost layer of the artery) are significantly different [31]. The media is much stiffer than the adventitia. In particular, in the unloaded configuration the mean value of Young's modulus for the media, for several pig thoracic aortas, is about an order of magnitude higher than that of the adventitia [32]. In addition, the arterial layers have different physiological tasks, and hence the artery is modeled as a thick-walled elastic circular tube consisting of two layers corresponding to the media and adventitia. In a young nondiseased artery the intima (innermost layer of the artery) exhibits negligible wall thickness and mechanical strength.

Each tissue layer is treated as a composite reinforced by two families of collagen fibers which are symmetrically disposed with respect to the cylinder axis. Hence, each tissue layer is considered as *cylindrically orthotropic* (already postulated in the early work [20]) so that a tissue layer behaves like a so-called balanced angle-ply laminate. We use the *same* forms of strain energy functions (Eqs. 4 and 5) for each tissue layer (each layer responds with similar mechanical characteristics) but use a *different* set of material parameters. Hence, Eq. 2 takes on the specified form

$$\bar{\Psi}_M = \frac{c_M}{2}(\bar{I}_1 - 3) + \frac{k_{1M}}{2k_{2M}} \sum_{\alpha=1}^{2} \left\{ \exp\left[ k_{2M} (\bar{I}^*_{\alpha M} - 1)^2 \right] - 1 \right\} \tag{10}$$

$$\bar{\Psi}_A = \frac{c_A}{2}(\bar{I}_1 - 3) + \frac{k_{1A}}{2k_{2A}} \sum_{\alpha=1}^{2} \left\{ \exp\left[ k_{2A} (\bar{I}^*_{\alpha A} - 1)^2 \right] - 1 \right\} \tag{11}$$

We end up with a two-layer model incorporating *six* material parameters, three for the media $M$, i.e., $c_M$, $k_{1M}$, $k_{2M}$, and three for the adventitia $A$, i.e., $c_A$, $k_{1A}$, $k_{2A}$.

The invariants associated with the anisotropic parts of the two tissue layers are defined by $\bar{I}^*_{1j} = \bar{\mathbf{C}} : \mathbf{A}_{1j}$ and $\bar{I}^*_{2j} = \bar{\mathbf{C}} : \mathbf{A}_{2j}$, $j = M, A$. The structure tensors $\mathbf{A}_{1j}$, $\mathbf{A}_{2j}$ are given by

$$\mathbf{A}_{1j} = \mathbf{a}_{01j} \otimes \mathbf{a}_{01j}, \quad \mathbf{A}_{2j} = \mathbf{a}_{02j} \otimes \mathbf{a}_{02j}, \quad j = M, A. \tag{12}$$

Employing a cylindrical coordinate system, the components of the unit (direction) vectors $\mathbf{a}_{01j}$ and $\mathbf{a}_{02j}$ read in matrix notation

$$[\mathbf{a}_{01j}] = \begin{bmatrix} 0 \\ \cos\beta_j \\ \sin\beta_j \end{bmatrix}, \quad [\mathbf{a}_{02j}] = \begin{bmatrix} 0 \\ \cos\beta_j \\ -\sin\beta_j \end{bmatrix}, \quad j = M, A, \tag{13}$$

and $\beta_j$, $j = M, A$, are the angles between the collagen fibers and the circumferential direction in the media and adventitia (see Figure 10.11.3). Small components of the (collagen) fiber orientation in the radial direction, as, for example, reported for human brain arteries [5], are neglected.

### 10.11.5.1 Residual Stresses

It has been known for some years that arteries which are excised from the body and not subjected to any loads are *not* stress-free (or strain-free) [28]. If, for example, the media and adventitia are separated and cut in a radial direction, the two arterial layers will spring open to form open (stress-free) sectors, which, in general, have different opening angles (see, for example, the experimental studies [29] for bovine specimens). In general, the residual stress state is very complex, and residual stresses (strains) in the axial direction may also occur. Note that residual stresses result from growth and remodeling mechanisms [21, 24].

By considering the arterial layers as circular cylindrical tubes, we may characterize the reference (stress-free) configuration of one arterial layer as a circular sector, as shown in Figure 10.11.4. For each arterial layer of the blood vessel a certain opening angle $\alpha$ can be found by experimental methods. The importance of incorporating residual stresses associated with the load-free

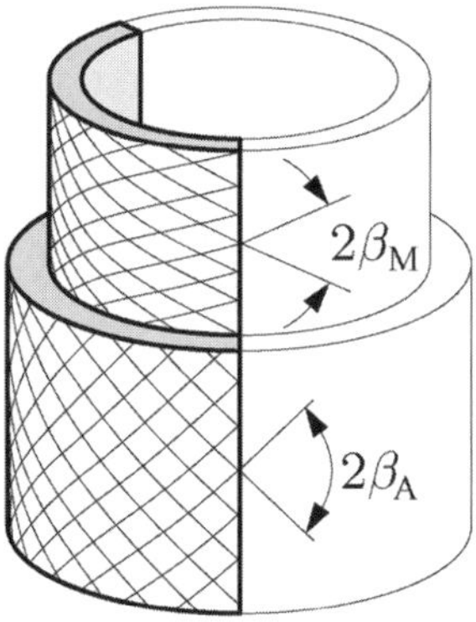


FIGURE 10.11.3 Load-free configuration of an idealized artery modeled as a thick-walled circular tube consisting of two layers, the media and adventitia.

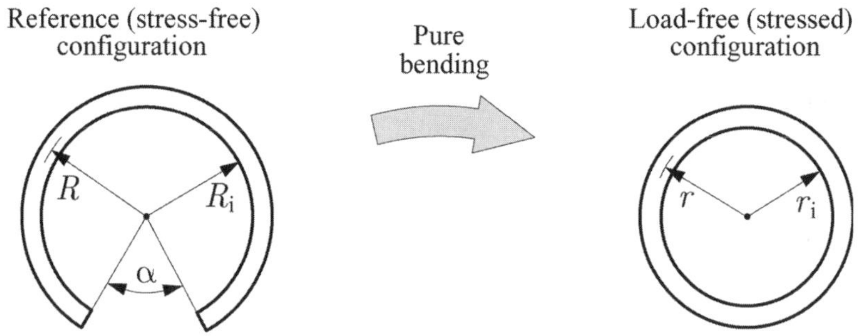


FIGURE 10.11.4 Cross-sectional representation of one arterial layer at the reference (stress-free) and load-free (stressed) configurations.

(but stressed) configuration into the computation has been emphasized in, for example, References [4, 12]. Consideration of residual strains has a strong influence on the global pressure and radius response of arteries and also on the stress and strain distributions across the deformed arterial wall. For analytical studies of residual stresses see, for example, the works [14, 22], which contain further references.

Therefore, it is essential to incorporate the residual stresses inherent in many biologic tissues. One possible approach to consideration of the influence of residual stresses on the overall three-dimensional stress behavior is to measure the strain energy from the load-free (stressed) configuration and to include the residual stresses [19]. Another approach is to start with the energy function relative to the stress-free (and fixed) configuration, as assumed in the presented models, and determine the deformation required to reach the load-free (stressed) configuration. Figure 10.11.4 shows the cross-sectional respresentation of one arterial layer at the load-free configuration obtained from the reference configuration by pure bending.

With the condition of incompressibility, the radius $r$ of an arterial layer in the load-free configuration may be computed from the radius $R$ of the associated reference configuration as [12]

$$r = \sqrt{\frac{R^2 - R_i^2}{k\lambda_z} + r_i^2}, \quad k = \frac{2\pi}{2\pi - \alpha} \tag{14}$$

where $r_i$, $R_i$ are the internal radii associated with the two configurations. The (constant) axial stretch is denoted by $\lambda_z$, and the parameter $k$ is a convenient measure of the tube opening angle in the stress-free configuration.

## 10.11.6 IDENTIFICATION OF THE MATERIAL PARAMETERS

Preferred directions in soft tissues are well specified by the orientation of prolate cell nuclei. They can be identified in microphotographs of appropriately stained histological sections. By visual inspection there exists a high directional correlation between smooth muscle cells and collagen fibers. Hence, the bell-shaped distribution of collagen fiber orientations may be obtained from an image-processing analysis of stained histological sections. The angle $\beta$ (and thus the unit vectors $\mathbf{a}_{01}$, $\mathbf{a}_{02}$) may be identified as the mean value of the corresponding statistical distribution.

Values of the material parameters associated with the model for soft tissues are then obtained by fitting the equations to the experimental data of the soft tissue of interest by using standard nonlinear fitting algorithms, such as the

Levenberg–Marquardt algorithm. If the mean values of the orientation of cell nuclei (collagen fiber) may not be identified experimentally, it is suggested to treat the collagen fiber orientations as additional (phenomenological) "material" parameters.

## 10.11.7 HOW TO USE THE MODEL

The energy functions are well suited for use in nonlinear finite element software, which enables complex boundary-value problems to be solved. Aspects of finite element implementation and numerical analysis of the model are presented in Reference [11]. Furthermore, computations may be carried out with some of the commercially available mathematical software packages such as *Mathematica* or *Maple*, which allow symbolic computation. Based on *Mathematica*, in Reference [12] a numerical technique for solving the bending, axial extension, inflation, and torsion problem of an artery is described.

## 10.11.8 TABLE OF PARAMETERS

Values of the parameters correspond to the Eqs. 10 and 11 and are given for a representative carotid artery from a rabbit (experiment no. 71 in Reference [7]). The material parameters $c$, $k_1$, $k_2$ and the (mean) values of collagen fiber angles $\beta$ are summarized in Table 10.11.2.

In the adventitia many collagen fibers run closer to the axial direction of the artery, whereas in the media the collagen fibers tend to run around the circumference. The fiber angles $\beta$ are meant to be associated with the reference (stress-free) configuration, and we assume that they are the same in the load-free (stressed) configuration. The difference in fiber angle which occur due to bending to the load-free configuration (see Figure 10.11.4) is small, so that it has a negligible influence on the stress–strain analysis of arteries.

**TABLE 10.11.2 Parameters for a Carotid Artery from a Rabbit (Experiment no. 71 in Reference [7]) in reference to Eqs. 10 and 11**

| Media | Adventitia |
|---|---|
| $c_M = 3.0000$ (kPa) | $c_A = 0.30000$ (kPa) |
| $k_{1M} = 2.3632$ (kPa) | $k_{1A} = 0.5620$ (kPa) |
| $k_{2M} = 0.8393$ (–) | $k_{2A} = 0.7112$ (–) |
| $\beta_M = 29.0°$ | $\beta_A = 62.0°$ |

By using a wall thickness of 0.39 mm (adopted from Reference [3]) and assuming that the media occupies two-thirds of the arterial wall thickness, the parameters in Table 10.11.2 predict the characteristic orthotropic behavior of a carotid artery under combined bending, inflation, axial extension, and torsion, as documented in Reference [12].

## REFERENCES


1. Betsch D. F., and Baer, E. (1980). Structure and mechanical properties of rat tail tendon. *Biorheology* **17**: 83–94.
2. Carew, T. E., Vaishnav, R. N., and Patel, D. J. (1968). Compressibility of the arterial wall. *Circ. Res.* **23**: 61–68.
3. Chuong, C. J., and Fung, Y. C. (1983). Three-dimensional stress distribution in arteries. *ASME J. Biomech. Engr.* **105**: 268–274.
4. Chuong, C. J., and Fung, Y. C. (1986). Residual stress in arteries, in *Frontiers in Biomechanics*, pp. 117–129, Schmid-Schönbein, G. W., Woo, S. L.-Y., and Zweifach, B. W., eds., New York: Springer-Verlag.
5. Finlay, H. M., McCullough, L., and Canham, P. B. (1995). Three-dimensional collagen organization of human brain arteries at different transmural pressures. *J. Vasc. Res.* **32**: 301–312.
6. Fung, Y. C. (1993). *Biomechanics: Mechanical Properties of Living Tissues*, 2nd ed., New York: Springer-Verlag.
7. Fung, Y. C., Fronek, K., and Patitucci, P. (1979). Pseudoelasticity of arteries and the choice of its mathematical expression. *Am. Physiological Soc.* **237**: H620–H631.
8. Gasser, T. C., and Holzapfel, G. A. (2001). Rate-independent elastoplastic constitutive modeling of biological soft tissues: Part I. Continuum basis, algorithmic formulation and finite element implementation. *Int. J. Solids and Structures* (in press).
9. Hoeve, C. A. J., and Flory, P. J. (1958). The elastic properties of elastin. *J. Am. Chem. Soc.* **80**: 6523–6526.
10. Holzapfel, G. A. (2000). *Nonlinear Solid Mechanics: A Continuum Approach for Engineering*, Chichester: John Wiley & Sons.
11. Holzapfel, G. A., and Gasser, T. C. (2001). A viscoelastic model for fiber-reinforced composites at finite strains: Continuum basis, computational aspects and applications. *Comput. Methods Appl. Mech. Engr.* **90**: 4379–4403.
12. Holzapfel, G. A., Gasser, T. C., and Ogden, R. W. (2001). A new constitutive framework for arterial wall mechanics and a comparative study of material models. *J. Elasticity* (in press).
13. Humphrey, J. D. (1995). Mechanics of the arterial wall: Review and directions. *Crit. Rev. Biomed. Engr.* **23**: 1–162.
14. Johnson, B. E., and Hoger, A. (1998). The use of strain energy to quantify the effect of residual stress on mechanical behaviour. *Math. Mech. of Solids* **3**: 447–470.
15. Martin, R. B., Burr, D. B., and Sharkey, N. A. (1998). *Skeletal Tissue Mechanics*, New York: Springer-Verlag.
16. Minns, R. J., Soden, P. D., and Jackson, D. S. (1973). The role of the fibrous components and ground substance in the mechanical properties of biological tissues: A preliminary investigation. *J. Biomech.* **6**: 153–165.
17. Nimni, M. E., and Harkness, R. D. (1988). Molecular structure and functions of collagen, *Collagen*, pp. 3–35, Nimni, M. E., ed., Boca Raton, FL: CRC Press.

18. Ogden, R. W. (1997). *Non-linear Elastic Deformations*, New York: Dover.
19. Ogden, R. W., and Schulze-Bauer, C. A. J. (2000). Phenomenological and structural aspects of the mechanical response of arteries. In *Mechanics in Biology*, Casey, J., and Bao, G. eds, New York, The American Society of Mechanical Engineers (ASME). AMD-Vol. 242/BED-Vol. 46, 125–140.
20. Patel, D. J., and Fry, D. L. (1969). The elastic symmetry of arterial segments in dogs. *Circ. Res.* **24**: 1–8.
21. Rachev, A. (1997). Theoretical study of the effect of stress-dependent remodeling on arterial geometry under hypertensive conditions. *J. Biomech.* **30**: 819–827.
22. Rachev, A., and Hayashi, K. (1999). Theoretical study of the effects of vascular smooth muscle contraction on strain and stress distributions in arteries. *Ann. Biomed. Engr.* **27**: 459–468.
23. Ramachandran, G. N. (1967). Chemistry of collagen, *Treatise on Collagen*, pp. 103–183, Ramachandran, G.N., ed., New York: Academic Press.
24. Rodriguez, E. K., Hoger, A., and McCulloh, A. D. (1994). Stress-dependent finite growth in soft elastic tissues. *J. Biomech.* **27**: 455–467.
25. Silver, F. H., Christiansen, D. L., and Buntin, C. M. (1989). Mechanical properties of the aorta: A review. Crit. Rev. Biomed. Engr. **17**: 323–358.
26. Spencer, A. J. M. (1984). Constitutive theory for strongly anisotropic solids, *Continuum Theory of the Mechanics of Fibre-Reinforced Composites*, pp. 1–32, Spencer, A.J.M., ed., Wien: Springer-Verlag, CISM Courses and Lectures No. 282, International Centre for Mechanical Sciences.
27. Tanaka, E., Yamada, H., and Murakami, S. (1996). Inelastic constitutive modeling of arterial and ventricular walls, in *Computational Biomechanics*, pp. 137–163, Hayasi, K., and Ishikawa, H. eds., Tokyo: Springer-Verlag.
28. Vaishnav, R. N., and Vossoughi, J. Estimation of residual strains in aortic segments, in *Biomedical Engineering II: Recent Developments*, pp. 330–333, Hall, C.W., ed., New York: Pergamon Press.
29. Vossoughi, J., Hedjazi, Z., and Borris, F. S. (1993). Intimal residual stress and strain in large arteries, in *Bed-Vol. 24, 1993 Bioengineering Conference ASME*, pp. 434–437.
30. Woo, S. L. Y., Gomez., M. A., and Akeson, W. H. (1985). Mechanical behaviors of soft tissues: Measurements, modifications, injuries and treatment, in *The Biomechanics of Trauma*, pp. 107–133, Nahum, A. M., and Melvin, J., eds., Norwalk: Appleton Crofts.
31. Xie, J., Zhou, J., and Fung, Y. C. (1995). Bending of blood vessel wall: Stress-strain laws of the intima-media and adventitia layers. *J. Biomech. Engr.* **117**: 136–145.
32. Yu, Q., Zhou, J., and Fung, Y. C. (1993). Neutral axis location in bending and Young's modulus of different layers of arterial wall. *Am. J. Physiol.* **265**: H52–H60.

# CHAPTER 11

# *Geomaterials*

SECTION **11.1**

# Introduction to Geomaterials

JEAN LEMAITRE
*Université Paris 6, LMT-Cachan, 61 avenue du Président Wilson, 94235 Cachan Cedex, France*

*Geomaterials* are "natural" materials such as **soils**, **rocks**, **sand**, **clay**, **salt**, etc., which are used without any technological transformation. A specific chapter is devoted to them because their behavior simultaneously involves many characteristics which do not follow the elementary classification: pressure dependency, dilatancy, time effects, internal friction, water saturation or not, fabric dependency, etc., and also because there is a specific community devoted to geomaterials.

The method of modeling does not differ from that for other materials, but specific variables are introduced to represent specific effects and the identification of material parameters often needs "*in situ*" tests and three axial tests.

A ***general background*** is given in Section 11.2 with special emphasis on the use of two potentials for two plastic mechanisms and an incrementally nonlinear model. Section 11.3 shows how to identify ***nonassociated elastic viscoplastic constitutive equations*** taking into account closing and opening of microcracks. Micro- and mesomechanical considerations are described in Section 11.4 for ***granular materials*** in which elastoplastic behavior induces shear banding and failure. Micromechanics allows for modeling the effects of friction and induced anisotropy which are beyond the classical normality rule (Section 11.5).

The general framework of ***linear poroelasticity*** may be found in Section 11.6, and ***nonlinear poroelasticity*** is described in Section 11.7 for liquid nonsaturated porous materials. ***Elastoplasticity*** for partially saturated soils is discussed in Section 11.8, taking into account the capillary effect of ***suction***. ***Strain-hardening*** models for ***sand***, ***clay***, and ***rocks*** are developed in Section 11.9. Extension to ***dynamic behavior***, including liquefaction, is given in Section 11.10. Bounding surface of plasticity may be established for sandy soils which are either drained or not, in monotonic and cyclic loading conditions (Section 11.11). Finally, Section 11.12 discusses a lattice model; i.e., a network of linear elements such as springs, trusses, or beams captures the main properties of fracture of concrete with a relatively small number of parameters.

*Handbook of Materials Behavior Models.* ISBN 0-12-443341-3.

SECTION 11.2

# Background of the Behavior of Geomaterials

FÉLIX DARVE
*l'INP Grenoble, L3S–BP 53 38041 Grenoble, France*

Contents



## 11.2.1 GENERAL BACKGROUND

Specific features of the behavior of geomaterials are as follows:

- pressure dependency, because of a failure criterion of the Mohr–Coulomb type, which is, roughly speaking, a generalization of the solid friction law for continuous media;
- nonassociated plastic strains, because the normality condition to a Mohr–Coulomb yield surface would imply too large a dilatancy under shear stress;
- incrementally nonlinear properties, because the experiments have not allowed us to exhibit finite domains of linearity for the relation between $\dot{\sigma}$ (or $d\sigma$) and $\dot{\varepsilon}$ (or $d\varepsilon$).

Here lie the basic reasons why nonlinear elasticity, associated elastoplasticity, and nonassociated elastoplasticity with a single plastic potential are today generally considered restricted frameworks for developing realistic constitutive models for geomaterials. Indeed, many constitutive relations for soils and rocks are incorporating two plastic potentials (usually one is linked to the


*Handbook of Materials Behavior Models.* ISBN 0-12-443341-3.

compressibility properties and the second to those of shearing), and few are thoroughly incrementally nonlinear.

Thus we propose first a global overview of the widely used constitutive relations for geomaterials and then a schematic presentation of one incrementally nonlinear model.

The general formalism of incremental constitutive relations is given by

$$F_h(d\varepsilon, d\sigma, dt) = 0$$

where $F$ is a tensorial function which depends on the material and on its stress-strain history by means of state variables and memory parameters $h$. For rate-independent materials, we can consider:

$$d\varepsilon = G_h(d\sigma) \text{ or } d\sigma = G_h^{-1}(d\varepsilon)$$

where $G$ and $G^{-1}$ are homogeneous functions of degree 1, because of the rate-independency condition.

According to Euler's identity,

$$d\varepsilon = \frac{\partial G}{\partial(d\sigma)} d\sigma = \mathbf{M}_h\left(\frac{d\sigma}{|d\sigma|}\right) d\sigma$$

or

$$d\sigma = \frac{\partial \mathbf{G}^{-1}}{\partial(d\varepsilon)} d\varepsilon = \mathbf{P}_h\left(\frac{d\varepsilon}{|d\varepsilon|}\right) d\varepsilon$$

where the elastoplastic tensors $M$ and $P$ are homogeneous functions of degree 0 with respect to $d\sigma$ and $d\varepsilon$, respectively, and where

$$|d\sigma| = \sqrt{\boldsymbol{d\sigma}_{ij}\boldsymbol{d\sigma}_{ij}}, \ |d\varepsilon| = \sqrt{\boldsymbol{d\varepsilon}_{ij}\boldsymbol{d\varepsilon}_{ij}}$$

Let us note:

$$u = \boldsymbol{d\sigma}/|\boldsymbol{d\sigma}| \text{ and } v = d\varepsilon/|d\varepsilon|$$

We call now "tensorial zone Z" any domain of incremental linearity in $d\sigma$ space (or, equivalently, in $d\varepsilon$ space) [1]:

$$\forall u \in Z : M(u) \equiv \mathbf{M}^z$$

While any elastic constitutive relation is characterized by a unique tensorial zone (these relations are so-called incrementally linear), conventional elastoplasticity with one unique plastic potential has two tensorial zones (the so-called loading domain and unloading domain). The behavior of geomaterials is often described through four tensorial zones, which correspond to two plastic potentials and two loading–unloading criteria:

- plastic mechanism 1 and related loading-unloading rule 1:

$$\frac{\partial f_1}{\partial \sigma} : d\sigma \langle 0 \text{ or} \rangle 0$$

- plastic mechanism 2 and related loading-unloading rule 2:

$$\frac{\partial f_2}{\partial \sigma} : d\sigma \langle 0 \text{ or} \rangle 0$$

The four tensorial zones are defined by the intersections of both hyperplanes (in $d\sigma$ space) whose equations are given by

$$\frac{\partial f_1}{\partial \sigma} : d\sigma = 0 \quad \text{and} \quad \frac{\partial f_2}{\partial \sigma} : d\sigma = 0$$

On these hyperplanes a continuity condition must be fulfilled by the elastoplastic tensors linked to the adjacent tensorial zones, as, for example,

$$\forall d\sigma \text{ such that} : \frac{\partial f_1}{\partial \sigma} : d\sigma = 0 \quad \text{and} \quad \frac{\partial f_2}{\partial \sigma} : d\sigma > 0$$

$$(\mathbf{M}^{++} - \mathbf{M}^{-+})d\sigma \equiv 0$$

where $M^{++}$ is associated with the tensorial zone defined by

$$\frac{\partial f_1}{\partial \sigma} : d\sigma > 0 \quad \text{and} \quad \frac{\partial f_2}{\partial \sigma} : d\sigma > 0$$

and $M^{-+}$ to the one defined by

$$\frac{\partial f_1}{\partial \sigma} : d\sigma < 0 \quad \text{and} \quad \frac{\partial f_2}{\partial \sigma} : d\sigma > 0$$

The introduction of such two plastic mechanisms is an attempt to describe the directional variation of $M$ (or equivalently $P$) with respect to $d\sigma$ (respectively, $d\varepsilon$), which is only roughly taken into account in conventional elastoplasticity with one elastic tensor $M^e$ for "unloading conditions" and one elastoplastic tensor $M^{ep}$ for "loading conditions."

If now we continue to increase the number of tensorial zones from one in elasticity, two in conventional elastoplasticity, four with two plastic potentials, etc., we obtain for an infinite number of tensorial zones the case of the incrementally nonlinear constitutive relations where $M$ (or $P$) is varying in a continuous manner with $d\sigma$ (or, respectively, $d\varepsilon$). This class of incrementally nonlinear models includes as particular cases endochronic relations [2] and hypoplasticity [3–5] We present in the next paragraph an incrementally nonlinear relation of second order [6].

Another point of interest in modeling geomaterial behavior is the question of hardening parameters. Here also the specific complexity of the irreversible behavior of geomaterials has led us to try to generalize the usual isotropic and kinematic hardening variables by taking into account a rotational hardening (for example, Lade [7]). This rotational hardening allows us to describe, as an example, the rotation of the yield surface in $(q,\ p)$ plane when the consolidation process is not isotropic but of the oedometric type.

Finally the third point which must be briefly evoked here is the modeling of the viscous behavior, which is quite an important phenomenon for clays and which cannot be negligible for rocks in certain situations. There are two possible ways to take into account viscous strains by considering two different decompositions of the incremental strains: $(1): d\varepsilon = d\varepsilon^{ep} + d\varepsilon^{v}$, where there is a distinction between the elastoplastic, instantaneous strains $(d\varepsilon^{ep})$ and the viscous, delayed strains $(d\varepsilon^{v})$; or $(2): d\varepsilon = d\varepsilon^{e} + d\varepsilon^{vp}$, where the distinction is made between the elastic, reversible strains $(d\varepsilon^{e})$ and the viscoplastic, irreversible strains $(d\varepsilon^{vp})$. For this second class of models the viscoplastic strains are usually treated by a viscoplastic potential as proposed by Perzyna [8].

## 11.2.2 INCREMENTALLY NONLINEAR CONSTITUTIVE RELATIONS

Returning to the relations:

$$d\varepsilon = M_h(u)d\sigma \text{ or } d\sigma = P_h(v)d\varepsilon$$

and considering the six-dimensional associated spaces, we develop the components of $M$ or $P$ in polynomial series expansions as follows:

$$M_{\alpha\beta} = M^1_{\alpha\beta} + M^2_{\alpha\beta\gamma}u_\gamma + M^3_{\alpha\beta\gamma\delta}u_\gamma u_\delta + \cdots$$

$$P_{\alpha\beta} = P^1_{\alpha\beta} + P^2_{\alpha\beta\gamma}v_\gamma + P^3_{\alpha\beta\gamma\delta}v_\gamma v_\delta + \cdots$$

Finally,

$$d\varepsilon_\alpha = M^1_{\alpha\beta}d\sigma_\beta + \frac{1}{|d\sigma|}M^2_{\alpha\beta\gamma}d\sigma_\beta d\sigma_\gamma + \cdots$$

$$d\sigma_\alpha = P^1_{\alpha\beta}d\varepsilon_\beta + \frac{1}{|d\varepsilon|}P^2_{\alpha\beta\gamma}d\varepsilon_\beta d\varepsilon_\gamma + \cdots$$

The first terms of the right-hand parts of these equations correspond to hypoelastic relations, and both the first terms are the general expression of incrementally nonlinear models of second order.

In order to determine the expression of $M^1, M^2, P^1$, and $P_2$, one assumes three supplementary hypotheses:

- The incremental relations are orthotropic.
- There is not any crossed terms:

$$\forall\beta \neq \gamma : M^2_{\alpha\beta\gamma} = 0 \text{ and } P^2_{\alpha\beta\gamma} = 0.$$

- The shear behavior is incrementally linear:

$$\forall\beta \geq 4 \text{ or } \forall\gamma \geq 4 : M^2_{\alpha\beta\gamma} = 0 \text{ and } P^2_{\alpha\beta\gamma} = 0$$

In orthotropy axes the first model (so-called direct) and the second model (so- called dual) are thus taking the following expressions. For the direct model:

$$\begin{cases} \begin{bmatrix} d\varepsilon_{11} \\ d\varepsilon_{22} \\ d\varepsilon_{33} \end{bmatrix} = A \begin{bmatrix} d\sigma_{11} \\ d\sigma_{22} \\ d\sigma_{33} \end{bmatrix} + \frac{1}{|d\sigma|} B \begin{bmatrix} (d\sigma_{11})^2 \\ (d\sigma_{22})^2 \\ (d\sigma_{33})^2 \end{bmatrix} \\ d\varepsilon_{12} = d\sigma_{12}/2G_3 \\ d\varepsilon_{23} = d\sigma_{23}/2G_1 \\ d\varepsilon_{31} = d\sigma_{31}/2G_2 \end{cases}$$

and for the dual model:

$$\begin{cases} \begin{bmatrix} d\sigma_{11} \\ d\sigma_{22} \\ d\sigma_{33} \end{bmatrix} = C \begin{bmatrix} d\varepsilon_{11} \\ d\varepsilon_{22} \\ d\varepsilon_{33} \end{bmatrix} + \frac{1}{|d\varepsilon|} D \begin{bmatrix} (d\varepsilon_{11})^2 \\ (d\varepsilon_{22})^2 \\ (d\varepsilon_{33})^2 \end{bmatrix} \\ d\sigma_{12} = 2G_3\, d\varepsilon_{12} \\ d\sigma_{23} = 2G_1\, d\varepsilon_{23} \\ d\sigma_{31} = 2G_2\, d\varepsilon_{31} \end{cases}$$

$A$, $B$, $C$, $D$ are $3 \times 3$ matrices which depend on state variables and memory parameters, and $G_1$, $G_2$, $G_3$ are the shear moduli.

In order to determine $A$ and $B$ (direct model), and $C$ and $D$ (dual model), we proceed to an identification, respectively, for the direct model with "generalized triaxial paths" and for the dual model with "generalized oedometric paths." "Generalized triaxial paths" refer to triaxial compressions (index" + ", $d\sigma_i > 0$) or extensions (index" − ", $d\sigma_i < 0$)with two constant and distinct lateral stresses (denoted by $\sigma_j$ and $\sigma_k$).

Let us notice for the direct model:

$$E_i = \left(\frac{\partial\sigma_i}{\partial\varepsilon_i}\right)_{\sigma_j,\sigma_k} \text{ and } V_i^j = -\left(\frac{\partial\varepsilon_j}{\partial\varepsilon_i}\right)_{\sigma_j,\sigma_k}$$

"Generalized eodometric paths" refer to oedometric compressions (index" + ", $d\varepsilon_i > 0$) or extensions (index" − ", $d\varepsilon_i < 0$) with two constant and distinct lateral strains (denoted by $\varepsilon_j$ and $\varepsilon_k$).

Let us notice for the dual model:

$$O_i = \left(\frac{\partial \sigma_i}{\partial \varepsilon_i}\right)_{\varepsilon_j,\varepsilon_k} \quad \text{and } K_i^j = \left(\frac{\partial \sigma_j}{\partial \sigma_i}\right)_{\varepsilon_j,\varepsilon_k}$$

Finally, let us consider

$$N^+ = \begin{bmatrix} \frac{1}{E_1^+} & -\frac{V_2^{1+}}{E_2^+} & -\frac{V_3^{1+}}{E_3^+} \\ -\frac{V_1^{2+}}{E_1^+} & \frac{1}{E_2^+} & -\frac{V_3^{2+}}{E_3^+} \\ -\frac{V_1^{3+}}{E_1^+} & -\frac{V_2^{3+}}{E_2^+} & \frac{1}{E_3^+} \end{bmatrix} \text{and } N^- \text{in the same way}$$

and for the dual model:

$$\mathcal{Q}^+ = \begin{bmatrix} O_1^+ & K_2^{1+}O_2^+ & K_3^{1+}O_3^+ \\ K_1^{2+}O_1^+ & O_2^+ & K_3^{2+}O_3^+ \\ K_1^{3+}O_1^+ & K_2^{3+}O_2^+ & O_3^+ \end{bmatrix} \text{and} \mathcal{Q}^- \text{in the same way}$$

It is now possible to prove that

$$\begin{cases} A = \frac{1}{2}(N^+ + N^-) \text{and } B = \frac{1}{2}(N^+ - N^-) \\ C = \frac{1}{2}\left(\mathcal{Q}^+ + \mathcal{Q}^-\right) \text{and } D = \frac{1}{2}\left(\mathcal{Q}^+ - \mathcal{Q}^-\right) \end{cases}$$

Besides these thoroughly incrementally nonlinear models, it is also possible to consider incrementally piecewise linear relations which are based on eight tensorial zones in $d\sigma$ space (or, respectively, in $d\varepsilon$ space) and which have the same tensorial structure as an elastoplastic model with three plastic potentials. These so-called "octolinear" models are given by

$$\begin{bmatrix} d\varepsilon_{11} \\ d\varepsilon_{22} \\ d\varepsilon_{33} \end{bmatrix} = A \begin{bmatrix} d\sigma_{11} \\ d\sigma_{22} \\ d\sigma_{33} \end{bmatrix} + B \begin{bmatrix} |d\sigma_{11}| \\ |d\sigma_{22}| \\ |d\sigma_{33}| \end{bmatrix}$$

and for the dual model by

$$\begin{bmatrix} d\sigma_{11} \\ d\sigma_{22} \\ d\sigma_{33} \end{bmatrix} = C \begin{bmatrix} d\varepsilon_{11} \\ d\varepsilon_{22} \\ d\varepsilon_{33} \end{bmatrix} + D \begin{bmatrix} |d\varepsilon_{11}| \\ |d\varepsilon_{22}| \\ |d\varepsilon_{33}| \end{bmatrix}$$

## 11.2.3 IDENTIFICATION OF THESE CONSTITUTIVE RELATIONS

As we have just seen, the identification of the direct model is performed on triaxial paths and for the dual model on oedometric paths.

We have developed sophisticated expressions of $E_i$ and $V_i^j$ (and, respectively, $O_i$ and $K_i^j$) [9], and but also very simple expressions that are particularly useful for inverse analyses in geomechanics. We are only presenting here these last simple formulations [10] for the direct model:

- The plastic limit surface is given by a Mohr–Coulomb criterion which is characterized by cohesion C and friction angle $\varphi$
- Inside the limit surface, the tangent Young's moduli are equal to

$$E_i = E_0\sqrt{\frac{p}{p_0}}$$

($p$, mean pressure and $p_0$ reference pressure) for sands and to $E_i = E_0\, p$ for clays.

The tangent Poisson's ratios $V_i^j$ are equal to: $V_0$.

- On the limit surface, the moduli are taken as arbitrarily small and the Poisson's ratios are deduced from the dilatancy angle: $\psi$
- After a stress reversal the tangent moduli are mutiplied by $k$.

For the dual model:

- The plastic limit surface is of the Mohr–Coulomb type ($C$, $\varphi$) as for the direct model.– Inside the limit surface, the tangent oedometric moduli are equal to

$$O_i = E_0^{oedo}\sqrt{\frac{p}{p_0}}\text{for sands}$$

  and to

$$E_0^{oedo}\, p \text{ for clays}$$

  The tangent lateral pressure coefficients $K_{ij}$ are equal to $k_O$ for oedometric compression and to $k_d$ for extension.
- On the limit surface, the tangent oedometric moduli are divided by $\alpha$ and the tangent lateral pressure coefficients are equal to the plastic passive pressure coefficient.
- After a stress reversal, the tangent moduli are multiplied by $k$.

### 11.2.4 PARAMETERS

Table 11.2.1 and Table 11.2.2. present typical values of the parameters which have been defined in 11.2.3

TABLE 11.2.1 Table of parameters for the Direct Model

| Materials | C(MPa) | $\varphi(°)$ | $E_0$(MPa) | $V_0$ | $\psi(°)$ | $k$ |
|---|---|---|---|---|---|---|
| Sand | 0 | $30 \rightarrow 40$ | $50 \rightarrow 120$ | 0.25 | $-3 \rightarrow 10$ | $3 \rightarrow 6$ |
| Clay | $0 \rightarrow 0.3$ | $20 \rightarrow 35$ | $10 \rightarrow 80$ | 0.30 | $-5 \rightarrow 15$ | $3 \rightarrow 6$ |

TABLE 11.2.2 Table of parameters for the Dual Model

| Materials | C(MPa) | $\varphi(°)$ | $E_0^{oed}$(MPa) | $k_0$ | $k_d$ | $\alpha$ | $k$ |
|---|---|---|---|---|---|---|---|
| Sand | 0 | $30 \rightarrow 40$ | $60 \rightarrow 150$ | $0.3 \rightarrow 0.6$ | $0.1 \rightarrow 0.4$ | $3 \rightarrow 12$ | $3 \rightarrow 6$ |
| Clay | $0 \rightarrow 0.3$ | $20 \rightarrow 35$ | $20 \rightarrow 100$ | $0.4 \rightarrow 0.7$ | $0.2 \rightarrow 0.5$ | $2 \rightarrow 7$ | $3 \rightarrow 6$ |

## REFERENCES


1. Darve, F. (1990). The expression of rheological laws in incremental form and the main classes of constitutive equations, in *Geomaterials: Constitutive Equations and Modelling*, pp. 123–148. Darve, F., ed., Elsevier applied Science.
2. Valanis, K.C. (1971). A theory of viscoplasticity without a yield surface. *Arch. Mech.* **23**: 517–551.
3. Dafalias, Y. F. (1986). Bounding surface plasticity: I. Mathematical foundation and hypoplasticity. *J. Eng. Mech.* **112**(9): 966–987.
4. Kolymbas, D. (1984). A constitutive law of the rate-type for soils. Position calibration and prediction, in *Constitutive Relations for Soils*, pp. 419–437, Gudehus G., Darve, F., and Vardoulakis, I., eds., Balkema.
5. Chambon, R., Desrues, J., Hammad, W., and Charlier, R. (1994). Cloe a new rate-type constitutive model for geomaterials. *Int. J. Num. Anal. Geom.* **18**(4): 253–278.
6. Darve, F (1990). Incrementally non-linear constitutive relationships, in *Geomaterials; Constitutive Equations and Modelling*, pp. 213–238,. Darve, F., ed., Elsevier Applied Science
7. Lade, P., and Inel, S. (1997). Rotational kinematic hardening model for sand. Part I. Concept of rotating yield and plastic potential surfaces. *Comp. Geotech.* **21**(4): 183–216.
8. Perzyna, P. (1963). The constitutive equations for work-hardening and rate-sensitive plastic materials. *Proc. Vibrational Problems* 4(3): 281–290.
9. Darve, F., and Dendani, H. (1988). An incrementally non-linear constitutive relation and its predictions, in *Constitutive Equations for Granular Soils*, pp. 237–254, Saada, A.S., and Bianchini, G., eds., Balkema.
10. Darve, F., and Pal, O. (1998). A new incrementally non-linear constitutive relation with 5 material constants, in *Computer Methods and Advances in Geomechanics*, pp. 2445–2454, Yuan, ed., Balkema.

SECTION 11.3

# Models for Compressible and/or Dilatant Geomaterials

N.D. CRISTESCU
*231 Aerospace Building, University of Florida, Gainesville, Florida*

Contents



## 11.3.1 VALIDITY

The model to be used for geomaterials must describe the main mechanical properties exhibited by such materials, such as instantaneous response, time effects (creep, relaxation, rate effects), and irreversible volumetric deformation (compressibility and/or dilatancy). The main mechanisms governing the deformability of Geomaterials are closing and opening and microcracks. This article shows what kind of testing is needed to formulate a nonassociated elastic-viscoplastic constitutive equation for such materials. One starts by defining the compressibility/dilatancy boundary and that of instantaneous failure, both to be incorporated into the constitutive equation. Then triaxial tests are used to determine the energy of deformation, the yield function, and finally the viscoplastic potential. The elastic parameters are determined from small unloading/reloading cycles which follow a short period of creep. The models can describe transient and steady-state creep, irreversible damage developed in time or irreversible compaction in time, convergence of the walls in underground excavations, landslides, etc.


*Handbook of Materials Behavior Models.* ISBN 0-12-443341-3.

The models can be used for any kind of rock or geomaterial to describe the deformation in time (creep) and irreversible volumetric changes, i.e., compressibility and/or dilatancy. Also, the model can be used to describe compaction and/or creep flow of powder like materials in bulk, such as ceramic powders, pharmaceuticals, food, etc.

## 11.3.2 FORMULATION

The model can be formulated using the data obtained in triaxial testing of geomaterials. For instance Figure 11.3.1 shows the volumetric behavior of rock salt in true triaxial tests, where $\sigma$ is the mean stress and $\tau$ is the octahedral shear stress. The locus in the $\sigma\tau$-plane where one is passing from compressibility to dilatancy is the ***compressibility/dilatancy boundary***. For rock salt it is

$$X(\sigma, \tau) := -\frac{\tau}{\sigma_*} + f_1\left(\frac{\sigma}{\sigma_*}\right)^2 + f_2\frac{\sigma}{\sigma_*} \tag{1}$$

where $f_1$ and $f_2$ are material constants and $\sigma_*$ is the unit stress. In reality this boundary is a strip of incompressibility which, for convenience, is approximated by a curve (Fig. 11.3.2). The upper curve shown in Figure

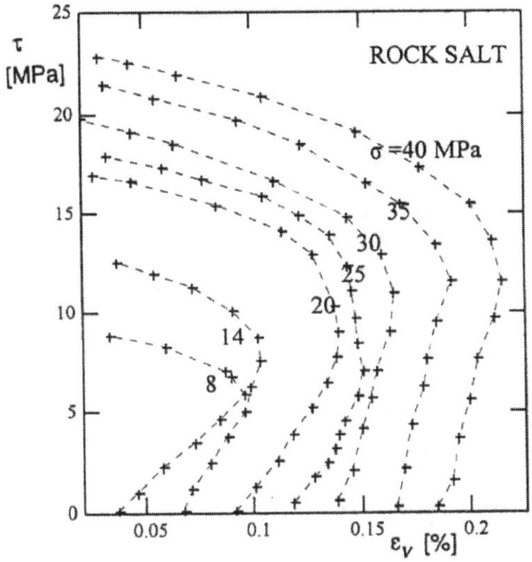


FIGURE 11.3.1 Dependency of volumetric strain on octahedral shear stress for rock salt in true triaxial tests.

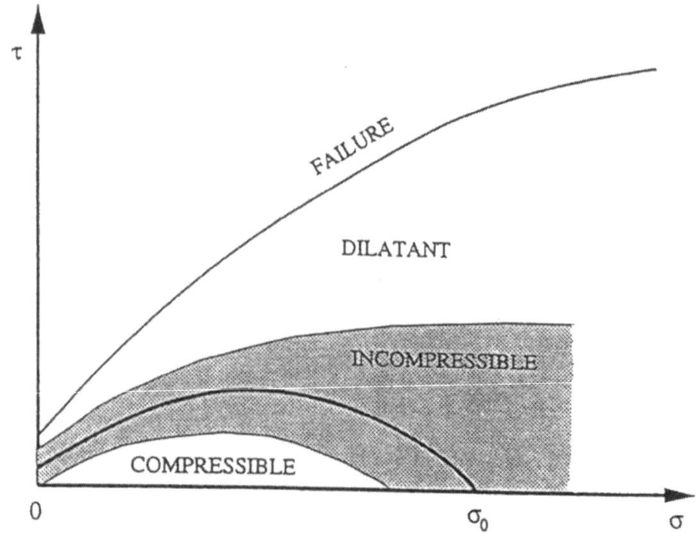


FIGURE 11.3.2 Domains of compressibility and dilatancy, separated by a strip of incompressibility.

11.3.2 is the *short-term failure surface*, which for rock salt is

$$Y(\sigma, \tau) := -r\frac{\tau}{\sigma_*} - s\left(\frac{\tau}{\sigma_*}\right)^6 + \frac{\sigma}{\sigma_*} \tag{2}$$

where $r$ and $s$ are material constants.

The model is based on the following assumptions: the material is homogeneous and isotropic; the material displacements and rotations are small; the geomaterial exhibits instantaneous response (i.e., the two extended body elastic (seismic) waves can propagate in that material); the yield stress can be close to zero; the material exhibits time effects (transient and steady-state creep as well as loading rate effect) for any stress state; and the constitutive domain is bounded by the *short-term failure surface*. With these assumptions the constitutive equation is of the form

$$\boldsymbol{D} = \frac{\dot{\boldsymbol{T}}}{2G} + \left(\frac{1}{3K} - \frac{1}{2G}\right)\dot{\sigma}\boldsymbol{1} + k_T\left\langle 1 - \frac{W(t)}{H(\boldsymbol{T})}\right\rangle\frac{\partial F}{\partial \boldsymbol{T}} + k_S\frac{\partial S}{\partial \boldsymbol{T}} \tag{3}$$

where $\boldsymbol{D}$ is the strain rate tensor, $\boldsymbol{T}$ is the stress tensor, $\boldsymbol{1}$ is the unit stress tensor, $H(\boldsymbol{T})$ is the yield function, $F(\boldsymbol{T})$ is the viscoplastic potential for transient creep, and $S(\boldsymbol{T})$ is the potential for the steady-state creep. All these functions depend on stress invariants. Further, $k_T$ and $k_S$ are two viscosity coefficients, and $K$ and $G$ are the elastic parameters (generally variable) which are defining the "instantaneous response" of the geomaterial.

$$W(t) = \int_0^t \sigma \dot{\varepsilon}_V^I \, dt + \int_0^t T' : D'^I dt \tag{4}$$

is the irreversible stress work per unit volume, with "prime" standing for deviator and

$$H(T) = W(t) \tag{5}$$

is the stabilization boundary for transient creep. The term $\partial F/\partial T$ can be replaced by an irreversible strain rate orientation tensor $N(T)$, which is easier to formulate from tests.

## 11.3.3 DESCRIPTION OF THE MODEL

The yield function is determined by performing triaxial tests from which $W(t)$ is obtained following various loading paths, and thus the boundary (Eq. 5) is determined.

The functions $F$ (or $N$ ) and $S$ are determined starting from the expression for the irreversible volumetric strain rate deformation

$$\dot{\varepsilon}_V^I = k_T\left\langle 1 - \frac{W(t)}{H(T)}\right\rangle \frac{\partial F}{\partial \sigma} + k_S\frac{\partial S}{\partial \sigma}\left(\text{or } k_T\left\langle 1 - \frac{W(t)}{H(T)}\right\rangle tr(N) + k_S\frac{\partial S}{\partial \sigma}\right) \tag{6}$$

where the right-hand-side derivatives are with respect to the mean stress $\sigma$. These functions must satisfy the inequalities

$$\begin{aligned} &\frac{\partial F}{\partial \sigma} > 0, \ \frac{\partial S}{\partial \sigma} = 0, \ \text{in the compressibility domain} \\ &\frac{\partial F}{\partial \sigma} = 0, \ \frac{\partial S}{\partial \sigma} = 0, \ \text{in the C/D boundary strip} \\ &\frac{\partial F}{\partial \sigma} < 0, \ \frac{\partial S}{\partial \sigma} < 0, \ \text{in the dilatancy domain} \end{aligned} \tag{7}$$

This is obtained by incorporating both expressions $X$ and $Y$ (see (Eqs. 1 and 2) into the functions $F$ and $S$. For instance, for $F$ one can use

$$k_T\frac{\partial F}{\partial \sigma} = \frac{X(\sigma, \bar{\sigma})\Psi(\sigma)}{Y(\sigma, \bar{\sigma})}[\bar{\sigma}G(\bar{\sigma}) + \sigma] \tag{8}$$

where $\Psi(\sigma)$ is determined from the hydrostatic tests and $G(\bar{\sigma})$ from the deviatoric part of the triaxial tests (in many cases $G$ can be disregarded). Here $\bar{\sigma}$ is the equivalent stress $\tau = (2/3)\,\bar{\sigma} = [(2/3)II_{T'}]^{1/2}$ and $II_{T'}$ is the second invariant of the stress deviator. With this procedure dilatancy and/or compressibility, as well as short-time failure, is described exactly by the constitutive equation.

## 11.3.4 IDENTIFICATION OF MATERIAL PARAMETERS

First to be determined are the elastic parameters $K$ and $G$. The procedure is described in this volume in "Elasticity of Porous Materials," Section 2.3.

The yield function $H$ is determined in the two stages of triaxial tests: hydrostatic and deviatoric; thus $H(\sigma, \bar{\sigma}) := H_H(\sigma) + H_D(\sigma, \bar{\sigma})$, where generally

$$H_H(\sigma) := \begin{cases} h_0 \sin\left(\omega \dfrac{\sigma}{\sigma_*} + \varphi\right) + h_1 & \text{if } \sigma \leq \sigma_0 \\ h_0 + h_1 & \text{if } \sigma \geq \sigma_0 \end{cases} \tag{9}$$

where $h_0$, $h_1$, $\omega$ are material constants, $\sigma_* = 1$ (unit stress), and $\sigma_0$ is the pressure closing all pores and microcracks. Further, $H_D$ is a polynomial in $\tau$ with coefficients depending on mean stress. For rock salt:

$$H_D(\sigma, \tau) := A(\sigma)\left(\frac{\tau}{\sigma_*}\right)^{14} + B(\sigma)\left(\frac{\tau}{\sigma_*}\right)^{3} + C(\sigma)\left(\frac{\tau}{\sigma_*}\right) \tag{10}$$

The viscoplastic potential $F$ is determined by starting from Eq. 8, integrating with respect to $\sigma$, and then differentiating with respect to $\bar{\sigma}$ (or $\tau$), and using the formula

$$k_T \frac{\partial F}{\partial \bar{\sigma}} = \frac{\sqrt{2}}{3} \frac{1}{\left\langle 1 - \dfrac{W(t)}{H(T)} \right\rangle} \sqrt{\left(\dot{\varepsilon}_1^I - \dot{\varepsilon}_2^I\right)^2 + \left(\dot{\varepsilon}_2^I - \dot{\varepsilon}_3^I\right)^2 + \left(\dot{\varepsilon}_3^I - \dot{\varepsilon}_1^I\right)^2} \tag{11}$$

and the triaxial data for the irreversible strain rate components.

The potential $S$ is determined starting from volumetric deformation in long-term creep tests performed for various stress states. From these tests follow

$$\dot{\varepsilon}_V^I|_S = k_S \frac{\partial S}{\partial \sigma} := \begin{cases} b\left(\dfrac{\tau}{\sigma_*}\right)^m \left(\dfrac{\sigma}{\sigma_*}\right)^n & \text{if } X < 0 \\ 0 & \text{if } X \geq 0 \end{cases} \tag{12}$$

from where follows for rock salt in the dilatancy domain

$$k_S \frac{\partial S}{\partial \tau} = \frac{bm}{n+1}\left(\frac{\tau}{\sigma_*}\right)^{m-1}\left(\frac{\sigma}{\sigma_*}\right)^{n+1} + \frac{p}{\sigma_*}\left(\frac{\tau}{\sigma_*}\right)^{5} \frac{1}{\sigma_*} \tag{13}$$

Here $b$, $m$, $n$, and $p$ are material constants.

## 11.3.5 HOW TO USE THE MODEL

Two boundary value problems can be described: (a) creep under constant stresses, and (b) a general variation of both stresses and strains in long-term intervals. In the first case, an "instantaneous" loading results in a stress and strain distribution at time $t_0$ (the elastic solution), and afterwards the stresses are held constant (or nearly constant) and the variation of the strains are obtained from

$$\varepsilon_{ij}(t) = \varepsilon_{ij}^{E} + \frac{\left\langle 1 - \dfrac{W_T(t_0)}{H} \right\rangle \dfrac{\partial F}{\partial T_{ij}}}{\dfrac{1}{H}\dfrac{\partial F}{\partial T_{kl}} T_{kl}} \left\{ 1 - \exp\left[ \frac{1}{H} \frac{\partial F}{\partial T_{mn}} T_{mn} [k_T(t_0 - t)] \right] \right\} + \frac{\partial S}{\partial T_{ij}} k_S (t - t_0) \tag{14}$$

where $\varepsilon_{ij}^{E}$ are the elastic strains. This formula can be used for creep tests, gravitational compaction, and any deformation by creep when stresses are nearly constant. For instance, for a radial convergence $u$ of a vertical borehole of initial radius $a$ at time $t_0$, we have

$$\frac{u}{a} = -\frac{p - \sigma_h}{2G} \frac{a}{r} + \frac{r}{a} \left\{ \frac{\left\langle 1 - \dfrac{W_T(t_0)}{H} \right\rangle \dfrac{\partial F}{\partial T_\theta}}{\dfrac{1}{H}\dfrac{\partial F}{\partial T_{ij}} T_{ij}} \left\{ 1 - \exp\left[ \frac{k_T}{H} \frac{\partial F}{\partial T_{mn}} T_{mn} \right] \right\} + k_S \frac{\partial S}{\partial T_\theta} (t - t_0) \right\} \tag{15}$$

where $p$ is the internal pressure, $\sigma_h$ is the far-field stress and $r$ is the cylindrical coordinate.

For the second kind of problem, when, because of the change of geometry, or for other reasons, the stresses are also varying during creep, one has to introduce the constitutive equation in a code (see examples in Cristescu [1] and Cristescu and Hunsche [4]).

Another example is compaction of a geomaterial (generally particulate material) under its own weight. The volumetric deformation describing compaction in time, as obtained from

$$\varepsilon_V^I(t) = \frac{\left\langle 1 - \frac{W(t_0)}{H(T)} \right\rangle \frac{\partial F}{\partial \sigma}}{\frac{1}{H}\frac{\partial F}{\partial T_{ij}} T_{ij}} \left\{ 1 - \exp\left[ \frac{1}{H} \frac{\partial F}{\partial T_{mn}} T_{mn} [k_T(t - t_0)] \right] \right\} \tag{16}$$

## 11.3.6 PARAMETERS

For rock salt (tests performed by Hunsche),

$$A(\sigma) := a_1 + \frac{a_2}{\left(\frac{\sigma}{\sigma_*}\right)^6},\; B(\sigma) := b_1 \frac{\sigma}{\sigma_*} + b_2,\; C(\sigma) := \frac{c_1}{\left(\frac{\sigma}{\sigma_*}\right)^3 + c_3} + c_2$$

$a_1 = 7\times 10^{-21}$ MPa, $a_2 = 6.73\times 10^{-12}$ MPa, $b_1 = 1.57\times 10^{-6}$ MPa, $b_2 = 1.7\times 10^{-5}$ MPa, $c_1 = 26.12$ MPa, $c_2 = -0.00159$ MPa, $c_3 = 3134$, and $\sigma_* = 1$ MPa.

The viscoplastic potential for rock salt is

$$\begin{aligned}
k_T F(\sigma, \tau) := \sigma_* \Bigg\{ & \frac{f_1 p_1}{4} [Y(\sigma, \tau)]^4 + \left[ -\frac{4}{3} f_1 p_1 Z(\tau) + \frac{f_2 p_1 + f_1 p_2}{3} \right] [Y(\sigma, \tau)]^3 \\
& + \left[ 3 f_1 p_1 [Z(\tau)]^2 - \frac{3}{2}(f_2 p_1 + f_1 p_2) Z(\tau) + \frac{1}{2}\left( f_2 p_2 + f_1 p_3 - \frac{\tau}{\sigma_*} p_1 \right) \right] [Y(\sigma, \tau)]^2 \\
& + \left[ -4 f_1 p_1 [Z(\tau)]^3 + 3 (f_2 p_1 + f_1 p_2) [Z(\tau)]^2 - 2 \left( f_2 p_2 + f_1 p_3 - \frac{\tau}{\sigma_*} p_1 \right) Z(\tau) \right] Y(\sigma, \tau) \\
& + \Bigg[ f_1 p_1 [Z(\tau)]^4 - (f_2 p_1 + f_1 p_2) [Z(\tau)]^3 + \left( f_2 p_2 + f_1 p_3 - \frac{\tau}{\sigma} p_1 \right) [Z(\tau)]^2 \\
& - \left( f_2 p_3 - \frac{\tau}{\sigma_*} p_2 \right) Z(\tau) - \frac{\tau}{\sigma_*} p_3 \Bigg] \ln Y(\sigma, \tau) + \left( f_3 p_3 - \frac{\tau}{\sigma_*} p_2 \right) \frac{\sigma}{\sigma_*} \Bigg\} (G(\tau) + 1) + g(\tau)
\end{aligned}$$

with

$$G(\tau) := u_1 \frac{\tau}{\sigma_*} + u_2 \left(\frac{\tau}{\sigma_*}\right)^2 + u_3 \left(\frac{\tau}{\sigma_*}\right)^3 + u_4 \left(\frac{\tau}{\sigma_*}\right)^8$$

$$Z(\tau) := -r \frac{\tau}{\sigma_*} - s \left(\frac{\tau}{\sigma_*}\right)^6 + \tau_o = Y(\sigma, \tau) - \frac{\sigma}{\sigma_*}$$

with $u_1 = 0.036$, $u_2 = -0.00265$, $u_3 = 5.256\times 10^{-5}$, and $u_4 = 1.57\times 10^{-12}$.

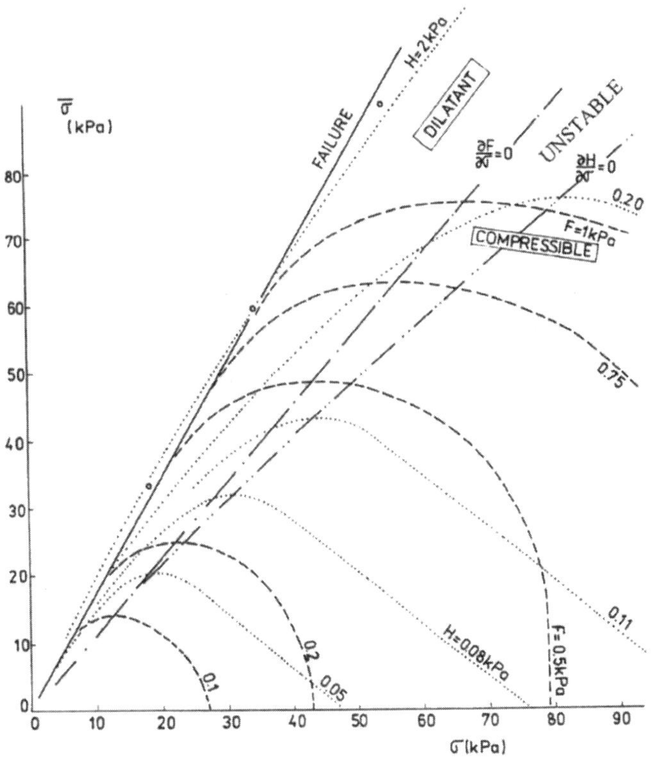


FIGURE 11.3.3 Yield surfaces (dotted lines) and viscoplastic potential surfaces (interrupted lines) for saturated sand. In the domain $\partial F/\partial\sigma > 0$, $\partial H/\partial\sigma < 0$ the sand is instable.

The potential $S$ for rock salt is given by Eqs. 12 and 13 with $b = -1\times10^{-14}\,s^{-1}$, $m = 5$, $n = -0.1$, $p = 3\times10^{-13}$ MPa $s^{-1}$.

For other geomaterials, such as dry or saturated sand, granite, andesite, coal or cement concrete, see Cristescu [1] and Cristescu and Hunsche [4], and for ceramic powder see O. Cazacu *et al.* [2], and Cristescu *et al.* [3].

As another example, in Figure 11.3.3 is shown the yield surface (dotted lines) for saturated sand, with

$$H(\sigma,\,\bar{\sigma}) : a\frac{\bar{\sigma}^7}{(\sigma-\bar{\sigma}/3)^5} + b\bar{\sigma} + c\sigma$$

and $a = 4.8\times10^{-7}(kPa)^{-1}$, $b = 0.0013$. The viscoplastic potential surfaces are shown as dotted lines with

$$\frac{\partial F}{\partial\sigma} = h_1\frac{(-\bar{\sigma}+2f\sigma)\sqrt{\sigma}}{(2f+\alpha)\sigma-(1+\alpha/3)\bar{\sigma}}$$

where the expression at the denominator is just that involved in the short-term failure and $-\bar{\sigma} + 2f\sigma = 0$ is the equation of compressibility/dilatancy boundary ($\alpha = 1.34$, $f = 0.56$). Sand is dilatant when $\partial F/\partial\sigma < 0$, compressible when $\partial F/\partial\sigma > 0$, and unstable if $\partial F/\partial\sigma > 0$, $\partial H/\partial\sigma < 0$.

## REFERENCES

1. Cristescu, N. (1989). *Rock Rheology*, Kluwer Academic.
2. Cazacu, O., Jin, J., and Cristescu, N.D. (1997). A new constitutive model for alumina powder compaction, *KONA. Powder and Particle*, **15**: 103–112.
3. Cristescu, N.D., Cazacu, O., and Jin, J. (1997). Constitutive equation for compaction of ceramic powders, in *IUTAM Symposium on Mechanics of Granular and Porous Materials*, pp. 117–128, Fleck, Norman, ed., Kluwer Academic.
4. Cristescu, N.D., and Hunsche, U. (1998). *Time Effects in Rock Mechanics*, Chichester-New York-Weinheim-Brisbane-Singapore-Toronto: John Wiley and Sons.

SECTION **11.4**

# Behavior of Granular Materials

IOANNIS VARDOULAKIS
*National Technical University of Athens, Greece*

Contents



## 11.4.1 MICRO- AND MACROMECHANICAL CONSIDERATIONS

Granular materials are random assemblies of small grains in very large numbers. Here situations are considered where the grains are always in contact with their neighbors. In a first approximation the grains are assumed to be spherical and granular materials are classified on the basis of their *grain size distribution curve* and its statistical moments, like the mean grain diameter. The *geometric fabric* of a granular medium at a given configuration in space is the three-dimensional truss with its nodes located at grain centers [1] (Fig. 11.4.1).

The property which differentiates granular materials from other solids, like, e.g., metals, is their pronounced *pressure sensitivity*, which is attributed to the existence of *internal friction* and which is developing at grain contacts. The corresponding *static fabric* is a three-dimensional truss with its bars along intergranular grain contact forces. Intergranular forces must always lie inside

*Handbook of Materials Behavior Models.* ISBN 0-12-443341-3.

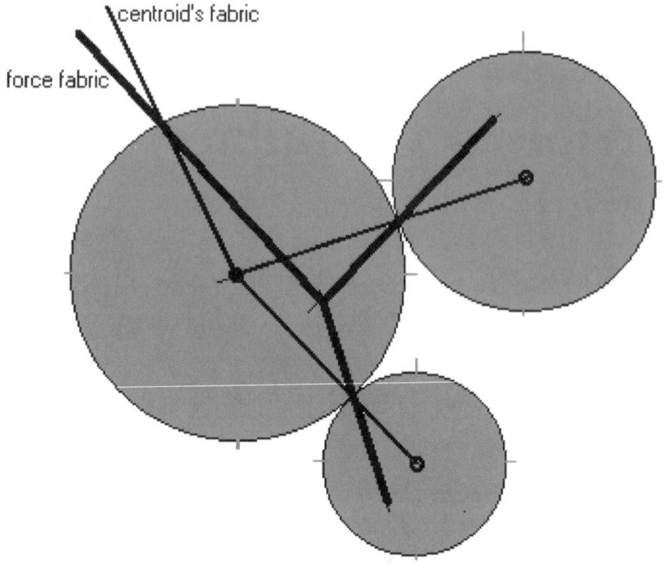


FIGURE 11.4.1 Force and centroid's fabric.

or be tangential to the local friction cone of the corresponding grain contact. Thus, because of friction at grain contacts, the centroid's and force's lattices do not coincide. Accordingly, granular kinematics and granular statics are not linked to each other in a fashion that is amenable to simple analysis. This inherent difficulty of the micromechanical description of granular media with internal friction explains why until now phenomenological as opposed to micromechanical models seem to be more useful for engineering applications. However, in the last 20 years considerable progress has been made in computer-aided micromechanical simulations of granular assemblies [2–4].

Granular materials are porous media. From a continuum mechanics point of view, their porous or granular structure is described by macroscopic properties of the granular assembly such as the porosity $n$ and bulk density $\rho = (1 - n)\rho_s$, where $\rho_s$ is the density of the grain. Grain displacements are averaged over a representative elementary volume (REV), and the macroscopic velocity $v_i$ and the rate of deformation $D_{ij}$ fields can be computed [5];

$$D_{ij} = \frac{1}{2}\left(\frac{\partial v_i}{\partial x_j} + \frac{\partial v_j}{\partial x_i}\right) \tag{1}$$

Since grains are practically incompressible ($\rho_s = \text{const.}$) and mass convection is in most cases negligible, mass balance yields that volume changes are simply due to changes in porosity

$$D_{kk} = -\frac{\partial}{\partial t}\ln(1-n) \tag{2}$$

The intergranular forces, $t_i$, acting at grain contacts are linked to a local Cauchy stress tensor $\tilde{\sigma}_{ij}$, which is assumed to fulfill equilibrium with negligible body forces. Through this local stress tensor a macroscopic $\sigma_{ij}$ stress can be computed as a volume average assigned to the center of a considered REV:

$$\sigma_{ij} = \frac{1}{V_{REV}}\int_{(V_{REV})} \tilde{\sigma}_{ij}\, dV = \frac{1}{V_{REV}}\int_{(\partial V_{REV})} x_i t_j\, dS \tag{3}$$

In the case of a granular medium which is fully saturated with a pore fluid of low viscosity (e.g., water), an '*effective*' stress is defined as a measure of the intergranular forces. We notice that in granular media the mixtures theory decomposition of the total stress in partial stresses (as is the case in gases) is not justified experimentally. Experimental evidence [6] and micromechanical considerations [7] support the idea of the total stress decomposition according to Terzaghi's [8] (empirical) effective stress principle into an effective stress and into a pore-water pressure

$$\sigma_{ij} = \sigma'_{ij} - p_w\delta_{ij} \quad (p_w > 0)^1 \tag{4}$$

The macroscopic (effective) stress tensor is decomposed in deviatoric and isotropic part,

$$\sigma_{ij} = s_{ij} + \frac{1}{3}\sigma_{kk}\delta_{ij} \tag{5}$$

For a spherical REV and statistically uniform contact-normal distribution, it can be shown [9] that the *mean normal traction* $p$ over all intergranular contacts is given by the first invariant of the stress tensor, $I_{1\sigma} = \sigma_{kk}$, and the *mean shear traction* $\tau_m$ or the second invariant of the deviator, $T = \sqrt{J_{2s}} = \sqrt{s_{ij}s_{ji}/2}$,

$$p = I_{1\sigma}/3; \;\; \tau_m = \sqrt{\frac{2}{5}}T = \sqrt{\frac{2}{5}J_{2s}} \tag{6}$$

A simple micromechanical interpretation of the third stress invariant is not known [9]. However, the most lucid selection of a third stress invariant follows if one considers the ratio of maximum to mean shear stress. Let, for example, $\sigma_2 \le \sigma_3 \le \sigma_1$ be the principal stresses for a given state of stress and $\tau_{3,\max} = |\sigma_1 - \sigma_2|/2$; then one can show that this ratio is a simple function of

[1] Compression negative.

the so-called *stress invariant angle of similarity* $\alpha_s$:

$$\frac{\tau_{3,\max}}{\tau_m} = \sqrt{\frac{5}{2}}|\sin(\pi/3 + \alpha_s)| \tag{7}$$

where

$$\cos 3\alpha_s = \frac{3\sqrt{3}}{2}\frac{J_{3s}}{J_{2s}^{3/2}} \quad (0 \leq \alpha_s \leq \pi/3) \tag{8}$$

## 11.4.2 ELASTOPLASTIC BEHAVIOR OF GRANULAR MATERIALS

Typical examples of *granular geomaterials* are sands, silts, and psammitic rocks (sandstones). For slow deformation processes, granular materials exhibit predominantly nonlinear stress-strain behavior upon loading as well as upon unloading and reloading. Deformations are mostly irreversible and practically strain-rate-insensitive. Granular geomaterials are good examples of deformation-rate-insensitive *plastic* solids.

The behavior of granular materials is strongly history-dependent. Thus finite constitutive laws are excluded, and constitutive equations can only be written in the form of evolution-type equations that relate an objective (effective) stress rate $\dot{\sigma}_{ij}$ to the rate of deformation tensor $D_{ij}$. The response function depends strongly on the direction of the strain rate, resulting generally in nonlinear rate-type equations. The assumption of rate independence restricts $\dot{\sigma}_{ij}$ to being a first-degree homogeneous function of $D_{ij}$. As an example, we refer here to Chambon *et al.* [10], who suggested recently the following simple *incrementally nonlinear* model:

$$\dot{\sigma}_{ij} = T_{ij}(\sigma_{ij}, D_{ij}, \ldots) = A_{ij}||D_{kl}|| + B_{ijkl}D_{kl} + \cdots \tag{9}$$

In the frame of classical *elastoplasticity theory*, we distinguish roughly among two such "directions," namely, between *loading and unloading*. Accordingly, the rate of deformation tensor is decomposed into an elastic (reversible) and a plastic (irreversible) part,

$$D_{ij} = D_{ij}^e + D_{ij}^p \tag{10}$$

Upon virgin loading the response is relatively soft, since both elastic and plastic strains are generated. Elastic strains are relatively small, and they reflect reversible, nonlinear deformations of grain contacts. The elastic stiffness that relates the stress rate to the elastic deformation rate,

$$\dot{\sigma}_{ij} = C_{ijkl}^e D_{kl}^e \tag{11}$$

is stress-dependent, and accordingly it is derived from a suitably constructed elastic strain energy function [11, 12].

In most cases plastic deformations dominate, since they reflect irreversible grain rearrangements, which in turn include grain contact detachment and intergranular slip. The state of stress of a given granular material point (REV) during plastic yielding is restricted by the so-called *plastic yield condition*, which has the form

$$F(\sigma_{ij}, \Psi, \ldots) = 0 \tag{12}$$

The parameter $\Psi$ in the argument list of $F$, and possibly other scalar- and tensor-valued variables, is used to describe, to some degree of approximation, the history of plastic deformation, as this is reflected in the evolving force fabric. The yield condition of granular materials is expressing the constraint that is imposed by internal friction. Macroscopically, this results in *pressure-sensitive behavior*, which is expressed by linking the shear stress intensity $T$ to the mean (effective) normal stress though a constraint of the form

$$F(\sigma_{ij}, \Psi) = T - f_C(q - p) = 0 \tag{13}$$

where $f_C(\Psi, \ldots)$ is the macroscopic or *Coulomb friction coefficient*.

The yield condition is visualized in stress space by the corresponding *yield surface* $F|_{\Psi=\text{const.}} = 0$ (Fig. 11.4.2). With the concept of yield surface one can generalize the one-dimensional concepts of loading and unloading to more complicated stress paths. This is done by assuming a) an elastic domain "inside" the yield surface and b) that plastic strains are generated only if the stress increment points in the direction of growth of the yield surface.

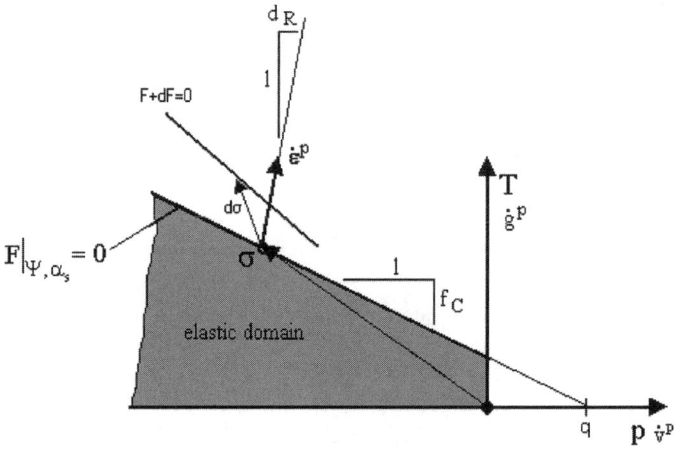


FIGURE 11.4.2 Coulomb-type yield surface and dilatancy flow rule in $(T, p)$ stress subspace.

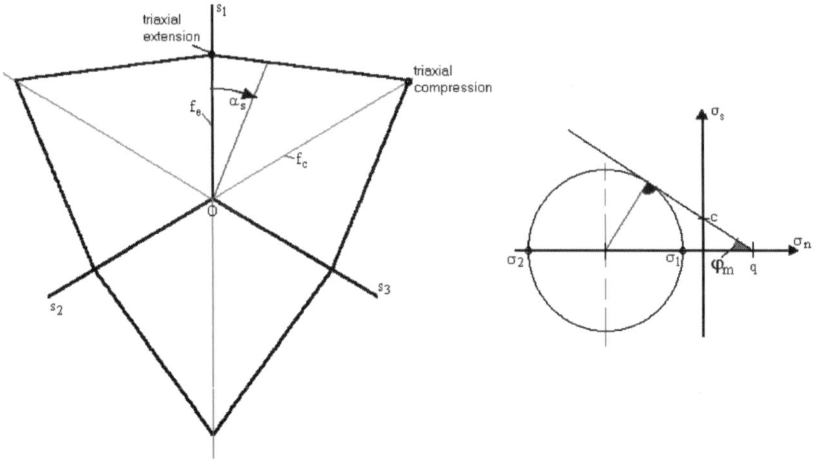


(a) (b)

FIGURE 11.4.3 a. Deviatoric trace of the Mohr-Coulomb yield surface in principal stress space. b. Mohr circle of stresses with linear yield envelope and the definition of $\varphi_m$ and c.

In granular materials the response under continued loading is practically independent of the intermediate principal stress, as is shown in Figure 11.4.3. Mathematically this is expressed by the so-called *Mohr-Coulomb* yield condition [9, 13], with

$$f_C = \frac{\sin \varphi_m}{\sin(\alpha_s + \pi/3) + \frac{1}{\sqrt{3}} \cos(\alpha_s + \pi/3) \sin\varphi_m} \tag{14}$$

In Eq. 14 $\varphi_m$ is called the *mobilised Mohr-Coulomb angle of internal friction*, and reflects an average value of the internal friction, which at any instant of the deformation process is mobilised at grain contacts inside the REV. The parameter $q$ is related to the cohesion $c$ of the geomaterial, $c = q \tan \phi_m$. Both mobilised friction and cohesion are assumed to be functions of the plastic history parameter, $\sin \phi_m = \mu(\Psi)$, $c = c(\Psi)$.

Besides internal friction, granular media are distinguished from other plastic solids by the phenomenon of *dilatancy*. Dilatancy means plastic volume increase under shear and is observed in relatively dense packings. Loose packings undergo instead plastic volume decrease during shear (*contractancy*). Dilatancy or contractancy are measured by the plastic volumetric strain-rate $\dot{v}^p = D^p_{kk}$ and are attributed to grain rearrangements due to grain slip, which in turn is measured by the second invariant of the deviator of the rate of deformation tensor, $\dot{g}^p = \sqrt{2D'^p_{ij}D'^p_{ji}}$ ($D^p_{ij} = \dot{v}^p\delta_{ij}/3 + D'^p_{ij}$). Since elastic deformations are subordinate to plastic ones, Eq. 2 tells us that dilatancy or contractancy will be directly evident in the experiment through visible (detectable) changes in porosity [14]. Dilatancy is an *internal constraint* for the plastic deformation of the form

$$\dot{v}^p = d_R \dot{g}^p \Rightarrow \frac{dv^p}{dg^p} = d_R(\Psi) \tag{15}$$

The parameter $d_R$ in this relation is called the *Reynolds* [15] dilatancy coefficient. The dilatancy constraint is depicted in Figure 11.4.2 as the corresponding vector of the plastic deformation rate, attached to the current stress state. By selecting an appropriate *plastic potential function* $\mathcal{Q} = \mathcal{Q}(\sigma_{ij}, \Psi)$, the dilatancy constraint, Eq. 15, follows from the so-called *plastic flow rule*

$$D^p_{ij} = \dot{\Psi}\frac{\partial \mathcal{Q}}{\partial \sigma_{ij}} \tag{16}$$

Introducing the *mobilized dilatancy angle* $\psi_m$ (Fig. 11.4.4), and in analogy to Eqs. 13 and 14, we have

$$\mathcal{Q}(\sigma_{ij}, \Psi) = T + d_R p + \text{const.} \tag{17}$$

$$d_R = \frac{\sin \psi_m}{\sin(\alpha_s + \pi/3) + \dfrac{1}{\sqrt{3}}\cos(\alpha_s + \pi/3)\sin \psi_m} \tag{18}$$

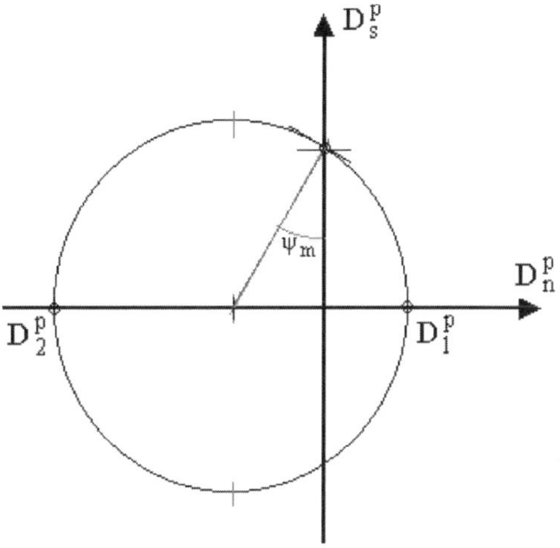


FIGURE 11.4.4 Mohr-circle of plastic strain rates and the definition of $\psi_m$.

The phenomenological parameters, $\varphi_m$ and $\psi_m$, in the definitions of the yield function $F$ and plastic potential $Q$, are assumed to describe the essential future of the structure of the force- and centroid's fabrics, respectively. As already mentioned in Section 11.4.1, these two fabrics do not coincide, and the corresponding plasticity theory must require distinct functions that specify the yield condition and the flow rule ($F \neq Q$).

The way to relate plastic yielding to plastic flow in granular materials is to consider the fact that plastic work,

$$\dot{w}^p = \sigma_{ij} D_{ij}^p \tag{19}$$

should after all express the work done by intergranular forces during grain slip. Let $f_c$ be the "*true*" *interparticle friction* coefficient; then the previous requirement is fulfilled if

$$d_R = f_c + f_C \tag{20}$$

This is *Taylor's friction-dilatancy rule* [16–18] and is interpreted as follows: From the point of view of energy dissipation, granular materials behave like incompressible frictional materials, with constant friction coefficient $f_c$ (Fig. 11.4.5).

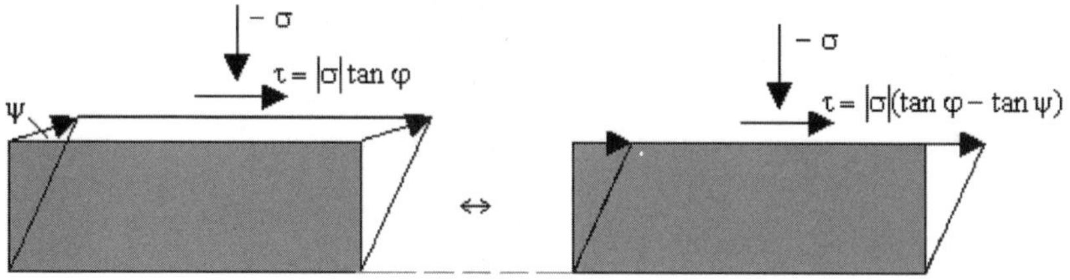


**FIGURE 11.4.5** Taylor's friction-dilatancy rule in simple shear conditions [16].

Equation 20 is the simplest generalization of the so-called *normality condition*[2] of the classical (associated) flow theory of plasticity. Indeed, for ideally smooth particles $f_c = 0$, $d_R = f_C$ form Eq. 20, and yield and plastic potential functions coincide. This is consistent with the micromechanics of granular media, since the condition $f_c = 0$ also means that the granular medium is nondissipative and that the centroid's- and force fabrics coincide (cf. Fig. 11.4.1).

The flow rule, (Eq. 16) and the corresponding flow theory of plasticity are called *nonassociated*, because no further restriction is imposed on the

[2]The normality condition of the flow theory of plasticity is stating that (a) stress and plastic strain rate are coaxial, and (b) in stress space representations the plastic strain rate vector is normal to the yield surface at the considered state of stress.

dilatancy coefficient, except that energy dissipation is never negative:

$$\dot{w}^p \geq 0 \Rightarrow d_R \leq f_C \tag{21}$$

Taylor's rule (Eq. 20) satisfies the previous constitutive inequality and provides a micromechanically motivated constitutive law for the dilatancy coefficient. Dense sands are dilatant, since $f_C > f_c$ and $d_R > 0$; loose sands are contractant, since $f_C < f_c$ and $d_R < 0$. For $f_C = f_c$, the granular material is said to be at a *critical state* and it deforms isochorically, $d_R = 0$.

With distinct yield and plastic potential functions, one arrives at the following constitutive equations of the so-called *Mroz* [19,20] nonassociate, elastoplasticity theory for frictional-dilatant solids,

$$\dot{\sigma}_{ij} = C^{ep}_{ijkl} D_{kl} = \left( C^e_{ijkl} - \left\langle C^p_{ijkl} \right\rangle \right) D_{kl} \tag{22}$$

with:

- *Plastic stiffness tensor:*

$$C^p_{ijkl} = \frac{\langle 1 \rangle}{H} \frac{\partial Q}{\partial \sigma_{mn}} C^e_{mnij} \frac{\partial F}{\partial \sigma_{pq}} C^e_{pqkl}$$

- *Plastic modulus:*

$$H = H_0 + H_t > 0$$

- *Hardening (softening) modulus:*

$$H_t = -\frac{\partial F}{\partial \Psi} \begin{cases} > 0 \ : \ \text{hardening} \\ < 0 \ : \ \text{softening} \end{cases}$$

- *Snap-back threshold value for the softening modulus:*

$$H_0 = \frac{\partial F}{\partial \sigma_{kl}} C^e_{klmn} \frac{\partial Q}{\partial \sigma_{mn}} > 0$$

- $\langle \bullet \rangle$ *Foeppl-Macauley brackets:*

$$\begin{cases} 1 \text{ if } : F = 0 \text{ and } B_{kl} \dot{\varepsilon}_{kl} > 0 \\ 0 \text{ if } : F < 0 \text{ or } F = 0 \text{ and } B_{kl} \dot{\varepsilon}_{kl} \leq 0 \end{cases}; B_{ij} = \frac{\partial F}{\partial \sigma_{kl}} C^e_{klij}$$

$C_{ijkl}^{ep}$ is a quasi-linear operator and in the case of associative plasticity is satisfying major symmetry conditions. We notice that for granular materials nonassociativity of the flow rule usually holds only for the volumetric component of the plastic strain rate. At the same time, the *deviatoric normality* is assumed to hold [21]; i.e.,

$$\frac{\partial F}{\partial \sigma_{ij}} - \frac{\partial Q}{\partial \sigma_{ij}} = \lambda \delta_{ij}$$

## 11.4.3 SHEAR-BANDING AND POSTFAILURE BEHAVIOR

Otto Mohr [22] published in the year 1900 the original strength theory of cohesive-frictional or *Mohr-Coulomb materials*. These materials fail under shear by forming a set of conjugate slip lines. Granular materials are good examples of Mohr-Coulomb materials, since failure is manifested in these materials in the form of conjugate shear-bands (Fig. 11.4.6).

A shear-band is a narrow zone of intense shear with a thickness that is a small multiple of the mean grain diameter [23]. In the past 25 years extensive work on shear-banding in granular media has been initiated by the works of

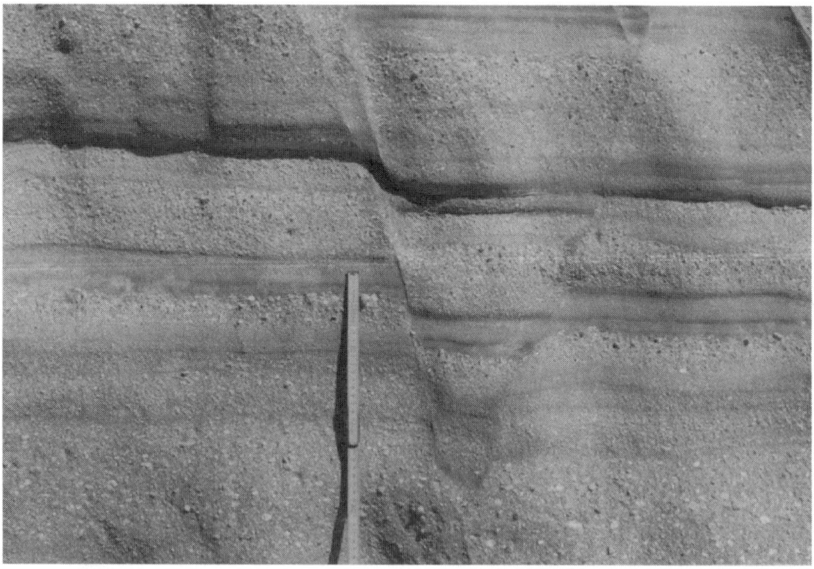

FIGURE 11.4.6 Conjugate shear-bands in perlite (volcanic soft rock in Melos island, Greece).

Vardoulakis [24] and Desrues [25]; see Reference [9] for an extensive literature review. Shear localization induces intense intergranular slip and dilatancy of the material inside the localized zone [14], which is due to grain rearrangement, grain slip, and rotation [26]. Increasing porosity naturally reduces the coordination number of the granular assembly (i.e., the number of contacts per grain), yielding progressively to macroscopic *material softening* inside the localized zone. For equilibrium reasons the material outside the localized zone is unloading (Fig. 11.4.7).

From the micromachanical point of view, an important structure that appears to dominate localized deformation is the formation and collapse (buckling) of grain columns. We notice that in order to account for these effects, higher moments concerning the grain geometry must be accounted for, such as their ellipticity, angularity, etc., which in turn lead to a basic asymmetry of shear stress and to micropolar effects [27, 28]. At any rate, localization of deformation leads to a change of scale of the problem so that phenomena occurring at the scale of the grain cannot be ignored anymore in the modeling process of the macroscopic behavior of the material. Under these circumstances, it appeared necessary to resort to continuum models with a *microstructure*, which allow us to some degree to describe localization phenomena. These generalized continua contain additional kinematical degrees of freedom and/or consider higher deformation gradients. These observations have prompted the extension of classical continuum mechanical descriptions for granular media past the softening regime by resorting to the so-called *Cosserat* [9, 29] or *gradient models* [9, 30, 31].

In a recent paper, Zervos *et al.* [32] presented a new unified gradient elastoplasticity theory for cohesive/frictional, dilatant materials, where

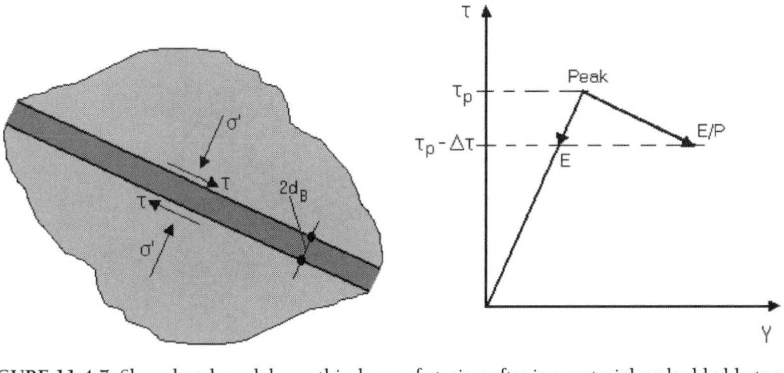


FIGURE 11.4.7 Shear-band model as a thin layer of strain-softening material embedded between elastically unloading half-spaces.

gradient terms were considered in the elastic regime as well, and the stress rate equation reads as follows:

$$\dot{\sigma}_{ij} = \left(C^e_{ijkl} - \left\langle C^p_{ijkl}\right\rangle\right)D_{kl} - \left(\ell_e^2 C^e_{ijkl} + \ell_c^2\left\langle C^p_{ijkl}\right\rangle\right)\nabla^2 D_{kl} \tag{23}$$

As a result, the order of the governing equations remains the same everywhere in the deforming solid throughout the loading history. The consistency condition of the flow theory of plasticity, which in this case is a differential equation, is solved analytically in an approximate fashion [9]. Therefore, only displacements need to be discretized in a finite element formulation, where a $C^1$ three-noded triangle with quintic interpolation for the displacement field was implemented. The ability of such a theory to model progressive

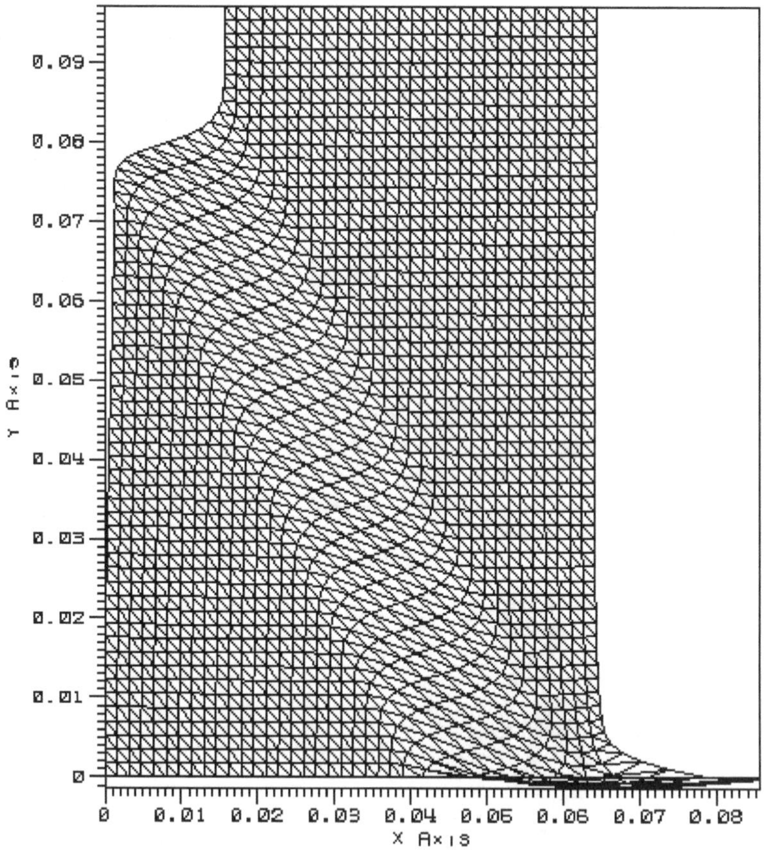


FIGURE 11.4.8 Finite-element postlocalization analysis of the biaxial test, using second-gradient elastoplasticity theory [31].

localization of deformation is demonstrated by simulating the biaxial test on a weak sandstone. After the critical load level is reached, a shear-band is formed (Fig. 11.4.8). The material inside the band is in the softening regime and continues to deform plastically. The material outside unloads elastically. The inclination and thickness of the shear-band and the load displacement response of the sample are not affected by mesh refinement.

## ACKNOWLEDGEMENTS

This article is a partial result of research supported by funds of GSRT of Greece through the program PENED 99 ED 642.

## REFERENCES


1. Satake, M. (1982). Fabric tensor in granular materials, in *IUTAM Conf. on Deformation and Failure of Granular Materials*, pp. 63–68, Delft, Balkema.
2. Cundall, P. A., and Strack, O. D. L. (1979). A discrete numerical model for granular assemblies. *Géotechnique* **29**: 47–65.
3. Jean, M. (1995). Frictional contact in collections of rigid or deformable bodies: A numerical simulation of geomaterial motion, in *Mechanics of Geomaterial Interfaces*, Selvadurai, A. P. S., ed., Elsevier.
4. Emeriault, F., and Cambou, B. (1996). Micromechanical modelling of anisotropic non-linear elasticity of granular medium. *Int. J. Solids Structures* **33**: 2591–2607.
5. Bagi, K. (1996). Stress and strain in granular assemblies. *Mech. Materials* **22**: 165–177.
6. Bishop, A. W., and Skinner, A. E. (1977). The influence of high pore-pressure on the strength of cohesionless soils. *Phil. Trans. Roy. Soc. London* **284**: 91–130.
7. Oka, F. (1996). Validity and limits of the effective stress concept in geomechanics. *Mech. Coh. Frict. Mat.* **1**: 219–234.
8. Terzaghi, K. V. (1936). The shearing resistance of saturated soils, in *Proc. 1st ICSMFE Cambridge*, pp. 54-56. vol. 1.
9. Vardoulakis, I., and Sulem, J. (1995). *Bifurcation Analysis in Geomechanics*, Blackie Academic and Professional.
10. Chambon, R., Desrues, J., Hammad, W., and Charlier, R. (1994). A new rate-type constitutive model for geomaterials. *Int. J. Num. Anal. Meth. Geomechanics* **18**: 253–278.
11. Loret, B. (1985). On the choice of elastic parameters for sand. *Int. J. Num. Anal. Meth. Geomechanics* **9**: 285–287.
12. Lade, P. V., and Nelson, B. (1987). Modelling the elastic behavior of granular materials. *Int. J. Num. Anal. Meth. Geomechanics* **11**: 521–554.
13. Chen, W. F., and Han, D. J. (1988). *Plasticity for Structural Engineers*, Springer.
14. Desrues, J., Chambon, R., Mokni, M., and Mazerolle, F. (1996). Void ratio evolution inside shear bands in triaxial sand specimens studied by computed tomography. *Géotechnique* **46**: 529–546.
15. Reynolds, O. (1885). On the dilatancy of media composed of rigid particles in contact. With experimental illustrations. *Phil. Mag.* **2(20)**: 469–481. Also: Truesdell, C., and Noll, W.

(1948). *The Non-Linear Field Theories of Mechanics*, Handbuch der Physik Band III/3, section 119, Springer.

16. Taylor, D. W. (1948). *Fundamentals of Soil Mechanics,* John Wiley.
17. Rowe, P. W. (1962). The stress-dilatancy relation for static equilibrium of an assembly of particles in contact. *Proc. Roy. Soc.* **269**: 500–527.
18. De Josselin de Jong (1976). Rowe's stress-dilatancy relation based on friction. *Géotechnique* **26**: 527–534.
19. Mroz, Z. (1963). Non-associate flow laws in plasticity. *Journal de Mécanique* **2**: 21–42.
20. Nguyen, Q. S., and Bui, H. D. (1974). Sur les matériaux élastoplastiques à écrouissage positif ou négatif. *Journal de Méchanique* **3**: 322–432.
21. Gudehus, G. (1972). Elastic-plastic constitutive equations for dry sand. *Arch. Mech. Stosowanej* **24**: 395–402.
22. Mohr, O. (1900). Welche Umstände bedingen die Elastizitätsgrenze und den Bruch eines Materials? *Zeitschrift des Vereines deutscher Ingenieure* **44**: 1–12.
23. Roscoe, K. H. (1970). The influence of strains in soil mechanics. *Géotechnique* **20**: 129–170.
24. Vardoulakis, I. (1977). *Scherfugenbildung in Sandkoerpern als Verzweigungsproblem*, Doktorarbeit, Universitaet Karlsruhe.
25. Desrues, J. (1984). *La Localization de la Déformation dans les Matériaux Granulaires*. Thése de Doctorat et Science, USMG & INPG, Grenoble.
26. Zervos, A., Vardoulakis, I., Jean, M., and Lerat, P. (2000). Numerical investigation of granular kinematics, in *Mechanics of Cohesive-Frictional Materials* (in press).
27. Oda M. (1997). A micro-deformation model for dilatancy of granular materials, in *ASCE/ ASME Symposium on Mechanics of Deformation and Flow of Particulate Materials*, pp. 24–37, Chang, C.S., Misra, A., Liang, Ry., and Babic, M. eds.
28. Bardet, J.-P., and Vardoulakis, I. (2000). The asymmetry of stress in granular media. *Int. J. Solids Struct.* **38**: 353–367.
29. Mühlhaus, H.-B., and Vardoulakis, I. (1987). The thickness of shear bands in granular materials. *Géotechnique* **37**: 271–283.
30. Vardoulakis, I., and Aifantis, E. C. (1991). A gradient flow theory of plasticity for granular materials. *Acta Mechanica* **87**: 197–217.
31. Vardoulakis, I., and Frantziskonis, G. (1992). Micro-structure in kinematic-hardening plasticity. *Eur. J. Mech./Solids* **11**: 467–486.
32. Zervos, A., Papanastasiou, P., and Vardoulakis, I. (2001). A finite element displacement formulation for gradient elastoplasticity, *Int. J. Numer. Meth. Engng.* **50 (6)**: 1369–1388.

SECTION **11.5**

# Micromechanically Based Constitutive Model for Frictional Granular Materials

SIA NEMAT-NASSER

*Center of Excellence for Advanced Materials, Department of Mechanical and Aerospace Engineering, University of California, San Diego, California*

Contents



## 11.5.1 INTRODUCTION

The response of frictional granules (i.e., granular masses which support the applied loads through contact friction) is strongly dominated by anisotropy or fabric which is invariably induced upon shearing under confinement. Concomitant features of this characteristic of frictional granules are their dilatancy in monotonic shearing and their densification upon subsequent unloading. A further remarkable property of

*Handbook of Materials Behavior Models*. ISBN 0-12-443341-3.

materials of this kind is that *they can actually undergo reverse inelastic shearing* **against** *an applied shear stress when a monotonic shearing under confining pressure is followed by unloading* [1]. *The energy required for plastic deformation against the applied shear stress is supplied by the work done by the confining pressure going through the accompanying shear-induced volumetric contraction.* Experiments show that, under relatively small confining pressures, the following features are among the essential characteristics which must be captured in modeling the inelastic deformation of granular masses:

- An initial densification (decrease in void volume), the magnitude of which decreases as the void ratio (the ratio of void volume to the volume of the solid) approaches a minimum value.
- If the sample is dense (i.e., its void ratio is close to the corresponding minimum value), then the initial (small) densification is followed by dilatancy (increase in void volume), which continues until a critical void ratio is attained asymptotically.
- If at a certain stage during the course of dilatancy, the shearing is reversed and the shear strain is gradually decreased to its initial zero value (completing half of a strain cycle), then there is always a net amount of densification, this amount decreasing as the void ratio approaches its minimum value.
- If the sample is loose (i.e., its void ratio is larger than a critical value), then the sample may densify continuously until the critical void ratio is reached asymptotically.

Based on micromechanics, a plasticity model which includes both isotropic and kinematic hardening has been developed [2] to describe the response of frictional granules under small confining pressures. The model includes pressure sensitivity and dilatancy, as well as the effect of *induced anisotropy* and the noncoaxiality of the inelastic strain rate and the stress tensors, and does capture the basic features listed previously. The basic hypothesis is that the deformation of frictional granular masses occurs through simple shearing accompanied by dilatation or densification (mesoscale), depending on the microstructure (microscale) and the loading conditions (continuum scale). The microstructure and its evolution are defined in terms of the fabric and its evolution, and this fabric is characterized in terms of a back stress (kinematic hardening). In this theory, the fabric affects both the inelastic and the elastic deformation, in a most profound way. An expression for the overall elasticity tensor is developed in terms of a fabric tensor.

## 11.5.2 BACK STRESS AND FABRIC

The term *fabric* is used to define the overall anisotropic distribution of the granules, their contact forces, the associated voids, and other microstructural parameters which are responsible for the anisotropic behavior of the granular mass. Assuming a uniform sample, here we measure the granular fabric by the *distribution of the contact unit normals*. Denote by $\boldsymbol{n}$ a typical contact unit normal, and let $E(\boldsymbol{n})$ be its *distribution-density function*. Hence, $E(\boldsymbol{n})\, d\Omega$ represents the fractional number of contact unit normals whose directions fall within the *solid angle* $d\Omega$. Various aspects of this representation of the fabric of a granular mass are discussed by Mehrabadi *et al.* [3], Kanatani [4], and Subhash *et al.* [5]. Here, we use the second-order approximation of $E(\boldsymbol{n})$, given by

$$E(\boldsymbol{n}) = \frac{1}{2\pi(r-1)}(\mathbf{1} + \boldsymbol{E}) : (\boldsymbol{n} \otimes \boldsymbol{n}) = \frac{1}{2\pi(r-1)}(1 + E_{ij} n_i n_j) \qquad \text{(1a)}$$

where $r = 2$ in two dimensions and $r = 3$ in three dimensions; $\mathbf{1}$ with components $\delta_{ij}$ is the identity tensor; $\boldsymbol{E}$ is a second-order symmetric and *deviatoric* tensor with components $E_{ij} = 4(< n_i n_j > -\frac{1}{2}\delta_{ij})$ in two dimensions, and $E_{ij} = 15/2(< n_i n_j > -1/3\delta_{ij})$ in three dimensions, where $< \ldots >$ denotes volume average; and the repeated indices are summed. The back stress is defined by

$$\boldsymbol{\beta} = \sqrt{2} p \boldsymbol{J},\ \boldsymbol{J} = -\hat{\mu} \boldsymbol{E} \qquad \text{(1b, c)}$$

where $p$ is the pressure, and $\hat{\mu}$ is an effective friction coefficient which is used as a normalizing factor. The minus sign in Eq. 1c ensures that the major principal direction of $\boldsymbol{\beta}$ coincides with the direction of the least density of the contact unit normals. The fabric tensor $\boldsymbol{J}$ is used as an internal variable to represent the effects of the microstructure. Tensors $\boldsymbol{E}$, $\boldsymbol{\beta}$, and $\boldsymbol{J}$ are all deviatoric.

The internal resistance to deformation in frictional granules can now be represented by a stress tensor $\boldsymbol{\tau}$ which consists of a hydrostatic pressure, $-p\mathbf{1}$, the (deviatoric) back stress $\boldsymbol{\beta}$ corresponding to the fabric tensor, and *a deviatoric tensor* $\boldsymbol{S}$ *representing the internal isotropic Coulomb friction*,

$$\boldsymbol{\tau} = -p\mathbf{1} + \boldsymbol{\beta} + \boldsymbol{S} \qquad \text{(2)}$$

The deviatoric tensor $\boldsymbol{S}$ is called the *stress difference*.

## 11.5.3 YIELD CRITERION

The yield condition is defined in terms of the isotropic Coulomb friction, using the stress difference tensor $S$, as follows:

$$f \equiv \tau - F(p, \Delta, \gamma) \leq 0, \; \tau = (\tfrac{1}{2}S : S)^{\frac{1}{2}} \tag{3a, b}$$

where $\Delta$ is the accumulated volumetric strain and $\gamma$ is the effective inelastic strain, defined in terms of the deviatoric part, $\mathbf{D}^{p}$, of the plastic deformation rate tensor $\mathbf{D}^{p}$, as follows:

$$\gamma = \int_0^t \dot{\gamma}\, dt, \; \dot{\gamma} = \sqrt{2}\boldsymbol{\mu} : \mathbf{D}^{p} \tag{3c, d}$$

where $\boldsymbol{\mu}$ is defined in the following text by Eq. 4c. As has been discussed in Nemat-Nasser and Shokooh [6], $\Delta$ can be related directly to the void ratio, $e$, and $M = \partial F/\partial p$ is the isotropic friction coefficient. Indeed, in two dimensions, $M = sin\,\phi_{\mu}$, where $\phi_{\mu}$ is the internal friction angle [2]. With this in mind, the yield condition is written as

$$f \equiv \tau - pM \leq 0 \tag{3e}$$

in what follows.

## 11.5.4 INELASTIC DEFORMATION RATE TENSOR

Generalizing the double sliding models proposed by Spencer [7, 8], de Josselin de Jong [9], Mehrabadi and Cowin [10], and Balendran and Nemat-Nasser [11, 12], it is shown by Nemat-Nasser [2] that the inelastic deformation rate tensor, $\mathbf{D}^{p}$, can be expressed as

$$\mathbf{D}^{p} = \dot{\gamma}\frac{\boldsymbol{\mu}}{\sqrt{2}} + \eta\alpha(\mathbf{1}^{(4s)} - \boldsymbol{\mu} \otimes \boldsymbol{\mu}) : \mathbf{D}' + \frac{1}{3}\dot{\gamma}B\mathbf{1} \tag{4a}$$

or, in component form, as

$$D^{p}_{ij} = \dot{\gamma}\frac{\mu_{ij}}{\sqrt{2}} + \eta\alpha(1^{(4s)}_{ijkl} - \mu_{ij}\mu_{kl})D'_{kl} + \frac{1}{3}\dot{\gamma}B\delta_{ij} \tag{4b}$$

where $\mathbf{1}^{(4s)}$ is the fourth-order symmetric identity tensor; $\mathbf{D}'$ is the deviatoric part of the deformation rate tensor $\mathbf{D}$; and

$$\boldsymbol{\mu} = \frac{S}{\sqrt{2}\tau} \tag{4c}$$

is a unit tensor coaxial with $S$. In Eq. 4a,b, $\eta = \pm 1$ with the sign chosen such that the associated rate of work is positive; $B$ is the dilatancy parameter which relates the *shear-induced* volumetric strain rate to the shear strain rate; and $\alpha \geq 0$ is the noncoaxiality coefficient. In the literature, the part of the plastic deformation rate tensor which is noncoaxial with the stress difference is expressed in terms of the Jauman rate $\overset{0}{\boldsymbol{\mu}}$, which is orthogonal to $\boldsymbol{\mu}$ i.e., $\overset{0}{\boldsymbol{\mu}} : \boldsymbol{\mu} = 0$; see References [13–15]. Since $[(\mathbf{1}^{(4s)} - \boldsymbol{\mu} \otimes \boldsymbol{\mu}) : \boldsymbol{D}'] : \boldsymbol{\mu} = 0$, a direct relation exists between the two representations; this is discussed in a forthcoming book by the author.

In general, the response of frictional granular masses is dominated by their fabric. Therefore, the noncoaxiality term in the plastic strain rate can be neglected in many cases, especially at small pressures; i.e., $\alpha = 0$ may be assumed.

## 11.5.5 DILATANCY, FRICTION, AND FABRIC

An expression for the dilatancy parameter $B$ is obtained by setting the rate of frictional loss equal to the rate of distortional plastic work. To arrive at an explicit result, assume that the rate of frictional loss (per unit volume) is given by

$$\dot{w}_f = \dot{\gamma} p M_f \tag{5a}$$

where $M_f$ is an effective friction factor, a function of the void ratio and the fabric. Then, neglecting the work associated with the noncoaxiality term in Eq. 4a, b, from $\boldsymbol{D}^p : \boldsymbol{\tau} = \dot{w}_f$, obtain

$$-B = M_f - (M + \mu_f), \; \mu_f = \boldsymbol{J} : \boldsymbol{\mu} \tag{5b, c}$$

In this equation, $\mu_f$ represents the effect of the fabric anisotropy on the dilatancy. Since $M_f > 0$ and $M = \tau/p$ is generally very small, the dilatancy parameter $B$ is initially negative (densification) for an initially isotropic virgin sample, until $\mu_f$ becomes suitably large, as $\boldsymbol{\mu}$ and $\boldsymbol{J}$ tend to become coaxial (in general, $0 < M \ll 1$, for small pressures; i.e., most geomaterials have a very small elastic range). When the direction of shearing is reversed, $\mu_f$ changes its sign and attains a negative value. Hence, in reversing the direction of shearing, the dilatancy parameter $B$ becomes negative, predicting that the sample would densify. This is in agreement with essentially all experimental observations. As the reversed shearing continues, the fabric anisotropy evolves and $\boldsymbol{J}$ changes with the applied stress. For the model to predict positive dilatancy in continued monotonic, say, shearing, the evolution of $\boldsymbol{J}$ must be such that $\boldsymbol{J}$ tends to become coaxial with $\boldsymbol{\mu}$, rendering $\boldsymbol{J} : \boldsymbol{\mu}$ positive again. The evolution equations for the fabric tensor $\boldsymbol{J}$, as well as for the overall effective friction coefficient, $M_f$, and the radius of the yield surface defined by $pM$, must thus reflect these basic physical requirements.

Since, from Eq. 4a, b, $D^p_{kk} = \dot{\gamma}B$, it follows that

$$\dot{\gamma}B = \frac{\dot{e}}{1+e} \tag{5d}$$

so that the void ratio $e$ is given by

$$e = (1+e_0)\exp\left(\int_0^{\gamma} B\, d\gamma\right) - 1 \tag{5e}$$

where $e_0$ is the initial void ratio. As is seen, the exponential dependency of the void ratio $e$ on the dilatancy parameter $B$ makes the result very sensitive to the variation of the parameter $B$ with the fabric parameter $\mu_f$ in Eq. 5b. Comparison with experiments has shown that Eq. 5b may have to be modified to reduce the effect of the fabric parameter on $B$, by replacing $\mu_f$ in Eq. 5b with $\zeta\mu_f$, where $0 < \zeta \leq 1$.

## 11.5.6 ELASTICITY RELATIONS

The elasticity of an assembly of frictional granules depends on the fabric and the applied pressure. Let $\boldsymbol{L}$ be the instantaneous elasticity tensor. Based on two-dimensional deformation of granular materials, it can be shown that the elasticity tensor is given by

$$\boldsymbol{L} = 2G(\mathbf{1}^{(4s)} - 1/3\mathbf{1} \otimes \mathbf{1}) + 2\bar{G}\boldsymbol{\mu}_\beta \otimes \boldsymbol{\mu}_\beta + \sqrt{2}\bar{K}(\boldsymbol{\mu}_\beta \otimes \mathbf{1} + \mathbf{1} \otimes \boldsymbol{\mu}_\beta) + K\mathbf{1} \otimes \mathbf{1} \tag{6a}$$

where $G$, $\bar{G}$, $K$, and, $\bar{K}$ are pressure-dependent moduli (all functions of the void ratio), and

$$\boldsymbol{\mu}_\beta = \frac{\boldsymbol{\beta}}{\sqrt{2}\beta}, \; \beta = (\tfrac{1}{2}\boldsymbol{\beta} : \boldsymbol{\beta})^{\frac{1}{2}} = pJ, \; J = (\boldsymbol{J} : \boldsymbol{J})^{\frac{1}{2}} \tag{6b--d}$$

In the present formulation, the fabric tensor, $\boldsymbol{\beta}$, affects both the elastic and inelastic response of the material.

## 11.5.7 RATE CONSTITUTIVE RELATIONS

The final constitutive relation is now given by

$$\overset{o}{\boldsymbol{\tau}} = \boldsymbol{L} : (\boldsymbol{D} - \boldsymbol{D}^p) \tag{7a}$$

or, in component form, by

$$\overset{0}{\tau}_{ij} = L_{ijkl}(D_{kl} - D^p_{kl}) \tag{7b}$$

For the evolution of the fabric, assume

$$\overset{0}{\boldsymbol{J}} = \Lambda\dot{\gamma}\boldsymbol{\mu}, \quad \overset{0}{\boldsymbol{\beta}} = \sqrt{2}p\overset{0}{J} \tag{8a, b}$$

where $\Lambda$ is a material function, depending on the fabric measure, say, $J$, and possibly its history (e.g., the last extreme state of the fabric, in cyclic loading). In monotonic loading, the induced fabric (anisotropy) will eventually saturate. Hence, $\Lambda$ must then tend to zero. Upon reverse loading, a jump in the value of $\Lambda$ is expected; see the example in Section 11.5.9.

The effective strain rate $\dot{\gamma}$ is obtained from the consistency condition; i.e., $\dot{f} = 0$ for continued plastic deformation. In the present case, we obtain

$$\dot{\gamma} = \frac{A_1 + A_2}{H + G + \bar{G}\mu_\beta^2 + \mu_\beta(B + M)\bar{K} + MKB} \tag{9a}$$

where

$$A_1 = \sqrt{2}(GD_\mu + \bar{G}\mu_\beta D_\beta) + (\bar{K}\mu_\beta + MK)D_{kk} + \sqrt{2}M\bar{K}D_\beta$$

$$A_2 = \sqrt{2}\eta\alpha(\bar{G}\mu_\beta + M\bar{K})(\mu_\beta D_\mu - D_\beta) \tag{9b, c}$$

In these equations, $\mu_\beta = \boldsymbol{\mu} : \boldsymbol{\mu}_\beta = \mu_f/J$, $D_\mu = \mathbf{D} : \boldsymbol{\mu}$, $D_\beta = \mathbf{D} : \boldsymbol{\mu}_\beta$, and $H$ is the workhardening parameter. To obtain an explicit expression for $H$, consider Eqs. 3e and 5c, and assume that $M$ in Eq. 3e is a function of the void ratio $e$ and the fabric parameter $J$, i.e., $M = M(e, J)$; for small pressures, the assumption $F(p, \Delta, \gamma) = pM(e, J)$ is appropriate, but not for large pressures, where the dependence on the pressure may be nonlinear. From the consistency relation $\dot{f} = 0$ at constant pressure, we thus obtain

$$H = \frac{dF}{d\gamma} + \frac{\overset{0}{\boldsymbol{\beta}} : \boldsymbol{\mu}}{\sqrt{2}\dot{\gamma}} = p\left[(1 + e)B\frac{\partial M}{\partial e} + \left(\mu_\beta\frac{\partial M}{\partial J} + 1\right)\Lambda\right] \tag{9d}$$

For a small elastic range, $M \approx 0$ and hence $H \approx p\Lambda$. In this case, the elastic volumetric strain rate may be neglected; i.e., $D_{kk} \approx D^p_{kk} = \dot{\gamma}B$. When, in addition, we set $\alpha = 0$, then Eq. 9a reduces to

$$\dot{\gamma} = \frac{\sqrt{2}(GD_\mu + \bar{G}\mu_\beta D_\beta)}{p\Lambda + G + \bar{G}\mu_\beta^2} \tag{9e}$$

## 11.5.8 MATERIAL FUNCTIONS

In this theory, there are three basic material functions which characterize the inelastic response of the granular mass. These are:

(1) $M$, which defines the elastic range of the material. At small pressures, $M \approx 0$.

(2) $M_f$, which defines the overall frictional resistance of the material and, in general, is a function of the void ratio $e$ and the fabric measure $\mu_f$. The following form for this function has been suggested by Nemat-Nasser and Zhang [16], based on experimental results of Okada [17]:

$$M_f = (a + b(e - e_m)^{n_1})(\mu_0 - \mu_f)^{n_2} \tag{10a}$$

where $e_m$ is the minimum value of the void ratio $e$, and $\mu_0$ is a reference value of the fabric parameter $\mu_f$.

(3) $\Lambda$, which defines the evolution of the fabric tensor through the back stress. In monotonic loading, the induced anisotropy (fabric) is eventually saturated. To simulate this, $\Lambda$ must tend to zero in such loadings. Then, upon unloading, $\Lambda$ must jump to a finite positive value [16].

In addition, there are four elastic moduli, $G$, $\bar{G}$, $K$, and, $\bar{K}$, all of which depend on the void ratio and the pressure. They may be approximated by the following general form:

$$A = A_0(1 + c(e_M - e)^{n_3}) \tag{10b}$$

where $A$ stands for any of the four elastic moduli.

In expressions Eq. 10a, b, the free parameters must be established based on experimental results, as has been illustrated by Nemat-Nasser and Zhang [16]. An example, given by these authors, is used in following text for illustration.

## 11.5.9 ILLUSTRATIVE EXAMPLE

Okada [17] has performed experiments on large hollow cylindrical samples of Silica 60, for which $e_m = 0.631$ and $e_M = 1.095$. Some of the results are examined and modeled by Nemat-Nasser and Zhang [16], where details of the modeling are given. The samples have 25-cm outside and 20-cm inside diameters, and 25-cm height. They are cyclically sheared under a constant confining pressure in a drained condition. A pure shearing with $\tau_{rr} = \tau_{\theta\theta} = \tau_{zz} = -p$ and $\tau_{r\theta} = \tau_{rz} = 0$ may be assumed, using a polar coordinate system. Setting $\theta = 1$ and $z = 2$, and assuming a negligible elastic

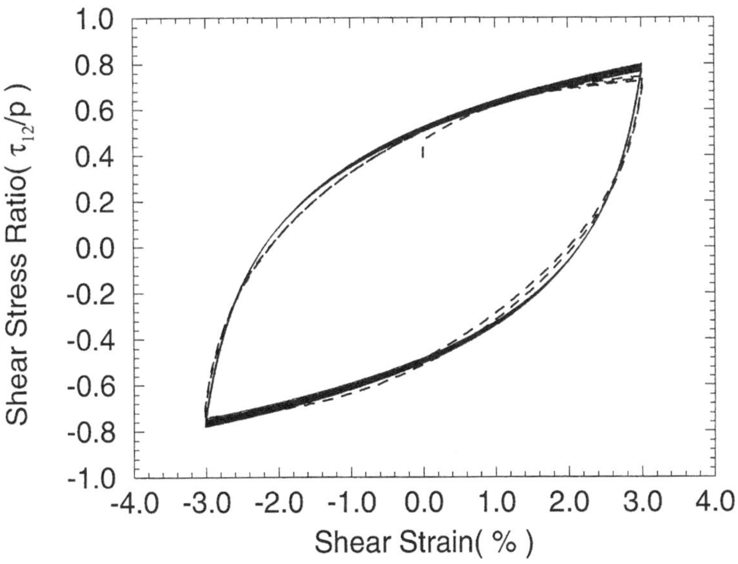


FIGURE 11.5.1.

range, we obtain

$$\dot{\tau}_{12} \approx \dot{\beta}_{12} = \pm p\Lambda\dot{\gamma},\ \dot{\gamma} = \pm\frac{2D_{12}}{1+\Lambda/\kappa} > 0,\ \kappa = (G+\bar{G})/p \qquad \text{(11a–c)}$$

The combined elastic modulus $\kappa$ has the functional form given by Eq. 10b, and for $\Lambda$ one may use

$$\Lambda(t) = \frac{\kappa}{1+d|\varepsilon(t)-\varepsilon_e|^{n_4}},\ \varepsilon(t) = \int_0^t D_{12}(t)\,dt \qquad \text{(11d, e)}$$

where $\varepsilon_e$ is the value of $\varepsilon(t)$ attained just before unloading. From Eqs. 11a–e, the shear stress $\tau_{12} = \tau_{\theta z}$ is obtained as a function of the strain $\varepsilon$. A typical example is given in Figure 11.5.1 , where the solid curve is the experimental result [17] and the dashed curve is the theoretical result. The material constants used are: $\kappa = 130$, $c = 1.42$, $d = 1.40 \times 10^3$, and $n_4 = 1.1$. The pressure is 195 Pa.

The void ratio is obtained from Eq. 5e with

$$B = -M_f + \zeta\mu_f,\ \mu_f = \pm\frac{\beta_{12}}{p} \qquad \text{(12a, b)}$$

The results for the same example are shown in Figure 11.5.2, where the experimental data are presented by open circles. The material constants are: $a = 1/4$, $b = 0.1$, $n_1 = 0.40$, $n_2 = n_3 = 1$, and $\zeta = 3/4$.

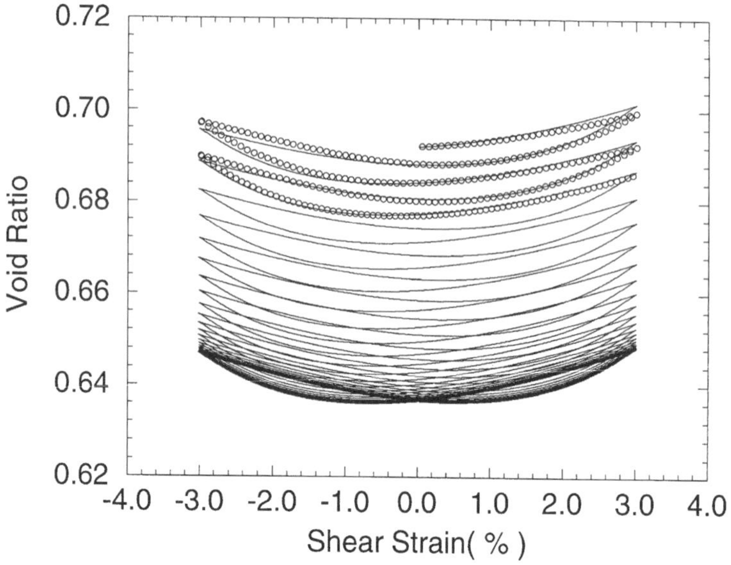


FIGURE 11.5.2.

## ACKNOWLEDGEMENTS

The work reported here has been supported by the U.S. National Science Foundation under Grant CMS-9729053, with the University of California, San Diego.

## REFERENCES

1. Okada, N., and Nemat-Nasser, S. (1994). Energy dissipation in inelastic flow of saturated cohesionless granular media. *Géotechnique* **44**(1): 1–19.
2. Nemat-Nasser, S. (2000). A micromechanically-based constitutive model for frictional deformation of granular materials. *J. Mech. Phys. Solids* **48**(6–7): 1541–1563.
3. Mehrabadi, M. M., Nemat-Nasser, S., and Oda, M. (1982). On Statistical description of stress and fabric in granular materials. *Int. J. Num. Anal. Methods in Geomechanics* **6**: 95–108.
4. Kanatani, K. (1984). Distribution of directional data and fabric tensors. *Int. J. Eng. Sci.* **22**(2): 149–164.
5. Subhash, G., Nemat-Nasser, S., Mehrabadi, M. M., and Shodja, H. M. (1991). Experimental investigation of fabric–stress relations in granular materials. *Mech. Mat.* **11**(2): 87–106.
6. Nemat-Nasser, S., and Shokooh, A. (1980). On finite plastic flow of compressible materials with internal friction. *Int. J. Solids Struct.* **16**(6): 495–514.
7. Spencer, A. J. M. (1964). A theory of the kinematics of ideal soils under plane strain conditions. *J. Mech. Phys. Solids* **12**: 337–351.

8. Spencer, A. J. M. (1982). Deformation of ideal granular materials, in *Mechanics of Solids*, pp. 607–652, Hopkins, H. G., and Sewell, M. J. eds., Oxford: Pergamon.
9. de Josselin de Jong, G. (1959). Statics and Kinematics in the Failable Zone of Granular Material. Ph.D. thesis, University of Delft.
10. Mehrabadi, M. M., and Cowin, S. C. (1978). Initial planar deformation of dilatant granular materials. *J. Mech. Phys. Solids* **26**: 269–284.
11. Balendran, B., and Nemat-Nasser, S. (1993). Double sliding model for cyclic deformation of granular materials, including dilatancy effects. *J. Mech. Phys. Solids* **41**(3): 573–612.
12. Balendran, B., and Nemat-Nasser, S. (1993). Viscoplastic flow of planar granular materials. *Mech. Mat.* **16**: 1–12.
13. Rudnicki, J. W., and Rice, J. R. (1975). Conditions for the localization of deformation in pressure-sensitive dilatant materials. *J. Mech. Phys. Solids* **23**: 371–394.
14. Nemat-Nasser, S. (1983). On finite plastic flow of crystalline solids and geomaterials. *Journal of Applied Mechanics* (50th Anniversary Issue) **50**(4b): 1114–1126.
15. Nemat-Nasser, S. (1992). Phenomenological theories of elastoplasticity and strain localization at high strain rates. *The Journal of Applied Mechanics Reviews* **45**(3, Part 2): 519–545.
16. Nemat-Nasser, S. and Zhang, J. Constitutive relations for cohensionless frictional granular materials. *J. of Plasticity* (in press).
17. Okada, N. (1992). Energy Dissipation in Inelastic Flow of Cohensionless Granular Media. Ph.D. thesis, University of California, San Diego.

SECTION **11.6**

# Linear Poroelasticity

J. W. Rudnicki
*Department of Civil Engineering, Northwestern University, Evanston, Illinois*

Contents



## 11.6.1 VALIDITY

Linear poroelasticity is a theory that includes the coupling between linear diffusion of a mobile species and the stress and deformation of a linear elastic porous solid. This theory has been widely applied not only to soils and rock masses infiltrated by groundwater but also to coupling of fluid flow and deformation in biological materials and diffusion of hydrogen in metals. Although departures from linear behavior are common in actual materials, linear poroelasticity is an improvement over idealizations that neglect either diffusion or deformation entirely. Because parameters describing nonlinear material behavior are difficult to obtain with accuracy, poroelasticity often offers a more practical approach than complex, nonlinear models. Furthermore, the insights gained from linear poroelasticity can be valuable for interpreting more complex models.

## 11.6.2 FORMULATION

The general development of linear poroelasticity was first given by Biot [1], but a revealing presentation of the constitutive relations by Rice and Cleary


*Handbook of Materials Behavior Models*. ISBN 0-12-443341-3.

[2] is followed here. Detournay and Cheng [3] have recently given a detailed discussion of the theory with a variety of solutions, especially for borehole problems, and an introduction to numerical formulations.

In addition to the total stress $\sigma_{ij}$ and the (infinitesimal) strain of the solid matrix $\varepsilon_{ij}$, two additional variables are needed to describe the fluid phase (or other diffusing species). One choice for these is the fluid mass per unit reference volume of porous solid $m$ and the pore fluid pressure $p$. The latter is defined as that pressure in an imaginary reservoir of fluid connected to the material point that equilibrates any mass flow to the porous solid. If a "point" of the porous solid is regarded as a representative volume element containing differently oriented fissures, then the assignment of a single scalar pore pressure to the point assumes that the time scale of deformation is slow enough to allow pressure equilibration within this element.

Linear relations for the stress and alteration of fluid mass content from a reference value $m_0$ in terms of the strain and pore pressure take the form

$$\sigma_{ij} = L_{ijkl}\varepsilon_{kl} - A_{ij}p \tag{1a}$$

$$m - m_0 = R_{ij}\varepsilon_{ij} + Qp \tag{1b}$$

where $L_{ijkl}$, $A_{ij}$, and $R_{ij}$ are arrays of factors reflecting the symmetries of $\sigma_{ij}$ and $\varepsilon_{ij}$. The first of these has the usual form of the linear elastic relation with the stress replaced by the *effective stress*, $\sigma_{ij} + A_{ij}p$. The fluid mass content can be written as the product of $\rho$, the density of homogeneous fluid, and $\upsilon$, the apparent volume fraction of pore space, $m = \rho\upsilon$. Linearizing this relation and substituting into Eq. 1b yields

$$\upsilon - \upsilon_0 = \frac{1}{\rho_0}R_{ij}\varepsilon_{ij} + \frac{p}{\rho_0}\left(Q - \frac{\rho_0\upsilon_0}{K_f}\right) \tag{2}$$

where the last term results from the linearized density change and $K_f = \rho_0 p/(\rho - \rho_0)$ is the bulk modulus of the pore fluid. An additional constraint on constitutive parameters arises from the relation for changes in the Helmholtz function per unit mass $\phi(\varepsilon_{ij}, m)$ for isothermal deformation :

$$d\phi = \sigma_{ij}d\varepsilon_{ij} + \mu dm \tag{3}$$

where the chemical potential $\mu$ is given by

$$\mu = \int_{p_0}^{p} \frac{dp}{\rho(p)} \tag{4}$$

(McTigue [4] has given a formulation for linearized porothermoelasticity.) A Legendre transformation to the potential $\psi = \phi - \mu m$ yields

$$d\psi = \sigma_{ij}d\varepsilon_{ij} - \upsilon dp \tag{5}$$

Because $L_{ijkl} = \partial^2\psi/\partial\varepsilon_{ij}\partial\varepsilon_{kl}$, these moduli satisfy the usual elastic symmetry with respect to interchange of the first and last pair of indices. In addition, the Maxwell relation that results from computing second derivatives of $\psi$ with respect to $\varepsilon_{ij}$ and $p$ in either order is

$$\frac{\partial\sigma_{ij}}{\partial p} = -\frac{\partial v}{\partial\varepsilon_{ij}} \tag{6}$$

Substituting Eqs. 1a and 2 yields

$$R_{ij} = \rho_0 A_{ij} \tag{7}$$

The remaining constitutive relation is Darcy's law:

$$q_i = -\rho_0\kappa_{ij}\left(\frac{\partial p}{\partial x_j} - \rho_0 g_j^f\right) \tag{8}$$

where $q_i$ is the fluid mass flow rate (relative to the solid phase) per unit area, $g_i^f$ is the body force per unit mass of fluid, and $\kappa_{ij}$ is a permeability tensor, required to be positive definite by the second law of thermodynamics. The permeability $\kappa_{ij}$ is more typically expressed as $k_{ij}/\eta$, where $\eta$ is the viscosity of the fluid and $k_{ij}$ is a permeability tensor with dimensions of $(\text{length})^2$, frequently measured in units of darcies (1 darcy=$10^{-12}\text{m}^2$). Alternatively, the permeability is expressed as an equivalent velocity $\bar{k} = \gamma_0\kappa$ where $\gamma_0$ is the weight density of the fluid.

If deformation occurs very slowly, fluid mass diffusion will equilibrate any alterations of pore fluid pressure. In this *drained* limit, the pore pressure change vanishes. Thus the moduli $L_{ijkl}$ are those appropriate for drained deformation. In the contrasting *undrained* limit, rapid deformation allows no time for fluid mass alteration. Setting $m = m_0$ in Eq. 1b and substituting into Eq. 1a yields

$$\sigma_{ij} = L_{ijkl}^u\varepsilon_{kl} \tag{9}$$

where

$$L_{ijkl}^u = L_{ijkl} + Q^{-1}\rho_0 A_{ij}A_{kl} \tag{10}$$

are the moduli appropriate for undrained deformation. The pore pressure induced by undrained deformation can be expressed as

$$p = -\frac{1}{3}B_{ij}\sigma_{ij} \tag{11}$$

where

$$B_{ij} = \frac{3\rho_0 A_{mn}C_{mnij}}{Q - \rho_0 A_{mn}C_{mnpq}A_{pq}} \tag{12}$$

and $C_{pqrs}$ is the inverse of $L_{ijkl}$.

### 11.6.2.1 Isotropic Response

Although Cheng [5] has recently given an advantageous formulation of anisotropic poroelasticity, applications assuming isotropy are much more common because of the difficulties of obtaining a complete set of material parameters. For isotropy, the tensors $A_{ij}$, $B_{ij}$, and $\kappa_{ij}$ reduce to $\alpha$, $B$, and $\kappa$ times the Kronecker delta, $\delta_{ij}$, respectively. The coefficients $\alpha$ and $B$ are commonly referred to as *Biot's parameter* and *Skempton's coefficient*, respectively. The modulus tensor $L_{ijkl}$ assumes the usual form

$$L_{ijkl} = G(\delta_{ik}\delta_{jl} + \delta_{il}\delta_{jk}) + (K - 2G/3)\delta_{ij}\delta_{kl} \tag{13}$$

with inverse

$$C_{pqrs} = \frac{1}{2G}\left\{\frac{1}{2}(\delta_{pr}\delta_{qs} + \delta_{ps}\delta_{qr}) - \frac{v}{1+v}\delta_{pq}\delta_{rs}\right\} \tag{14}$$

where $G$ and $K$ are the drained shear and bulk moduli, respectively; and $v$ is the Poisson's ratio for drained response, related to $G$ and $K$ by $v = (3K - 2G)/2(G + 3K)$. The undrained modulus tensor has the same form as Eq. 13. Evaluation of Eq. 10 indicates that the shear modulus is the same for drained and undrained deformation and that the bulk modulus for undrained deformation $K^u$ is given by

$$K^u = K + \alpha^2 \rho_0 Q^{-1} \tag{15}$$

The expression for $B$, obtained from Eq. 12, is

$$B = (K^u - K)/\alpha K^u \tag{16}$$

Thus, in the isotropic case, Eqs. 1a and 1b reduce to

$$\sigma_{ij} = 2G\varepsilon_{ij} + (K - 2G/3)\delta_{ij}\varepsilon_{kk} - \alpha p \delta_{ij} \tag{17a}$$

$$m - m_0 = \frac{\rho_0 \alpha}{K}\left[\frac{1}{3}\sigma_{kk} + \frac{p}{B}\right] \tag{17b}$$

and the expression for the apparent volume fraction (Eq. 2) reduces to

$$\upsilon - \upsilon_0 = \frac{\alpha}{K}\left\{\frac{1}{3}\sigma_{kk} + \frac{1}{B}p\right\} - \frac{\upsilon_0 p}{K_f} \tag{18}$$

Insight into the meaning of $\alpha$, $B$, and $K_u$ can be obtained by considering the special loading $\sigma_{ij} = -\sigma\delta_{ij}$, $p = \sigma$. For idealized circumstances delineated by Rice and Cleary [2] (following an observation of Reference [6]), the resulting volume strain and fractional change of pore space are equal to $-\sigma$ divided by the bulk modulus of the solid constituent $K_s$. More generally, the moduli entering these expressions, denoted here (and in Reference [2]) by $K_s'$ and $K_s''$, will differ from $K_s$. Equating these expressions to those obtained from

Eqs. 17a, 17b and 18 yields the following relations:

$$\alpha = 1 - K/K'_s \tag{19a}$$

$$B = \frac{1/K - 1/K'_s}{1/K - 1/K'_s + \upsilon_0(1/K_f - 1/K''_s)} \tag{19b}$$

$$K_u = \frac{1/K - 1/K'_s + \upsilon_0(1/K_f - 1/K''_s)}{(1/K - 1/K'_s)K/K'_s + \upsilon_0(1/K_f - 1/K''_s)} \tag{19c}$$

The undrained response can also be expressed in terms of the undrained Poisson's ratio, $\nu_u$, related to $K_u$ by $\nu_u = (3K_u - 2G)/2(G + 3K_u)$ or, in terms of $\nu$, $B$, and $\alpha$ as

$$\nu_u = \nu + \frac{(1+\nu)(1-2\nu)\alpha B}{3 - (1-2\nu)\alpha B} \tag{20}$$

## 11.6.3 FIELD EQUATIONS

The relevant field equations are the usual ones of solid mechanics, equilibrium and strain displacement, and, in addition, conservation of fluid mass. Equilibrium is expressed as

$$\sigma_{ij,\,i} + F_j(x,\, t) = 0 \tag{21}$$

where $F_j$ is the body force per unit volume and $(\ldots)_{,\,i}$ denotes $\partial(\ldots)/\partial x_i$. The (small) strain displacement relation is

$$\varepsilon_{ij} = (u_{i,\,j} + u_{j,\,i})/2 \tag{22}$$

Conservation of fluid mass is

$$q_{k,k} + \partial m/\partial t = H(x,\, t) \tag{23}$$

where $H$ is a fluid mass source.

Substitution of Eq. 22 into Eq. 17a and the result into Eq. 21 yields

$$(K + G/3)e_{,j} + Gu_{j,\,ii} + F_j - \alpha p_{,j} = 0 \tag{24}$$

where $e = \varepsilon_{kk} = u_{i,i}$ is the dilatation. Thus gradients in pore pressure act as a distribution of body forces. Consequently, if the distribution of pore pressure is known, the displacements and stresses can be determined by superposition of the solution of Eq. 24 for a Dirac singular distribution of body force (point force). In general, however, the pore pressure field is coupled to the stress field and cannot be determined independently. Alternatively, Eq. 17 can be used to eliminate $p$ in favor of $m$:

$$(K_u + G/3)e_{,j} + Gu_{j,\,ii} + F_j - \frac{(K_u - K)}{\alpha\rho_0} m_{,j} = 0 \tag{25}$$

Substituting Darcy's law (Eq. 8), specialized to isotropy and a uniform permeability, into Eq. 23 gives

$$-\rho_0 \kappa p_{,kk} + \partial m/\partial t = H(\mathbf{x},\, t) - \rho_0^2 \kappa g^F_{k,k} \tag{26}$$

The divergence of Eqs. 24 and 17b can be used to eliminate $p$ from Eq. 26. The result is a diffusion equation for the fluid mass content

$$\frac{\partial m}{\partial t} - c m_{,kk} = H(\mathbf{x},\, t) + \rho_0 \kappa \left\{ \frac{(K_u - K)}{\alpha(K_u + 4G/3)} F_{k,k} - \rho_0 g^F_{k,k} \right\} \tag{27}$$

where Eqs. 16 and 17a have also been used. The diffusivity $c$ is

$$c = \kappa \frac{(K_u - K)(K + 4G/3)}{\alpha^2 (K_u + 4G/3)} \tag{28}$$

or, as given by Rice and Cleary [2] in terms of $B$, $\nu$, and $\nu_u$,

$$c = \kappa \left[ \frac{2G(1-\nu)}{(1-2\nu)} \right] \left[ \frac{B^2 (1+\nu_u)^2 (1-2\nu)}{9(1-\nu_u)(\nu_u - \nu)} \right] \tag{29}$$

where, as they note, the first bracket is the drained elastic modulus for one-dimensional strain and the second is unity for incompressible constituents.

### 11.6.3.1 Plane Strain

Two-dimensional solutions can provide a reasonable idealization of a variety of geomechanical problems. An important simplifying feature of the plane strain formulation, noted by Rice and Cleary [2], is that the governing equations can be expressed entirely in terms of the stress and pore pressure. For plane strain deformation in the $xy$ plane, the compatibility equation, in the absence of body forces, can be expressed as

$$\nabla^2(\sigma_{xx} + \sigma_{yy} + 2\eta p) = 0 \tag{30}$$

and the fluid mass diffusion equation, without source terms, becomes

$$(c\nabla^2 - \partial/\partial t)[\sigma_{xx} + \sigma_{yy} + (2\eta/\mu)p] = 0 \tag{31}$$

where $\nabla^2(\ldots)$ is the two-dimensional Laplace operator and the material constants enter only in the combinations

$$\eta = 3(\nu_u - \nu)/2B(1+\nu_u)(1-\nu) \tag{32}$$

$$\mu = (\nu_u - \nu)/(1-\nu) \tag{33}$$

If the boundary conditions can be expressed in terms of the stress and pore pressure, then Eqs. 30 and 31 and two of the three equilibrium equations (Eq. 21) suffice to determine these quantities.

## 11.6.4 MATERIAL PARAMETERS

The drained elastic constants and two additional parameters are needed to characterize a linear poroelastic solid. Ideally, the drained elastic constants would be measured on a saturated sample deformed very slowly so that no pore pressure changes are induced by fluid flow. In practice, however, the drained elastic constants are often assumed to be equal to the values obtained on dry samples. An undrained test, in which fluid exchange between the sample and the surroundings is prevented, would, in principal, suffice to determine the remaining two porous media parameters, for example, $K_u$ or $\nu_u$ and $B$. Unfortunately, these and other tests to determine the porous media parameters are difficult and not yet standard. Consequently, it is frequently necessary to resort to the assumption that both $K_s'$ and $K_s''$ are equal to the bulk modulus of the solid constituents $K_s$ and then to calculate $\alpha$, $B$, etc., from $K_s$, the porosity $\upsilon_0$, and the pore fluid bulk modulus $K_f$ using expressions such as Eqs. 19a and 19b. Values of $\nu_u$ and $B$ determined in this way by Rice and Cleary [2] in their Table 1 for six rock types range from 0.29 to 0.34 and from 0.51 to 0.88, respectively. Values of $\alpha$ are not given but can be calculated from the values in the table and range from 0.2 to 0.7. Values for three additional rock types listed by Detournay and Cheng [3] also fall within this range.

Limiting ranges of $\alpha$, $B$, and $\nu_u$ are easily obtained from Eqs. 19a, 19b and 20. From the first, it is evident that $\alpha$ approaches unity if the drained bulk modulus is much less than the bulk modulus of the solid constituents. If, in addition, the value of $K$ is much less than the bulk modulus of the pore fluid, so that $K \ll K_f/\upsilon_0$, then $B$ also approaches unity and $\nu_u = 1/2$. These approximations are appropriate for most soils. If, on the other hand, the pore fluid is very compressible so that $\upsilon_0/K_f \gg 1/K > 1/K_s'$, $1/K's$, then $B \simeq K_f/\upsilon_0 K$ and approaches zero. In this limit $\nu_u = \nu$ and, thus, $\nu < \nu_u < 1/2$ and $0 < B < 1$.

A further complication is that, at least for geomaterials, values in the field may be different from those determined in the laboratory because of the presence of long, narrow fissures. Such fissures tend to decrease $K$ but, if they are saturated, have little effect on $K_u$ and consequently, tend to increase the value of $\nu_u$ relative to $\nu$.

The diffusivity $c$ (Eqs. 28 and 29) controls the time scale of fluid mass flow. The diffusivity is proportional to the permeability entering Darcy's law (Eq. 8) but also involves a combination of moduli. Again, because large fractures and joints serve as conduits for fluid flow *in situ*, field values of the diffusivity are typically much larger than those measured in the laboratory. Roeloffs [7] schematically summarizes values from both laboratory and field data ranging from nearly $10^5$ to $10^{-11}$ m$^2$/s. Laboratory values for the rocks tabulated by Rice & Cleary [2] are about $10^{-2}$ to $10^{-4}$ m$^2$/s except for the Berea sandstone, for which $c = 1.6\,\mathrm{m}^2/\mathrm{s}$. A variety of observations related to earthquakes suggest diffusivities in the range 0.1 to 1.0 m$^2$/s, a range that may be representative of conditions near many faults, but values an order of magnitude smaller or larger are not unusual.

## REFERENCES

1. Biot, M. A. (1941). General theory of three dimensional consolidation. *J. Appl. Phys.* **12**: 155–164.
2. Rice, J. R., and Cleary, M. P. (1976). Some basic stress diffusion solutions for fluid-saturated elastic porous media with compressible constituents. *Rev. Geophy. Space Phys.* **14**: 227–241.
3. Detournay, E., and Cheng, A. H-D. (1993). Fundamentals of poroelasticity, in *Comprehensive Rock Engineering: Principles, Practice and Projects*, pp. 113–171, Vol. 2, Fairhurst, C., ed., New York: Pergamon.
4. McTigue, D. F. (1986). Thermoelastic response of fluid-saturated porous rock. *J. Geophys. Res.* **91**: 9533–9542.
5. Cheng, A. H.-D. (1997). Material coefficients of anisotropic poroelasticity. *Int. J. Rock Mech. Min. Sci.* **34**: 199–205.
6. Nur, A., and Byerlee, J. D. (1971). An exact effective stress law for elastic deformation of rock with fluids. *J. Geophys. Res.* **76**: 6414–6419.
7. Roeloffs, E. A. (1988). Hydrologic precursors to earthquakes: A review. *Pure and Applied Geophysics (PAGEOPH)* **126**: 177–209.

SECTION **11.7**

# Nonlinear Poroelasticity for Liquid Nonsaturated Porous Materials

Olivier Coussy, Patrick Dangla
*Laboratoire Central des Ponts et Chaussées, Paris, France*

Contents



## 11.7.1 AN ENERGY APPROACH OF POROELASTICITY: THERMODYNAMIC POTENTIALS

The approach of poroelasticity presented here is principally an extension to nonsaturated situations of the pioneering work of M.A. Biot on the modeling of deformable liquid saturated porous media (see in the references some of his classical papers). The extension to partially saturated porous media is mainly based upon thermodynamics of deformable open continua [7, 8, 12].

The porous material considered here is formed of a porous solid skeleton saturated by two fluids. These fluids are a wetting liquid, index $\alpha = l$, and a nonwetting gas, index $\alpha = g$. Submitted to external actions, an element $d\Omega$ of


*Handbook of Materials Behavior Models*. ISBN 0-12-443341-3.

this porous material simultaneously deforms and undergoes variations of its fluid mass contents $m_\alpha$. Referring the deformation to that of the porous solid skeleton, let $\varepsilon_{ij}$ be the strain tensor components. Let also be the components of the (total) stress tensor supported by both the solid skeleton and the saturating fluids. For *reversible* (here also isothermal for sake of clarity) processes, the first two laws of thermodynamics applied between times $t$ and $t + dt$ to the open deformable system $d\Omega$ yield the balance of its free energy in the form:

$$\sigma_{ij}d\varepsilon_{ij} + g_l dm_l + g_g dm_g - d\Psi = 0 \tag{1}$$

In this equation $\Psi$ is the free energy per unit of volume $d\Omega$ of porous material, and $g_\alpha$ is the Gibbs potential of fluid $\alpha$ per mass unit. Recall that:

$$g_\alpha = \psi_\alpha + \frac{p_\alpha}{\rho_\alpha} \tag{2}$$

where $\psi_\alpha$ is the free energy per mass unit, $p_\alpha$ is the fluid pressure, and $\rho_\alpha$ is the volumetric mass. The term $g_\alpha dm_\alpha = \psi_\alpha dm_\alpha + p_\alpha/\rho_\alpha\, dm_\alpha$ accounts both for the free energy convectively supplied by fluid $\alpha$ (term $\psi_\alpha dm_\alpha$) and for the work required to make the infinitesimal mass $dm_\alpha$ actually enter the open porous material element $d\Omega$(term $p_\alpha/\rho_\alpha dm_\alpha$). The term $\sigma_{ij}d\varepsilon_{ij}$ accounts for the additional work required to produce the deformation of volume $d\Omega$. Equation 1 is similar to the Gibbs-Duhem equation encountered in physical chemistry [1].

Poroelasticity corresponds to the constitutive equations of a saturated porous material deforming reversibly due to the elasticity of its solid constituent. For nonsaturated materials, poroelasticity accounts in addition for reversible capillary effects, not considering the hysteresis associated with such effects. In these conditions Eq. 1 holds for any process and, consequently, $d\Psi$ has to be the exact differential of some function $\Psi(\varepsilon_{ij}, m_\alpha)$ of the independent state variables $(\varepsilon_{ij}, m_\alpha)$. As a result, the state equations of poroelasticity derive from the potential function $\Psi$ according to:

$$\sigma_{ij} = \frac{\partial\Psi}{\partial\varepsilon_{ij}} \quad g_\alpha = \frac{\partial\Psi}{\partial m_\alpha} \quad \alpha = 1,\ v,\ a \tag{3}$$

Let now $\Psi_s$ be the potential defined by [10]

$$\Psi_s = \Psi - m_\alpha\psi_\alpha \tag{4}$$

where $\psi_\alpha$ is the free energy of fluid $\alpha$ per mass unit. Since $\Psi$ represents the free energy of all matter currently contained in $d\Omega$, $\Psi_s$ represents the free energy of the skeleton, *including* the fluid–fluid and solid–fluid interfaces, the latter possessing their own surface free energy. Let us also introduce the (Lagrangian) partial porosity $\phi_\alpha$ related to fluid $\alpha$ and referring to the initial volume $d\Omega$ such that $\phi_\alpha d\Omega$ represents the current volume of $d\Omega$ occupied by fluid $\alpha$. Therefore, we write

$$m_\alpha = \rho_\alpha \phi_\alpha \tag{5}$$

Recall also that

$$\frac{\partial g_\alpha}{\partial p_\alpha} = \frac{1}{\rho_\alpha} \tag{6}$$

Combining Eqs. 1 and 4 to 6, one derives

$$\sigma_{ij} d\varepsilon_{ij} + p_l d\phi_l + p_g d\phi_g - d\Psi_s = 0 \tag{7}$$

Equation 7 turns out to be the free energy balance of the skeleton as just defined (i.e., including the interfaces). In fact, $p_l d\phi_l + p_g d\phi_g$ represents the external work supplied by the liquid and the gas through the pressure they exert on the interface with the skeleton and reversibly stored by the latter in the form of free energy $d\Psi_s$. As a result of Eq. 7 the state equations of poroelasticity can be alternatively written in the form

$$\sigma_{ij} = \frac{\partial \Psi_s}{\partial \varepsilon_{ij}} \quad p_l = \frac{\partial \Psi_s}{\partial \phi_l} \quad p_g = \frac{\partial \Psi_s}{\partial \phi_g} \tag{8}$$

By contrast to Eq. 3, Eq. 8 addresses separately the constitutive equations of the skeleton, while submitted to the internal pressures $p_l$ and $p_g$ of the liquid and the gas. In turn, Eq. 5 can then be viewed as a consistency continuity equation ensuring the fullness of the porous space.

The set of state variables $(\varepsilon_{ij}, p_l, p_g)$ often proves to be more convenient as primary variables than the set of state variables $(\varepsilon_{ij}, \phi_l, \phi_g)$. Let us then introduce the Legendre-Fenchel transform $H_s(\varepsilon_{ij}, p_l, p_g)$ of $\Psi_s(\varepsilon_{ij}, \phi_l, \phi_g)$, reading

$$H_s = \Psi_s - p_l\phi_l - p_g\phi_g \tag{9}$$

A combination of Eqs. 8 and 9 yields [10]

$$\sigma_{ij} = \frac{\partial H_s}{\partial \varepsilon_{ij}} \quad \phi_l = -\frac{\partial H_s}{\partial p_l} \quad \phi_g = -\frac{\partial H_s}{\partial p_g} \tag{10}$$

## 11.7.2 INCREMENTAL EQUATIONS OF NONLINEAR POROELASTICITY

For many porous materials, potential $\Psi_s$ cannot be approximated by quadratic functions of their arguments, since experimental evidence shows that constitutive equations (Eq. 10) do not rely on linearly the set of variables $(\sigma_{ij}, \phi_l, \phi_g)$ to the conjugate set $(\varepsilon_{ij}, p_l, p_g)$. That means that the second-order derivatives of potential function $\Psi_s$ involved in the differentiation of

Eq. 8 are not constant and do depend on the state variables $(\varepsilon_{ij}, p_l, p_g)$. However, a usual assumption consists of considering that the deviatoric behavior is linear and is not coupled with the nonlinear volumetric behavior. For an isotropic material, the potential function reduces then to the form

$$H_s = H_{s\varepsilon}(\varepsilon, p_l, p_g) + 2G\, e_{ij}e_{ji} \tag{11}$$

where $\varepsilon$ is the volumetric strain $\varepsilon = \varepsilon_{ii}$ and $e_{ij} = \varepsilon_{ij} - \frac{1}{3}\varepsilon\delta_{ij}$ is the deviatoric strain tensor. A differentiation of state equations Eq. 10 yields

$$d\sigma = K(\varepsilon, p_l, p_g)d\varepsilon - b_l(\varepsilon, p_l, p_g)dp_l - b_g(\varepsilon, p_l, p_g)dp_g \tag{12}$$

$$d\phi_l = N_{ll}(\varepsilon, p_l, p_g)dp_l + N_{lg}(\varepsilon, p_l, p_g)dp_g + b_l(\varepsilon, p_l, p_g)d\varepsilon \tag{13}$$

$$d\phi_g = N_{gg}(\varepsilon, p_l, p_g)dp_g + N_{lg}(\varepsilon, p_l, p_g)dp_l + b_g(\varepsilon, p_l, p_g)d\varepsilon \tag{14}$$

$$ds_{ij} = 2G\, de_{ij} \tag{15}$$

where $\sigma = \frac{1}{3}\sigma_{ii}$ is the overall mean stress, and $s_{ij} = \sigma_{ij} - \sigma\delta_{ij}$ is the deviatoric stress tensor. $K$ is the (drained) tangent bulk modulus, as coefficients $b_l$ and $b_g$ can be considered as tangent Biot's coefficients. Finally, $G$ is the standard shear modulus.

## 11.7.3 ASSUMPTION OF NEGLIGIBLE VOLUME CHANGE FOR THE SOLID CONSTITUENT

In many evolutions of usual porous materials, the volume of their solid part does not change significantly when compared to the change of the volume of the porous space. The volumetric strain $\varepsilon$ is then only due to the variation of the total porosity $\phi = \phi_l + \phi_g$, reading

$$\varepsilon = \phi - \phi_0 = \phi_l + \phi_g - \phi_0 \tag{16}$$

where $\phi_0$ stands for the initial overall porosity. Equation 16 implies that only two variables among the set $(\varepsilon, \phi_l, \phi_g)$ are independent. We choose $(\varepsilon, \phi_l)$ so that Eqs. 16 and 7 yield

$$(\sigma_{ij} + p_g\delta_{ij})d\varepsilon_{ij} - p_c d\phi_l - d\Psi_s = 0 \tag{17}$$

where $p_c = p_l - p_g$ is the capillary pressure. Under the assumption of small volume change for the solid constituent, Eq. 17 shows that $\Psi_s$ depends only on $(\varepsilon_{ij}, \phi_l)$.

$$\sigma_{ij} + p_g\delta_{ij} = \frac{\partial \Psi_s}{\partial \varepsilon_{ij}} \quad p_c = -\frac{\partial \Psi_s}{\partial \phi_l} \tag{18}$$

Let $W(\varepsilon_{ij}, p_c)$ be the Legendre-Fenchel transform of $\Psi_s(\varepsilon_{ij}, \phi_l)$:

$$W = \Psi_s + p_c\phi_l \tag{19}$$

The state equations (Eq. 10) reduce then to

$$\sigma_{ij} + p_g\delta_{ij} = \frac{\partial W}{\partial \varepsilon_{ij}} \quad \phi_l = \frac{\partial W}{\partial p_c} \tag{20}$$

Equation 11 becomes

$$W = W_{s\varepsilon}(\varepsilon, p_c) + 2G\, e_{ij}e_{ji} \tag{21}$$

Differentiating Eqs. 20 and 16, one finally obtains

$$d(\sigma + p_g) = K(\varepsilon, p_c)d\varepsilon + b(\varepsilon, p_c)dp_c \tag{22}$$

$$d\phi_l = -N(\varepsilon, p_c)dp_c + b(\varepsilon, p_c)d\varepsilon \tag{23}$$

$$d\varepsilon = d\phi = d\phi_l + d\phi_g \tag{24}$$

while Eq. 15 remains unchanged. A comparison between the set of Eqs. 12–14 with the set of Eqs. 22–24 shows that the assumption of negligible volume change for the solid constituent entails

$$b = b_l \quad b_l + b_g = 1 \quad N = N_{ll} = -N_{lg} = N_{gg} \tag{25}$$

In addition, note that Maxwell's relations associated with Eq. 20 are

$$\left(\frac{\partial K}{\partial p_c}\right)_\varepsilon = \left(\frac{\partial b}{\partial \varepsilon}\right)_{pc} \quad \left(\frac{\partial N}{\partial \varepsilon}\right)_{pc} = -\left(\frac{\partial b}{\partial p_c}\right)_\varepsilon \tag{26}$$

### 11.7.4 THE EQUIVALENT INTERSTITIAL FLUID PRESSURE $\pi$

It remains to identify $K(\varepsilon, p_c)$, $b(\varepsilon, p_c)$, and $N(\varepsilon, p_c)$. A combination of the definition of the liquid water saturation, $S_l = \phi_l/\phi$, with Eqs. 23 and 24 yields

$$\phi dS_l = -Ndp_c + (b - S_l)d\varepsilon \tag{27}$$

In reversible (i.e., hysteresis associated with capillary effects is neglected) and nondeformable case (i.e., $\varepsilon = 0$, $\phi = \phi_0$), the previous equation reduces to

$$\phi_0 dS_l = -Ndp_c \tag{28}$$

yielding

$$p_c = p_c(S_l) \tag{29}$$

A very common macroscopic assumption which can receive some support and also some limitations from microscopic considerations [6] consists in assuming that Eq. 29 still holds in the deformable case (i.e., $\varepsilon \neq 0$, $\phi = \phi_0 + \varepsilon$). In other words, provided that the current ratio $S_l$ of the porous space occupied by the wetting liquid does not change, the capillary pressure is not affected by the strain. Then, a comparison of Eq. 29 with Eq. 27 leads us to identify Biot's coefficient $b$ as the liquid water saturation $S_l$:

$$b = S_l(p_c) \tag{30}$$

Equation 30 and Maxwell's equation (Eq. 26) shows then that the bulk modulus $K$ depends only on $\varepsilon$:

$$K = K(\varepsilon) \tag{31}$$

A substitution of Eqs. 30 and 31 into Eq. 22 entails

$$d(\sigma + \pi) = K(\varepsilon)d\varepsilon \tag{32}$$

where $\pi$ is the pressure defined by

$$\pi = p_g - \int_0^{p_c} S_l(\varpi)d\varpi \tag{33}$$

For $S_l = 1$, the capillary pressure being always non-negative, Eq. 33 reduces to $\pi = p_l$. Thus, provided that later on the porous space remains saturated, i.e., $dS_l = 0$, we identify $d\pi = dp_l$ and relation (32) reduces to

$$d(\sigma + p_l) = K(\varepsilon)d\varepsilon \tag{34}$$

Referring to the saturated case, i.e. Eq. 34, pressure $\pi$ involved in relation 32 can be interpreted in the nonsaturated case as *an equivalent interstitial fluid pressure*. In fact, relation 32 altogether with definition 33 ensures the continuity with the relation holding in the saturated case. Indeed, for negligible volume change for the solid constituent, Eq. 25 implies $b_l + b_g = 1$ and Eq. 34 can be directly and exactly recovered by letting $dp_l = dp_g = dp$ in Eq. 25. When the volume changes for the solid constituent are not negligible, noting $b_l + b_g = \beta$, the latter procedure yields

$$d\sigma + \beta dp = Kd\varepsilon \tag{35}$$

In the linear isotropic poroelastic saturated case, provided that the solid constituent is homogeneous and isotropic, Biot's relation (See, for instance, Reference [8]) provides $\beta$ as a function

$$\beta = 1 - \frac{K}{K_s} \tag{36}$$

of the drained bulk modulus $K$ and the bulk modulus $K_s$ of the solid constituent. Based upon Biot's relation, Eq. 32 is extended in the form

$$d(\sigma + \beta\pi) = K(\varepsilon)d\varepsilon \tag{37}$$

According to Eq. 37, $\sigma + \beta\pi$ governs the deformation of the solid skeleton (now not including the interfaces) over the whole range of saturation. Indeed, in the case of negligible volume change for the solid part of the skeleton ($K/K_s \ll 1$, $\beta \simeq 1$) and saturated situations ($S_l = 1$), Eqs. 33 and 37 indicate that the so defined effective stress $\sigma + \beta\pi$ reduces to the celebrated Terzaghi effective stress $\sigma + p$ [13]. Hence, the stress $\sigma + \beta\pi$ can be viewed as a generalized effective stress for liquid nonsaturated porous materials.

However, by contrast to general constitutive Eqs. 12–14, the effective stress concept is based upon assumptions and has to be experimentally checked. Indeed, for negligible volume changes of the solid part of the skeleton, a saturated experiment (i.e., $S_l = 1$, $\pi = p_l$, $d[\sigma + p_l] = K[\varepsilon]d\varepsilon$) must lead to the same identification of function $K(\varepsilon)$ as the one given by a free swelling experiment (i.e. $\sigma = 0$, $d\pi = Kp[\varepsilon]d\varepsilon$) [11].

The previous constitutive relations concern only reversible evolutions. This approach can be extended to account for irreversible processes, in order to include both the hysteresis associated with capillary effects and the plastic deformation of the solid skeleton [9].

## REFERENCES


1. Atkins, P. W. (1990). *Physical Chemistry*, $4^{th}$ ed., Oxford University Press.
2. Biot, M. A. (1941). General theory of three dimensional consolidation. *Journal of Applied Physics* **12**: 155–164.
3. Biot, M. A. (1962). Mechanics of deformation and acoustic propagation in porous media. *Journal of Applied Physics* **12**: 155–164.
4. Biot, M. A. (1972). Theory of finite deformation of porous solids. *Indiana University Mathematical Journal* **33**: 1482–1498.
5. Biot, M. A. (1977). Variational Lagrangian-thermodynamics of non-isothermal finite strain: Mechanics of porous solid and thermomolecular diffusion. *International Journal of Solids and Structures* **12**: 155–164.
6. Chateau, X., and Dormieux, L. (1998). A micromechanical approach to the behaviour of unsaturated porous media, in *Poromechanics, A Tribute to M.A. Biot, Thimus,* J.-F. *et al.*, eds., Proceedings of the Biot conference on poromechanics, Balkema.
7. Coussy, O. (1989). Thermodynamics of saturated porous solids in finite deformation. *European Journal of Mechanics, A/Solids* **8**: 1–14.
8. Coussy, O. (1995). *Mechanics of Porous Continua*, John Wiley & Sons.
9. Dangla, P., Malinsky, L., and Coussy, O. (1997). Plasticity and imbibition-drainage curves for unsaturated soils: A unified approach, in *Numerical Models in Geomechanics, NUMOG VI*, Pietruszczak and Pande, eds., Balkema.
10. Dangla, P., and Coussy, O. (1998). Non-linear poroelasticity for unsaturated porous materials: An energy approach, in *Poromechanics, A Tribute to M. A. Biot, Thimus, J.-F. et al.*, eds., Proceedings of the Biot conference on poromechanics, Balkema.

11. Dangla, P., Coussy, O., Olchitzky, E., and Imbert, C. (1999). Non linear thermo-mechanical couplings in unsaturated clay barriers, in *Theoretical and Numerical Methods in Continuum Mechanics of Porous Materials*, Ehlers, W., ed., Proceedings of IUTAM Symposium, Kluwer Academic Publishers.
12. Sih, G. C., Michopoulos, J. G., and Chou, S. C. (1986). *Hygrothermoelasticity*, Martinus Nijhoff Publishers.
13. Terzaghi, K. (1925). *Principles of Soil Mechanics, A summary of Experimental Results of Clay and Sand*, Eng. News Rec., 3–98.

SECTION **11.8**

# An Elastoplastic Constitutive Model for Partially Saturated Soils

B.A. Schrefler and L. Simoni

*Department of Structural and Transportation Engineering, University of Padua, Italy*

Contents



## 11.8.1 VALIDITY OF THE MODEL

This model has been developed by Bolzon *et al.* [1] for partially saturated geomaterials where capillary effects are of importance. It is an extension of the generalized plasticity model for fully saturated soils dealt with in previous sections. In the presented version a yield surface is used to take into account matric suction (also called capillary pressure), which has to be treated as an independent variable.



*Handbook of Materials Behavior Models*. ISBN 0-12-443341-3.

The model accounts for the following typical situations:

- Dependence of soil stiffness on suction; i.e., unsaturated specimens exhibit a lower volume change than saturated ones when, e.g., subjected to the same increment of vertical stress. However, when the unsaturated samples are soaked and hence saturated, the soil exhibits a significant compression strain under constant stress.
- When suction decreases (wetting), the geomaterial may exhibit first swelling (in the elastic range) and then collapse (irreversible compression).

The model has been applied successfully to partially saturated clays and sands. It has the most important features of the well-known Barcelona model [2]. Moreover, because of the choice of the stress variables (see next section), it can easily be applied to the dynamic analysis of problems where both fully and partially saturated zones are present, as, for instance, in the seismic analysis of earth dams, dikes, and slopes.

## 11.8.2 BACKGROUND

Vectorial notation is used in this chapter. When dealing with the mechanics of partially saturated porous media, we have to define a suitable stress measure: it is commonly accepted that two independent stress dimension parameters (a combination of total stress tensor and water and gas pressure) are needed to describe the deformational behavior [3]. In the following, the assumed stress variables are soil suction $s$, say, the difference between gas and water pressure ($p_g$ and $p_w$, respectively), and *Bishop's effective stress* $\sigma'_{ij}$. The latter is a combination of total stress $\sigma_{ij}$ and pressure acting on the solid, represented by a weighted average of water and gas pressure. As a weighting function, water saturation $S_r$ is assumed [3]; hence the stress measure takes the form

$$\sigma'_{ij} = \sigma_{ij} - \left[S_r p_w + (1 - S_r) p_g\right] \delta_{ij} = \sigma_{ij} - \left[p_g - S_r s\right] \delta_{ij} \tag{1}$$

with soil suction $s$ defined as

$$s = p_g - p_w \tag{2}$$

Soil suction and water saturation may be mutually related by means of the following relationship [2]:

$$S_r = 1 - m \tanh(ls) \tag{3}$$

where $m$ and $l$ are material parameters ($S_r = 1 - m$ represents *irreducible saturation*, i.e., the limiting value of $S_r$ as suction approaches infinity). In Eq. 1, parameter $S_r$ represents a phenomenological measure of the capillary effects, through its experimental relationship with suction. Bishop's stress definition recovers the Terzaghi effective stress definition, usually assumed in fully saturated soil mechanics when saturation equals one; hence the consistency condition between stress measures is guaranteed.

When one is observing the mechanical behavior of a partially saturated geomaterial, different deformation histories are obtained depending on the water content (or saturation or relative humidity) and also the so-called *hydric path* (or suction path); i.e., a forced change in saturation results in a sample deformation. Figure 11.8.1 shows these effects in terms of volumetric strain for a chalk.

## 11.8.3 MODEL DESCRIPTION

The elastoplastic model is now defined in the space $(p', q, s)$ of the mean stress $p'$, the deviatoric stress $q$, invariants of Bishop's stress, and suction $s$:

$$p' = \frac{\sigma'_{ii}}{3} = p - S_r p_w - (1 - S_r) p_g = p - p_g + S_r s \tag{4}$$

$$q = \sqrt{3J'_2} = \sqrt{\frac{3}{2}\chi_{ij}\chi_{ij}}, \quad \chi_{ij} = \sigma'_{ij} - p'\delta_{ij} \tag{5}$$

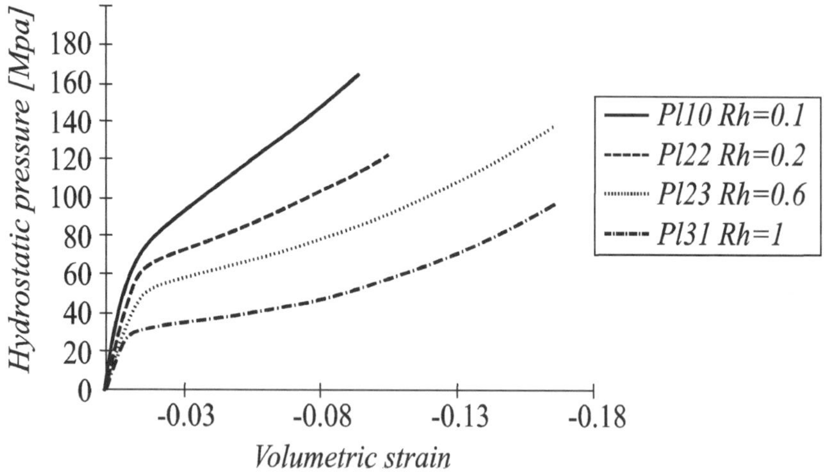


FIGURE 11.8.1 Volumetric strain of a chalk sample under isotropic compression at different levels of suction (*Rh*= relativehumidity). (Redrawn with permission from *Rock mechanics: Proceedings of the 35th US symposium*, Lake Taboe, 4–7 June 1995, Daeman, Jaak J.K. and Richard A. Schltz (eds.), 1995, Balkema.)

From Eq. 4 it results that changes in the mean stress $p'$ may be induced both by changes in gas pressure and changes in suction and saturation.

As shown by Pastor *et al.*, Section 11.10, this volume, generalized plasticity does not explicitly define the yield and potential surfaces but directly assumes the loading–unloading direction vector $\boldsymbol{n}$ and the direction vector defining the plastic flow $\boldsymbol{n}_{gL}$. When we refer to fully saturated conditions, i.e., when we use Terzaghi's definition of effective stress and invariants of Eqs. 4–5 must be evaluated for $S_r = 1$, these direction vectors can be given as functions of the stress ratio $\eta = q/p'$, e.g., following Nova [6], as

$$\boldsymbol{n} \equiv \frac{1}{\sqrt{1+d_f^2}}\{d_f; 1\}^{\mathrm{T}}, \; d_f = (1+\alpha)(M_f - \eta), \tag{6}$$

$$\boldsymbol{n}_{gL} \equiv \frac{1}{\sqrt{1+d_g^2}}\{d_g; 1\}^{\mathrm{T}}, \; d_g = (1+\alpha)(M_g - \eta), \tag{7}$$

where $\alpha$, $M_g$, and $M_f$ are material parameters; see Section 11.7. Whereas $\alpha$ can be assumed as independent of suction, parameters $M_g$ and $M_f$ must depend on suction in partially saturated problems (Laloui *et al.* [7], for $M_f$). Further assumptions for $\boldsymbol{n}$ and $\boldsymbol{n}_{gL}$ are, however, possible, for instance, $\boldsymbol{n} \equiv \boldsymbol{n}_{gL}$, which results in associative plasticity.

For the purpose of introducing the effects of suction, we define now a yield and potential surface, which are obtained by integration of Eqs. 6 and 7 as

$$f \equiv q - M_f p'\left(1+\frac{1}{\alpha}\right)\left[1-\left(\frac{p'}{p_f}\right)^{\alpha}\right] = 0 \tag{8}$$

$$g \equiv q - M_g p'\left(1+\frac{1}{\alpha}\right)\left[1-\left(\frac{p'}{p_g}\right)^{\alpha}\right] = 0 \tag{9}$$

where $p_f$ and $p_g$ are integration constants which determine the size of the surface but have no influence in defining the respective normals. Even though Eqs. 8 and 9 are formally the same as for fully saturated materials (Pastor *et al.*, this volume), they must be assumed in $(p', q, s)$ space when dealing with partially saturated problems (see Fig. 11.8.2). In addition to the aforementioned dependence of $M_g$ and $M_f$ on suction, parameters $p_g$ and $p_f$ also depend on suction, as suggested by experimental evidence [8]. This dependence will be discussed later.

Once the direction vectors have been defined, irreversible (or plastic) strains $d\boldsymbol{\varepsilon}^p$ are related to stress increments $d\boldsymbol{\sigma}'$ by the relationship

$$\mathrm{d}\boldsymbol{\varepsilon}^p = \frac{1}{H}\boldsymbol{n}^T d\boldsymbol{\sigma}' \boldsymbol{n}_{gL} \tag{10}$$

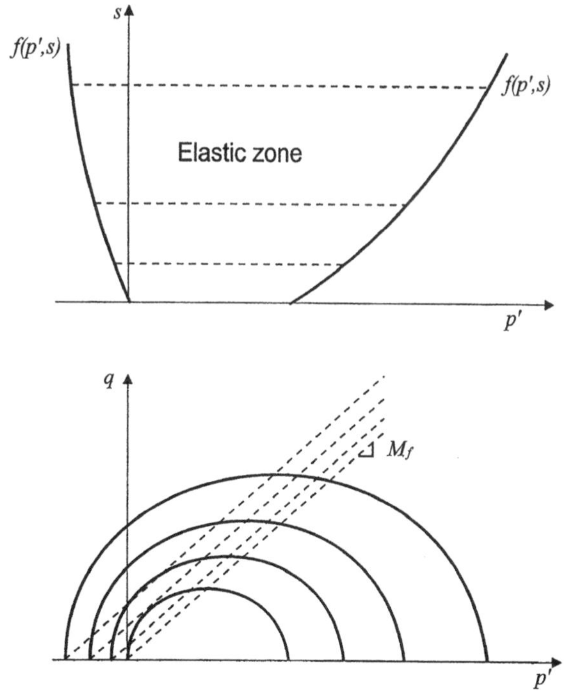


FIGURE 11.8.2 Yield surfaces in $(p', s)$ and $(p', q)$ planes.

where $H$ represents the plastic (or hardening) modulus. As usual, the strain increment $d\boldsymbol{\varepsilon}$ can be decomposed into an elastic part, $d\boldsymbol{\varepsilon}^e$, and an elastoplastic one:

$$\mathrm{d}\boldsymbol{\varepsilon} = \mathrm{d}\boldsymbol{\varepsilon}^e + \mathrm{d}\boldsymbol{\varepsilon}^p \tag{11}$$

Recalling that specific volume $v$ is the ratio between the sample volume and that of the solid, i.e., $v = 1 + e$, the total volumetric strain is defined as

$$\varepsilon_v = \frac{v - v_0}{v_0} = \frac{e - e_0}{1 + e_0} = \frac{n - n_0}{1 - n} \tag{12}$$

where $e$ is the void ratio, $n$ is the porosity, and subscript *0* refers to initial values. Following the critical state models [9], the total volumetric strain (recoverable and not) associated with the change in the hydrostatic component of the Bishop's stress tensor $p'$ can be linearized as

$$\mathrm{d}\varepsilon_v = \frac{\lambda(s)}{v_0}\frac{\mathrm{d}p'}{p'} \tag{13}$$

In this equation, $\lambda(s)$ is the soil compressibility, i.e., the slope of the virgin loading line in the ($e$, log $p'$) plane during a triaxial test. Its dependence on suction is assumed to be of the type

$$\lambda(s) = \frac{\lambda(0)}{1 + as} \tag{14}$$

where $\lambda(0)$ is the compressibility of the saturated sample and $a$ a material parameter. Other representations of the dependence of the compressibility parameter on suction can be found in the literature which have the advantage of giving a more correct shape of the yield surfaces, e.g., Reference [2]. They present, however, computational disadvantages because of their complexity and require at least two material parameters, which is more demanding for identification purposes. In a similar way, the elastic part of the volumetric strain can be written as

$$d\varepsilon_v^e = \frac{\kappa(s)}{v_0}\frac{\mathrm{d}p'}{p'} \tag{15}$$

$\kappa(0)$ being the slope of the elastic unloading line in the $(e, \ln p')$ plane during a triaxial test. A dependence on suction of $\kappa(s)$ similar to Eq. 14 is supposed in the following. Some authors (e.g., Laloui *et al.* [7]) assume the elastic compressibility modulus $\kappa$ independent of suction. This is, however, in contrast with the results of Figure 11.8.1 and probably depends on the range of applied forces, larger for rock-type materials.

For volumetric deformation during an isotropic virgin compression loading ($\eta = 0$), by substituting Eqs. 6–7 into Eq. 10, the plastic flow can be cast in the form

$$\mathrm{d}\varepsilon_v^p = \frac{1}{H_0 H_w}\frac{\mathrm{d}p'}{p'} \tag{16}$$

with the assumption that $H = H_0 H_w p'$: $H_w$ is related to partially saturated behavior, whereas term $H_0$ depends on the material characteristics in fully saturated conditions. Comparison of Eq. 16 with Eqs. 13 and 15 and accounting for Eqs. 11 and 14 gives the hardening modulus during isotropic compression and the hardening effects due to suction, respectively, as

$$H_0 = \frac{1 + e_0}{\lambda(0) - \kappa(0)}, \quad H_w = 1 + as \tag{17}$$

For generic (not isotropic) stress paths and for deviatoric effects, the reader is referred to the chapter by Pastor *et al.*, this volume. Using the same symbols, the plastic modulus results in

$$H = H_0 p' H_w H_f (H_v + H_s) \tag{18}$$

and can be further enhanced to account for memory effects and plastic unloading.

Once the dependence on suction of the plastic modulus has been defined, we have to introduce the same effects in the yield and potential surface equations (Eqs. 8 and 9). Experimental observations show that parameter $p_f$ is increasing with suction [3]. Given the initial yield stress $p'_{y0_i}$ for saturated conditions, the dependence of $p_f$ on suction is assumed as

$$p_f = p'_{y_i}(s) = p'_{y0_i} + is \tag{19}$$

Parameter $i$ has to be determined by interpolation of experimental data to obtain an increasing function of suction when water saturation is less than one.

Volumetric hardening is controlled by irreversible volumetric strain; hence, the evolution of the yield surfaces in the $(p', s)$ plane is given as

$$p'_y(s) = (p'_{y0_i} + is)\left(\frac{p'_{y0}}{p'_{y0_i}}\right)^{1+as} \tag{20}$$

This equation has been obtained by requiring that, when moving from one yield surface (e.g., the initial one) to another one, the related plastic volumetric strain must be the same independently of the starting point and the followed path.

The same variation with suction can be assumed for the parameter $p_g$, which means that yield and potential surface expand with the same law when suction is increasing. This is suggested by the possibility of assuming associative plasticity, in which case $f \equiv g$.

## 11.8.4 APPLICATIONS AND PARAMETERS

We show here two back calculations for kaolin and clay and one determination of parameters for a hydrocarbon reservoir sandstone by means of an inverse identification procedure. These examples present typical features of unsaturated soil behavior.

### 11.8.4.1 Tests on Compacted Kaolin

Data for a compacted kaolin have been obtained from the extensive experimental investigation conducted by Josa [10]. Some of these data are given in Table 11.8.1. Data of the present model not defined in Reference [10] are calculated assuming an intermediate value of 0.3 MPa for the mean net stress $\bar{p}$ (mean total stress in excess of air pressure) in the range of the applied pressures.

Volume changes resulting from the present model, due to isotropic compression at different but constant suctions, $s = 0.06\,\text{MPa}$ and $s = 0.09\,\text{MPa}$, are compared with experimental data in Figure 11.8.3. The initial mean stress $p'_{y0}$ is equal to 0.045 MPa, and bulk modulus $\kappa$ is equal to 0.015.

According to the experimental results, the specific volumes at the beginning of each loading path considered here have been assumed equal to 1.888 for a sample tested at $s = 0.06\,\text{MPa}$, and equal to 1.893 for $s = 0.09\,\text{MPa}$. Good agreement is achieved between model predictions and experimental data.

TABLE 11.8.1 Material Data for a Partially Saturated Kaolin [10]

| $s$ (MPa) | $S_r$ | $p'_y$ (MPa) | $\lambda(s)$ |
|---|---|---|---|
| 0.04 | 0.838 | 0.099 | 0.100 |
| 0.06 | 0.826 | 0.120 | 0.085 |
| 0.09 | 0.812 | 0.158 | 0.070 |

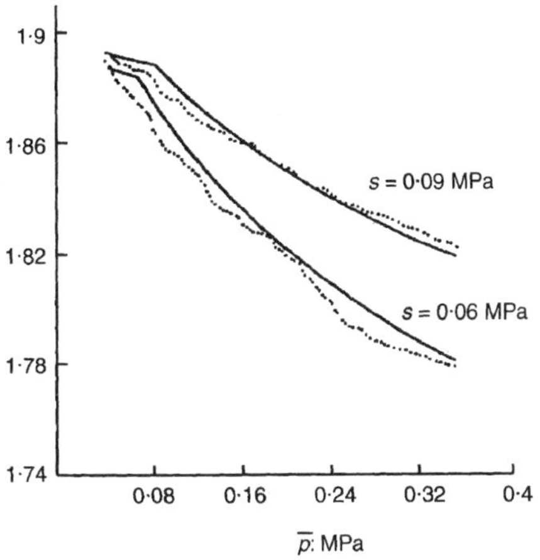


FIGURE 11.8.3 Specific volume plotted against mean net stress $(p - p_a)$ under isotropic compression at constant suction: comparison with experimental data (dots, reproduced from *Géotechnique*, vol. 40, 1990, Alonso, E.E., Gens, A., and Josa, A., A constitutive model for partially saturated soils, pages 405–430).

## 11.8.4.2 Tests on Clays

A characteristic behavior of partially saturated soils was indicated by the experiments performed by Escario and Sáez [11] on clay samples. As shown in Figure 11.8.4 (dots), along the loading path corresponding to suction decrease from 3.5 MPa to zero at constant vertical external pressure (0.2 MPa), the material exhibits first swelling and then collapse. The main features of this behavior can be reproduced as follows.

Data listed in the quoted paper and given in Table 11.8.3 are not enough to characterize the material completely according to the present model. Back calculation has yielded the following missing parameters: $v_o = 2.3$, $m = 0.8$, $l = 2\,\text{MPa}$, $a = 0.1\,\text{MPa}^{-1}$, $p'_{y0} = 0.07\,\text{MPa}$.

Only the contribution to volume changes due to the isotropic state of stress $p = 0.2\,\text{MPa}$ is considered, neglecting volume changes induced by soil dilatancy in the nonassociative plasticity model, depending on the values of the confining pressure. With these assumptions, the predicted volume changes plotted in Figure 11.8.4 are in good qualitative agreement with the experimental data.

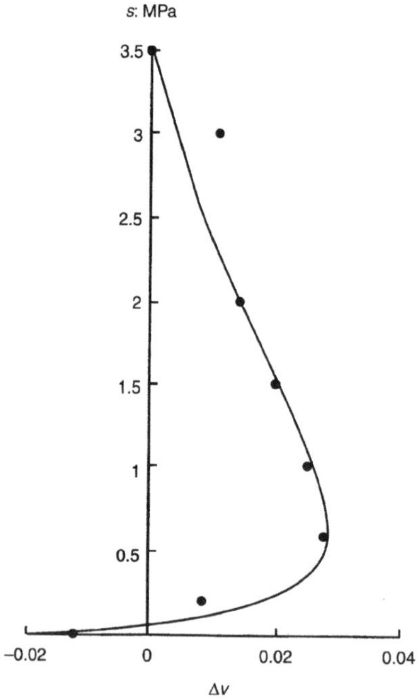


FIGURE 11.8.4 Specific volume plotted against suction under suction changes at constant net stress: comparison with experimental data (dots, reproduced from Escario, V., and Sáez, J., Measurement of the properties of swelling and collapsing soils under controlled suction, in *Proceedings 3rd I.C.E.S.*, 1973, Haifa, pages 195–200).

### 11.8.4.3 Material Parameters Identification in an Oedometric Test for Hydrocarbon Reservoir Sandstone

The last example deals with a laboratory experiment performed at IKU, SINTEF, Trondheim, on behalf of AGIP (Italian National Petroleum Company) [12] on a silty consolidated sandstone sample extracted from a gas bearing formation in the Northern Adriatic basin at a depth of 3400 m. Effective porosity, *in situ* water saturation, and irreducible saturation of the material were previously obtained, and then the specimen underwent an oedometric test. The loading was scheduled as follows: the sample at *in situ* saturation (0.38–0.45) was first stressed with an initial hydrostatic phase presenting $\sigma_r$-rate equal to 0.01 MPa/s until $\sigma_r = 0.5$ MPa. This was followed by an uniaxial phase with $\sigma_z$-rate of 0.004 MPa/s until $\sigma_z$ reached 35 MPa; the sample was then held at constant stress level and water was injected for 25 hours until full saturation was attained. During this period of time, volumetric changes of the specimen were recorded, as during the phases of stress changes. Once the full saturation was reached, a second uniaxial phase, at constant water content, with stress rate of 0.004 MPa/s till about 110 MPa was performed. The test also included unloading cycles to determine the elastic behavior and recoverable deformation. The water injection test (hydric path) simulated the behavior of the gas reservoir rock during artificial water injection or during the flooding associated with gas extraction.

The model response at the end of the identification procedure is shown in Figure 11.8.5. The agreement is very satisfactory, in particular for the increase of volumetric strain caused by water injection. The values of the identified parameters are shown in Table 11.8.2.

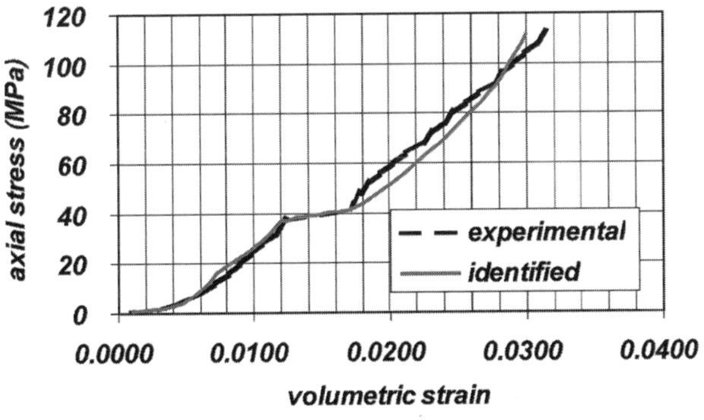


FIGURE 11.8.5 Results of the identification.

TABLE 11.8.2 Identified Material Parameters for a Reservoir Sandstone

| $l$ | $m$ | $a$ | $\lambda(0)$ | $i$ | $p_f$ | $\kappa(0)$ | $\beta_0$ | $\beta_1$ | $\eta_f$ |
|---|---|---|---|---|---|---|---|---|---|
| 0.9898 | 0.9867 | 0.9797 | 0.0122 | 1.0792 | 9.9490 | 0.0051 | 0.9675 | 1.0648 | 1.0016 |

TABLE 11.8.3 Material Parameters for Partially Saturated Materials

| Reference | $\kappa$ | $\lambda(0)$ | $p'_y$ (MPa) | $i$ |
|---|---|---|---|---|
| Escario and Saez [11] | 0.035 | 0.105 | 0.07 | 0.55 |
| Alonso *et al.*[2] | 0.015 | 0.14 | 0.055 | 1.18 |
| Cui and Delage [13] | 0.015 | 0.20 | 0.35 | 0.7 |

Finally, a set of parameters obtained from tests in the literature is listed in Table 11.8.3.

## ACKNOWLEDGMENTS

We thank AGIP for allowing us to use unpublished data concerning saturation experiments carried out for AGIP by IKU, SINTEF (Trondheim).

## REFERENCES

1. Bolzon, G., Schrefler, B.A., and Zienkiewicz, O.C. (1996). Elastoplastic soil constitutive laws generalised to partially saturated states. *Géotechnique* **46**: 279–289.
2. Alonso, E.E., Gens, A., and Josa, A. (1990). A constitutive model for partially saturated soils. *Géotechnique* **40**: 405–430.
3. Fredlund, D.G., and Morgenstern, N.R. (1977). Stress state variables for unsaturated soils. *J. Geotech. Engng. Div.*, ASCE, **103**: No. GT5, 447–466.
4. Schrefler, B.A., Simoni, L., Li, Xikui, and Zienkiewicz, O.C. (1990). Mechanics of partially saturated porous media, in *Numerical Methods and Constitutive Modelling in Geomechanics*, (CISM lecture notes), pp. 169–209, Desai, C.S., and Sioda, G., eds., Wein: Springer-Verlag.
5. Brignoli, M., Santarelli, F.J., and Papamichos, E. (1995). Capillary effects in sedimentary rocks: Application to reservoir water-flooding, in *Rock Mechanics*, Proc. 35th U.S. Symposium, pp. 619–625, Daemen, J.J.K., and Schltz, R.A., eds., Rotterdam: Balkema.
6. Nova, R. (1982). A constitutive model for soil under monotonic and cyclic loading, in *Soil Mechanics: Transient and Cycling loads*, Pande, G.N., and Zienkiewicz, O.C., eds., Chichester: Wiley.

7. Laloui, L., Geiser, F., Vulliet, L., Li, X.L., Bolle, A., and Charlier, R. (1997). Characterization of the mechanical behaviour of an unsaturated sandy silt, in *Proc XIVth Int. Conf. on Soil Mechanics and Foundation Engineering*, Hambourg, pp. 347–350.
8. Fredlund, D.G., Morgenstern, N.R., and Widger, R.A. (1978). The shear strength of unsaturated soils. *Canadian Geotech. J.* **15**: 313–321.
9. Wood, D.M. (1990). *Soil Behaviour and Critical State Soil Mechanics*, Cambridge University Press.
10. Josa, A. (1988). An Elastoplastic Model for Partially Saturated Soils. PhD thesis, ETSICCP, Barcelona.
11. Escario, V., and Sáez, J. (1973). Measurement of the properties of swelling and collapsing soils under controlled suction, *Proc. 3rd I.C.E.S.*, Haifa, pp. 195–200.
12. Papamichos, E., and Schei, G. (1998). Characterization of Adriatic soft weak sediments for subsidence studies, IKU, SINTEF, Tronfheim, Norway, Report Nr. 33.0693.00/01/01/98.
13. Cui, Y.J., and Delage, P. (1996). Yielding and plastic behaviour of unsaturated compacted silt. *Géotechnique* **46**: 291–311.

SECTION **11.9**

# "Sinfonietta Classica": A Strain-Hardening Model for Soils and Soft Rocks

Roberto Nova
*Milan University of Technology (Politecnico), Department of Structural Engineering, Milan, Italy*

Contents



## 11.9.1 A SHORT HISTORY OF THE MODEL

The model that will be presented here is the result of successive modifications of a constitutive law originally conceived to model virgin sand [1]. That law was characterized by a nonassociated flow rule and isotropic strain-hardening, depending on both deviatoric and volumetric plastic strains [2]. Two important features were the possibility of modeling dilatant behavior even for normally consolidated dense sand and the occurrence of static liquefaction for loose sand. The model was successively applied to clays [3]. A three-dimensional generalization with different mathematical expressions for loading function and plastic potential (with similar shape to the original version) was given in Reference [4]. The model was then extended to cover soft rock behavior [5]. The last is the version that will be presented here. More



*Handbook of Materials Behavior Models*. ISBN 0-12-443341-3.

recent developments concern mainly two aspects: the effect of induced anisotropy, taken into account by means of a mixed kinematic-isotropic hardening [6,7], and the degradation of the mechanical properties of soft rocks due either to bond crushing [8] or to chemical and temperature effects [9,10].

The unusual name given to the model is a homage to Prokofiev's *Sinfonia Classica*, op. 25, which the author composed while conjugating contemporary music and a classical framework.

## 11.9.2 VALIDITY

The model is applicable in geotechnical problems concerning a wide spectrum of soils (gravels, sands, silts, clays) and isotropic soft rocks (tuffs, marls, calcarenites, chalks) either in dry or fully saturated conditions. It can also be applied to describe the behavior of particulate materials of nongeological origin (e.g., rice, wheat grains, powders).

Either monotonically varying or oligocyclic loadings are considered. The model is appropriate, therefore, for describing neither ratcheting nor cyclic mobility of sand. It is also not suitable for describing highly anisotropic materials such as shales.

## 11.9.3 FORMULATION FOR UNCEMENTED MATERIALS

The model is elastic-plastic strain hardening. Hardening (or softening) is assumed to be isotropic. Stresses are intended as effective stresses and are taken as positive in compression as well as strains.
The plastic potential g and the loading function $f$ are defined as follows:

$$g = 9(\gamma - 3)\ln\frac{p'}{p_g} - \gamma J_{3\eta} + \frac{9}{4}(\gamma - 1)J_{2\eta} = 0 \qquad (1)$$

$$f = 3\beta(\gamma - 3)\ln\frac{p'}{p_c} - \gamma J_{3\eta} + \frac{9}{4}(\gamma - 1)J_{2\eta} \leq 0 \qquad (2)$$

where

$$p' \equiv \frac{1}{3}\sigma'_{ij}\delta_{ij} \qquad (3)$$

$$J_{2\eta} \equiv \eta_{ij}\eta_{ij} \qquad (4)$$

$$J_{3\eta} \equiv \eta_{ij}\eta_{jk}\eta_{ki} \qquad (5)$$

and the stress ratio tensor $\eta_{ij}$ is defined as

$$\eta_{ij} \equiv \frac{s_{ij}}{p'} \tag{6}$$

$p_c$ is the hidden variable which determines the size of the elastic domain and controls hardening or softening in the following way:

$$\dot{p}_c = \frac{p_c}{B_p}\left(\dot{\varepsilon}_v^p + \xi\left(\dot{e}_{hk}^p \dot{e}_{hk}^p\right)^{\frac{1}{2}} + \psi\left(\dot{e}_{ij}^p \dot{e}_{jk}^p \dot{e}_{ki}^p\right)^{\frac{1}{3}}\right) \tag{7}$$

where

$$\dot{\varepsilon}_v^p \equiv \dot{\varepsilon}_{ij}^p \delta_{ij} \tag{8}$$

$$\dot{e}_{hk}^p \equiv \dot{\varepsilon}_{hk}^p - \frac{1}{3}\dot{\varepsilon}_v^p \delta_{hk} \tag{9}$$

When $f < 0$ the behavior is assumed to be hypoelastic:

$$\dot{\varepsilon}_{ij}^e = \frac{1}{3} B_e \frac{\dot{p}'}{p'} \delta_{ij} + L\dot{\eta}_{ij} \tag{10}$$

Hypoelastic strain rates are added to the plastic ones to give total strain rates when $f = 0$ and $\dot{f} = 0$. It is assumed further that no principal stress can become negative (tension cut-off) and that there exists a small initial elastic nucleus delimited by a surface given by $f = 0$ with $p_c = p_{co}$.

It is also assumed that the elastic volumetric compliance cannot be larger than $B_e/p_{co}$.

## 11.9.4 PARAMETERS

In total, the constitutive law is characterized by seven nondimensional parameters and a reference pressure $p_{co}$. Only two constitutive parameters fully characterize plastic potential and loading function: $\beta$ and $\gamma$. The parameter $p_g$ is unessential since only the derivatives of g matter.

$\gamma$ is linked to the so-called characteristic state [11]. In axisymmetric (so-called triaxial) compression

$$\gamma = \frac{9 - M^2}{\frac{2}{9}M^3 + 3 - M^2} \tag{11}$$

where $M$ is in turn linked to the mobilized friction angle $\phi'_{cv}$ at the characteristic state (associated with zero plastic volumetric strain rate) by the relation

$$M = \frac{6 \sin \phi'_{cv}}{3 - \sin \phi'_{cv}} \tag{12}$$

$\phi'_{cv}$ typically ranges from 22° to 36°, the lower value being typical of plastic clays and the higher one of angular sands. Accordingly, $\gamma$ can vary between 3.44 and 4.41.

$\beta$ controls the deviation from normality, which is only due to the spherical part. If $\beta = 3$, normality holds. In general, $\beta$ is smaller than 3. Typical values are $\beta = 2$ for clays and $\beta = 1.2$ for sands.

The parameters $B_p$ and $B_e$ are the logarithmic volumetric compliances under isotropic loading, relative to plastic and elastic strains, respectively. The value of $B_p$ depends on the type and density of the material tested. For a dense silica sand the order of magnitude of $B_p$ is 0.01, for a loose sand the order of magnitude can be two or four times larger, and for a virgin clay it ranges from 0.02 to 0.1. The ratio $B_e/B_p$ is typically of the order of 0.2. However, the higher $B_p$ is, the lower the ratio between the two parameters tends to be.

The other elastic parameter, $L$, is linked to the shear modulus $G$. It can be derived, in fact, that in a test at constant isotropic pressure

$$G = \frac{p'}{2L}. \tag{13}$$

The apparent elastic shear modulus is therefore assumed to vary linearly with the isotropic pressure. The two parameters $\xi$ and $\psi$ are linked to the dilatancy at failure, $d_f$, in axisymmetric conditions

$$d_f = \frac{\dot{\varepsilon}_v^p}{\dot{\varepsilon}^p} = -3\left(\frac{\xi}{\sqrt{6}} \frac{\varepsilon^p}{|\varepsilon^p|} + \frac{\psi}{\sqrt[3]{36}}\right) \tag{14}$$

where

$$\varepsilon \equiv \frac{2}{3}(\varepsilon_1 - \varepsilon_3) \tag{15}$$

$\xi$ depends on the type and density of the soil. It is zero for normally consolidated kaolinitic clay and for loose sand (and can be negative for carbonate sands when particle crushing takes place). The denser the sand, the higher the $\xi$ value. A typical value for dense sand is 0.3, while a silty clay can have $\xi = 0.1$. When $\xi$ is positive, dilatancy at failure is negative (expansion) in drained tests, while in undrained tests the stress path has a characteristic hook across the phase transformation line [12]. Indeed, the model predicts that the phase transformation line and the characteristic state coincide. $\psi$ is generally ten times smaller than $\xi$ and can be put equal to zero for most practical purposes. In this case the absolute value of the dilatancy at failure is the same in compression and extension. Experimental data show, however, that there is actually a small difference that can be accounted for by $\psi$.

The value of $p_{co}$ is important only for small stress levels. It can be assumed to be of the order of a few dozen of kPa. In a boundary value problem, e.g., a shallow foundation, this means that the behavior of the elements close to the free surface is assumed to be elastic, for the first loading steps, at least.

## 11.9.5 PARAMETER DETERMINATION

All the constitutive parameters, with the exception of $p_{co}$, have a clear physical meaning. Their determination from experimental data is straightforward, and only few triaxial tests are necessary. From an isotropic test with a cycle of unloading–reloading it is possible to determine $B_e$ and $B_p$ (Fig. 11.9.1). In fact, their sum gives the slope of the straight line (in a semilog plot) connecting volumetric strains to isotropic effective pressure. $B_e$ is instead the slope of the unloading–reloading branch. These parameters can be conceptually linked to the traditional compressibility indices $C_c$ and $C_s$, respectively.

The other elastic parameter, $L$, can be obtained by the initial slope of the stress-strain curve in a $p'$ constant test (Fig. 11.9.2).

As already mentioned, $\xi$ and $\psi$ govern the plastic dilatancy at failure via Eq. 14. They can be determined from the results of a drained compression and a drained extension test, taking due care of the fact that sgn $\varepsilon^p$ is different in the two cases. If for the sake of simplicity $\psi$ is assumed to be zero, only the triaxial compression test is necessary to determine $\xi$, as shown in Figure 11.9.2.

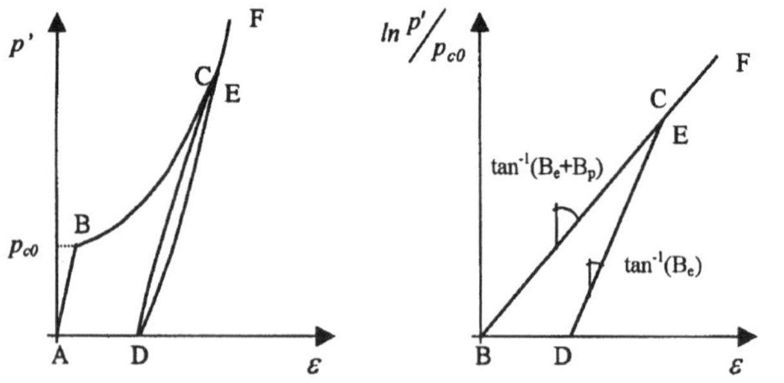


FIGURE 11.9.1 Schematic results in isotropic loading in arithmetic and semilogarithmic plot.

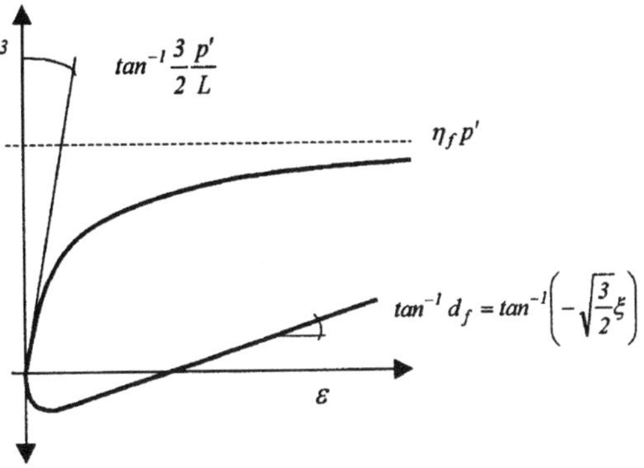


FIGURE 11.9.2 Schematic results of a $p'$ constant test on dense sand.

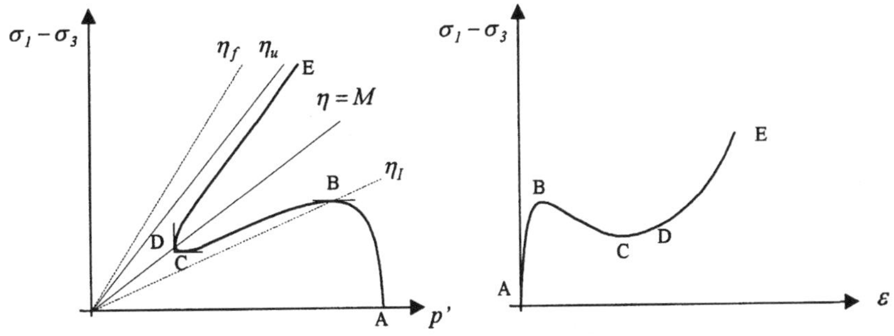


FIGURE 11.9.3 Schematic stress path and stress-strain behavior in an undrained test on a medium loose sand.

The stress level at failure $\eta_f$ depends on $d_f$ and on $\gamma$. Once $d_f$ has been determined, $\gamma$ can be derived from the measured value of $\eta_f$, which is directly linked to the friction angle $\phi'$ via a relationship formally identical to Eq. 12.

An alternative way to determine $\gamma$ is to perform an undrained test (Fig. 11.9.3). The inclination $M$ corresponding to $\gamma$ via Eq. 11 can be obtained by connecting the origin with point $D$ on the stress path curve, characterized by a vertical tangent. Note that if $\gamma$ is determined this way and $d_f$ as in Figure 11.9.2, the limit state is uniquely determined. The traditional limit value appears then to be determined by the inclination of the characteristic state and by the dilatancy, as in traditional soil mechanics. Note further that the asymptotic level $\eta_u$ reached by the stress path in the undrained test is slightly lower than $\eta_f$.

Finally, $\beta$ can be determined by fitting the calculated curve to the data of the drained test, since the hardening modulus is proportional to $\beta$, which therefore controls the specimen deformability. Because of such a proportionality, the value of $\beta$ can be calibrated also on the value of the coefficient of earth pressure at rest $K_0$ or the inclination of the "instability line" $\eta_I$ [13] (Fig. 11.9.3).

## 11.9.6 FORMULATION AND PARAMETER DETERMINATION FOR CEMENTED MATERIALS

The effect of bonding between grains is taken into account by the introduction of two parameters, $p_t$ and $p_m$, defined as in Figure 11.9.4. Formally, the equations of the plastic potential and yield function are the same as Eqs. 1 and 2, but the arguments of such functions are $p^*$ and $\eta^*_{ij}$ instead of $p'$ and $\eta_{ij}$, respectively, where

$$p^* \equiv p' + p_t \tag{16}$$

$$n^*_{ij} \equiv \frac{s_{ij}}{p^*} \tag{17}$$

Hardening is controlled by three parameters: $p_t$, $p_m$, and $p_s$, where the latter varies with plastic strains as does $p_c$ in Eq. 7, and

$$\dot{p}_t = -\rho p_t |\dot{\varepsilon}^p_v| \tag{18}$$

and $p_m$ is assumed to be proportional to $p_t$.

In total, three additional constitutive parameters are introduced: the initial value of $p_t$, $p_{t0}$, the proportionality constant $\alpha$, and a decay parameter of the bond strength, $\rho$.

Since in the tensile range failure occurs at the tension cut-off, $p_{t0}$ is nothing else than the absolute value of the tensile strength in uniaxial extension (which can be evaluated by performing a direct tension or a brasilian test). A typical value of $p_{t0}$ for soft rocks is between 100 and 200 kPa.

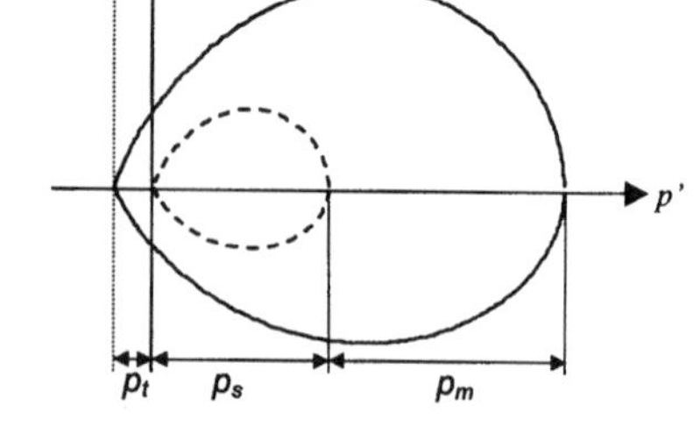


FIGURE 11.9.4 Yield locus for cemented and uncemented soil.

An isotropic compression test allows the yield value $p_{c0}$ to be determined. Since $p_{s0}$ is small, $\alpha$ can be evaluated as a first approximation as the ratio between $p_{c0}$ and $p_{t0}$. Such a value is usually between 10 and 20.

Finally, the parameter $\rho$ depends on the bond fragility, since it measures the rate at which the bonds are broken and the soft rock is transformed into a soil. It can be determined only by fitting theoretical results to experimental data in an isotropic compression test. The higher the $\rho$ value, the faster the compression curve will join the corresponding curve for an uncemented material. Such a parameter can usually vary between 0.5 and 5.

## 11.9.7 EXAMPLES

Examples of model performance can be found in Reference [4] (two types of sands in three-dimensional loading conditions), Reference [5] (sand, clay, grouted sand, tuff, and limestone in triaxial compression tests), and Reference [14] (natural and artificial calcarenite, marl in drained and undrained conditions).

## REFERENCES

1. Nova, R., and Wood, D. M. (1979). A constitutive model for sand in triaxial compression. *Int. J. Num. Anal. Meth. Geomech.* **3**: 255–278.
2. Nova, R. (1977). On the hardening of soils. *Archi. Mech. Stos.* **29** (3): 445–458.
3. Nova, R. (1982). A constitutive model for soil under monotonic and cyclic loading, in *Soil Mechanics: Transient and Cyclic Loading*, pp. 343–373, Pande, G .N., and Zienkiewicz, O. C., eds; Wiley.
4. Nova, R. (1988). Sinfonietta classica: An exercise on classical soil modelling, in *Constitutive Equations for Granular Non-cohesive Materials*, pp. 501–520, Saada and Bianchini, eds., Balkema.
5. Nova, R. (1992). Mathematical modelling of natural and engineered geomaterials. General lecture 1st E. C. S. M. Munchen, *Eur. J. Mech. A/Solids* **11**, (Special issue): 135–154.
6. di Prisco, C., Nova, R., and Lanier, J. (1993). A mixed isotropic-kinematic hardening constitutive law for sand, in *Modern Approaches to Plasticity*, pp. 83–124. Kolymbas, ed.
7. di Prisco, C., Matiotti, R., and Nova, R. (1995). Theoretical investigation of the undrained stability of shallow submerged slopes. *Géotechnique* **45** (3): 479–496.
8. Lagioia, R., and Nova, R. (1995). An experimental and theoretical study of the behaviour of a calcarenite in triaxial compression. *Géotechnique* **45** (4): 633–648.
9. Nova, R. (1986). Soil models as a basis for modelling the behaviour of geophysical materials. *Acta Mechanica* **64**: 31–44.
10. Nova, R. (2000). Modelling weathering effects on the mechanical behaviour of granite, in *Perspective and Developments of Hypoplasticity*, Kolymbas, D., ed.
11. Luong, M. P. (1980). Phénomènes cycliques dans les sols pulvèrulents. *Revue Française de Géotechnique* **10**: 39–53.

12. Ishihara, K., Tatsuoka, F., and Yasuda, S. (1975). Undrained deformation and liquefaction of sand under cyclic stresses. *Soils and Foundations* **15** (1): 29–44.
13. Lade, P. V. (1992). Static instability and liquefaction of loose fine sandy slopes. *Jour. of Geot. Engrg., ASCE* **118**, (1): 51–71.
14. Lagioia, R., and Nova, R. (1993). A constitutive model for soft rocks, *Int. Symp. Hard soils-soft rocks*, Athens, pp. 625–632.

SECTION **11.10**

# A Generalized Plasticity Model for the Dynamic Behavior of Sand, Including Liquefaction

M. Pastor[1], O.C. Zienkiewicz[2] and A.H.C. Chan[3]

[1] *Centro de Estudios y Experimentación de Obras Públicas and ETS de Ingenieros de Caminos, Madrid, Spain*
[2] *Department of Civil Engineering, University of Wales at Swansea, United Kingdom*
[3] *School of Engineering, University of Birmingham, United Kingdom*

Contents



## 11.10.1 RANGE OF APPLICATION OF MODEL

Classical plasticity models are not able to predict basic phenomena encountered in dynamic loading of saturated soils, such as the accumulation of plastic strain and the cycles, compaction, and pore pressure buildup under undrained conditions. To overcome such difficulties, a simple model was proposed by Pastor, Zienkiewicz, and Chan [1] in the framework of the generalized plasticity theory of Zienkiewicz and Mroz [3]. The model is able to reproduce the behavior of dense and loose sands under quasi-static and dynamic loading, and it can be applied to the following situations:

- static liquefaction of very loose sands and the instability phenomena found during undrained loading of medium-dense specimens;

*Handbook of Materials Behavior Models.* ISBN 0-12-443341-3.

- material softening of dense sands under drained conditions;
- zero volume change at the characteristic state line;
- densification of granular materials during drained cyclic loading;
- liquefaction and cyclic mobility of loose sands under cyclic undrained loading.

## 11.10.2 FRAMEWORK: GENERALIZED PLASTICITY THEORY

The relation between the increments of stress and strain for a material whose response does not depends on time can be written as

$$d\varepsilon = \mathbf{C} : d\boldsymbol{\sigma}$$

where the fourth-order constitutive tensor $\mathbf{C}$ depends on (i) the state of stress and strain, (ii) the past history, and (iii) the direction of the stress increment $d\boldsymbol{\sigma}$. In the case of soils, where coupling between skeleton and pore water exists, constitutive laws are formulated for the solid skeleton via the effective stresses $\boldsymbol{\sigma}' = \boldsymbol{\sigma} + p_w\mathbf{I}$, where $\boldsymbol{\sigma}$ is the total stress tensor in the mixture, $p_w$ is the pore pressure, and $\mathbf{I}$ is the second-order identity tensor. In what follows, and for the sake of simplicity, we will denote by $\boldsymbol{\sigma}$ the effective stress unless otherwise specified. Concerning sign conventions, compression is negative and pore pressure compression is negative [2].

As far as the last requirement is concerned, the dependency on the direction of $d\boldsymbol{\sigma}$ can be introduced in several ways, and the approaches of Darve and the Grenoble group [4], the hypoplastic law of Dafalias [5], and the hypoplastic model of Kolymbas *et al.* [6] are worth mentioning here.

The simplest approach consists of introducing a normalized direction $\mathbf{n}$ for any given state of stress such that all possible increments of stress are separated into classes, referred to as "loading" and "unloading":

$$d\boldsymbol{\varepsilon}_L = \mathbf{C}_L : d\boldsymbol{\sigma} \quad \text{for} \quad \mathbf{n} : \mathbf{d}\boldsymbol{\sigma} > 0 \text{ (loading)}$$

$$d\boldsymbol{\varepsilon}_U = \mathbf{C}_U : d\boldsymbol{\sigma} \quad \text{for} \quad \mathbf{n} : \mathbf{d}\boldsymbol{\sigma} < 0 \text{ (unloading)}$$

The limit case $\mathbf{n} : \mathbf{d}\boldsymbol{\sigma} = 0$ separating both classes of behavior will be called "neutral loading."

To ensure continuity between loading and unloading, it can be shown that tensors $\mathbf{C}_L$ and $\mathbf{C}_U$ have to be of the form

$$\mathbf{C}_{L/U} = \mathbf{C}^e + \frac{1}{H_{L/U}} \mathbf{n}_{gL/U} \otimes \mathbf{n}$$

where $\mathbf{n}_{gL}$ and $\mathbf{n}_{gU}$ are arbitrary tensors of unit norm and $H_{L/U}$ two scalar functions defined as loading and unloading plastic moduli. It can be very easily verified that both laws predict the same strain increment under neutral loading where both expressions are valid and hence nonuniqueness is avoided. As for such loading, the increments of strain using the expressions for loading and unloading are

$$d\boldsymbol{\varepsilon}_L = \mathbf{C}_L : d\boldsymbol{\sigma} = \mathbf{C}^e : d\boldsymbol{\sigma}$$

$$d\boldsymbol{\varepsilon}_U = \mathbf{C}_U : d\boldsymbol{\sigma} = \mathbf{C}^e : d\boldsymbol{\sigma}$$

This suggests that the strain increment can be decomposed into two parts:

$$d\boldsymbol{\varepsilon} = d\boldsymbol{\varepsilon}^e + d\boldsymbol{\varepsilon}^p$$

where the elastic and plastic components are

$$d\boldsymbol{\varepsilon}^e = \mathbf{C}^e : d\boldsymbol{\sigma}$$

$$d\boldsymbol{\varepsilon}^p = \frac{1}{H_{L/U}}\left(\mathbf{n}_{gL/U} \otimes \mathbf{n}\right) : d\boldsymbol{\sigma}$$

To account for softening behavior ($H_{L/U} < 0$), definitions of loading and unloading are modified as

$$d\boldsymbol{\varepsilon}_L = \mathbf{C}_L : d\boldsymbol{\sigma} \quad \text{for} \quad \mathbf{n} : d\boldsymbol{\sigma}^e > 0 \text{ (loading)}$$

$$d\boldsymbol{\varepsilon}_L = \mathbf{C}_L : d\boldsymbol{\sigma} \quad \text{for} \quad \mathbf{n} : d\boldsymbol{\sigma}^e < 0 \text{ (unloading)}$$

where $d\boldsymbol{\sigma}^e$ is given by

$$d\boldsymbol{\sigma}^e = \mathbf{C}^{e-1} : d\boldsymbol{\varepsilon}$$

All these relations can be cast in an equivalent vectorial form, substituting the fourth-order tensor by matrices and the second-order tensors by vectors. This simpler notation will be employed in the following sections.

## 11.10.3 MODEL DESCRIPTION

The proposed general law includes scalars $H_{L/U}$ and directions $\mathbf{n}$ and $\mathbf{n}_{gL/U}$, which have to be determined in order to fully characterize the material behavior.

First of all, the direction of plastic strain increment $\mathbf{n}_{gL}$ is obtained from triaxial tests, assuming that the ratio between total volumetric and deviatoric components is a good estimate of the ratio of the plastic components. The dilatancy law is approximated as

$$\frac{d\varepsilon_v^p}{d\varepsilon_s^p} \approx d_g = (1+\alpha)\left(M_g - \eta\right)$$

where $\alpha$ and $M_g$ are material parameters and $\eta$ is the stress ratio $\eta = q/p$, where $q = \sqrt{3J_2}$ and $p = -tr(\boldsymbol{\sigma})/3$.

Both parameters can be obtained from drained triaxial tests. It is interesting to note that $M_g$ characterizes the states at which no volume change occurs. This is the characteristic state line of Habib and Luong [7], which coincides with the projection of the critical state line on the plane ($p$–$q$). From this point, direction $\mathbf{n}_{gL}$ is obtained as

$$\mathbf{n}_{gL}^T = (n_{gv},\ n_{gs})$$

with

$$n_{gv} = \frac{d_g}{\sqrt{1+d_g^2}} \text{ and } n_{gs} = \frac{1}{\sqrt{1+d_g^2}}$$

Next, direction $\mathbf{n}$ will be assumed to be given by the law

$$\mathbf{n}^T = (n_v,\ n_s)$$

with

$$n_v = \frac{d_f}{\sqrt{1+d_f^2}} \text{ and } n_s = \frac{1}{\sqrt{1+d_f^2}}$$

where $d_f = d_g = (1+\alpha)M_f - \eta$.

In these equations, a new material parameter $M_f$ has been introduced. It can be seen that an associated plasticity model can be produced by choosing $M_f = M_g$. In the case of granular materials, it has been suggested to use the approximate relation $M_f/M_g = D_r$ to estimate $M_f$ [1].

The plastic modulus during virgin loading is postulated to be of the form [1]

$$H_L = H_0 p' H_f \{H_v + H_s\}$$

where

$$H_f = \left(1 - \frac{\eta}{\eta_f}\right)^4$$

$$\eta_f = \left(1 + \frac{1}{\alpha}\right) M_f$$

limit the possible states, and where

$$H_v = \left(1 - \frac{\eta}{M_g}\right)$$

$$H_s = \beta_0 \beta_1 \exp(-\beta_0 \xi)$$

$\xi$ being the accumulated plastic deviatoric strain $\xi = \int d\varepsilon_s$.

In these equations, the three new material parameters $H_0$, $\beta_0$, and $\beta_1$ have been introduced. Concerning $H_0$, it can be obtained from isotropic stress paths, but $\beta_0$ and $\beta_1$ have to be obtained by finding the values which better fit experiments.

The model so far developed needs to be completed by including plasticity during unloading and a memory function to describe past events. The latter aspect can be introduced by assuming that directions $\mathbf{n}$ and $\mathbf{n}_{gL}$ do not differ in loading and reloading, and by taking the plastic modulus $H_L$ as

$$H_L = H_0.p'.H_f(H_v + H_s)H_{DM}$$

where $H_{DM}$ is a *discrete memory factor* given by

$$H_{DM} = \left(\frac{\zeta\max}{\zeta}\right)^{\gamma}$$

where $\zeta$ was defined previously as

$$\zeta = p'.\left\{1 - \left(\frac{1+\alpha}{\alpha}\right)\frac{\eta}{M}\right\}^{1/\alpha}$$

and $\gamma$ is a new material constant which has to be chosen to provide the best fit to loading–reloading experiments.

Finally, we characterize plastic behavior during unloading by introducing the plastic modulus during unloading $H_u$ as

$$\begin{aligned} H_u &= H_{u0}\left(\frac{M_g}{\eta_u}\right)^{\gamma_u} && \text{for } \left|\frac{M_g}{\eta_u}\right| > 1 \\ &= H_{u0} && \text{for } \left|\frac{M_g}{\eta_u}\right| \leq 1 \end{aligned}$$

and assuming that direction $\mathbf{n}_{gU}$ is

$$\mathbf{n}_{gU} = \left(n_{guv}, n_{gus}\right)^T$$

where

$$n_{guv} = -abs\left(n_{gv}\right)$$

and

$$n_{gus} = +n_{gs}$$

In this way volumetric deformation during nature is of a contractive nature, as observed in experiments.

Elastic behavior is characterized by constants $K_{ev0}$ and $G_0$, the bulk and shear moduli of the material.

## 11.10.4 APPLICATIONS

We will show here some simple applications to the undrained behavior of sands of different relative densities under monotonic undrained loading and cyclic loading.

Figure 11.10.1 shows stress paths in the *p–q* space followed by samples of different densities during undrained triaxial loading. The predictions agree reasonably well with the experiments of Castro [8].

In the case of very loose specimens, it can be seen how liquefaction develops. It is interesting to note that even if there is a decrease in the deviatoric stress, the mobilized stress ratio, which in frictional materials is a measure of the mobilized strength, is always increasing. The material keeps during the whole process a tendency to densify, as the pore pressure is always increasing. The stress path approaches the origin, with an effective hydrostatic stress approaching zero. In this limit, the soil behaves no longer as a solid, but rather as a viscous fluid.

Under cyclic loading, liquefaction is the consequence of two factors, the accumulation of pore pressure during cyclic loading, and finally, liquefaction in a last cycle (in the same way that happens under monotonic loading).

Figure 11.10.2 shows a comparison between experimental data and model predictions for a very loose sand under undrained cyclic loading. Liquefaction develops in the last cycle.

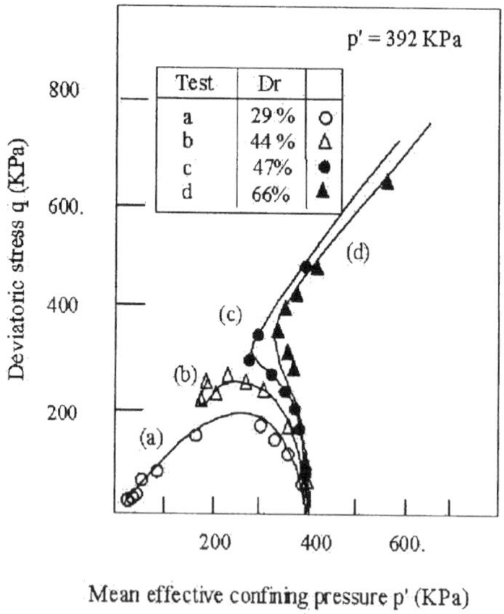


FIGURE 11.10.1 Undrained behavior of BANDING at different densities (data from Castro [8]).

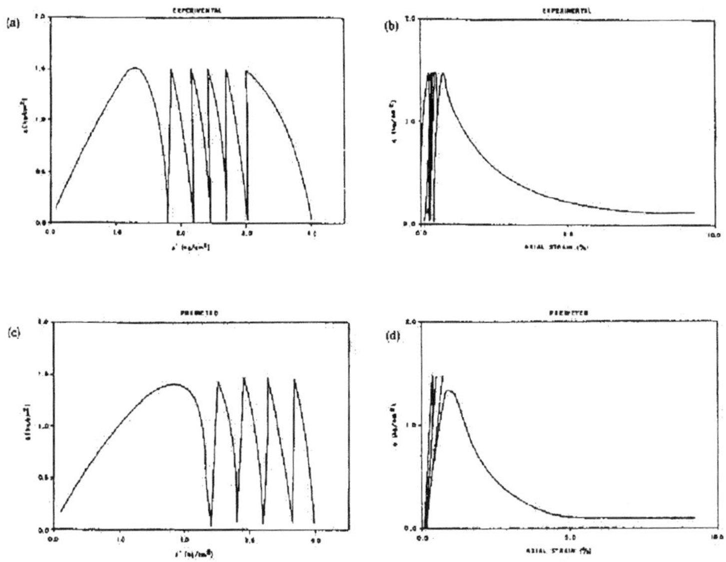

FIGURE 11.10.2 Liquefaction of a very loose sand under cyclic loading (experimental data from Castro [8]).

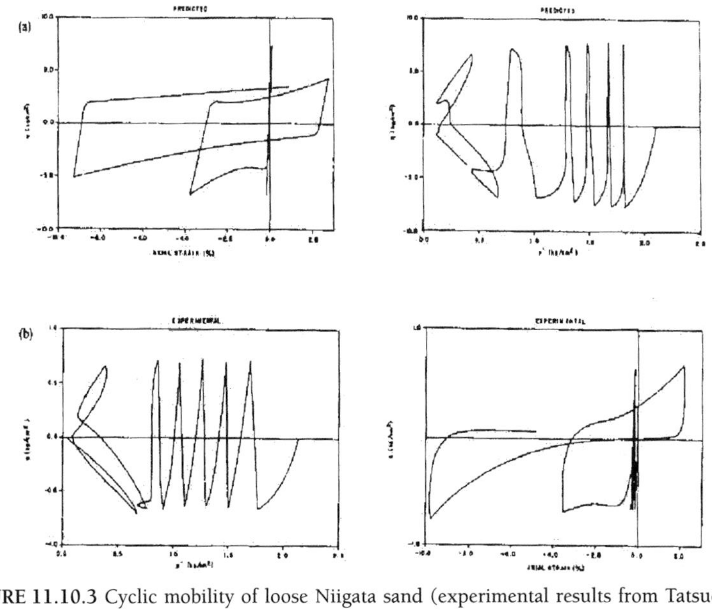

FIGURE 11.10.3 Cyclic mobility of loose Niigata sand (experimental results from Tatsuoka []).

TABLE 11.10.1 Material Properties

| | Fig. 11.10.1 and 11.10.2 ($D_r = 0.29$) | Fig. 11.10.1 ($D_r = 0.44$) | Fig. 11.10.1 ($D_r = 0.47$) | Fig. 11.10.1 ($D_r = 0.66$) | Fig. 11.10.3 |
|---|---|---|---|---|---|
| $K_{ev0}$ | 35,000 | 35,000 | 35,000 | 35,000 | 65,000 |
| $G_0$ | 52,500 | 52,500 | 52,500 | 52,500 | 30,000 |
| $M_f$ | 0.4 | 0.545 | 0.570 | 0.72 | 0.71 |
| $M_g$ | 1.5 | 1.32 | 1.12 | 1.03 | 1.5 |
| $H_0$ | 350 | 350 | 350 | 350 | 800 |
| $\beta_0$ | 4.2 | 4.2 | 4.2 | 4.2 | 3.8 |
| $\beta_1$ | 0.2 | 0.2 | 0.2 | 0.2 | 0.16 |
| $\gamma$ | 4 | 1.0 | — | — | 1 |
| $H_{u0}$ | 600 | 200 | — | — | 250 |
| $\gamma_u$ | 2 | 2 | — | — | 5 |

If density increases, the sand will exhibit cyclic mobility rather than liquefaction. Figure 11.10.3 shows cyclic mobility of loose Niigata sand. As density is higher, the stress path crosses the characteristic state line and, because of the tendency to dilate, turns towards the right. However, the accumulation of pore pressure shifts the cycles to the left, where the mean effective confining pressure is small. There, a moment arrives at which large deformations occur, as shown in the figure.

## 11.10.5 MATERIAL PROPERTIES

The material properties are shown in Table 11.10.1.

## REFERENCES


1. Pastor, M., Zienkiewicz, O. C., and Chan, A. H. C. (1990). Generalized plasticity and the modelling of soil behaviour. *Int. J. Numer. Anal. Methods Geomech.* **14**: 151–190.
2. Zienkiewicz, O. C., Chan, A. H. C., Pastor, M., Schrefler, B., and Shiomi, T. (1999). *Computational Geomechanics*, John Wiley and Sons.
3. Zienkiewicz O. C., and Mroz, Z. (1984). Generalized plasticity formulation and application to geomechanics, in *Mechanics of Engineering Materials*, Desia, C. S., and Gallaher R. H., eds., John Wiley and Sons.
4. Darve, F., ed. (1990). *Geomaterials: Constitutive Equations and Modelling*, Elsevier Applied Science.
5. Dafalias, Y. F. (1986). Bounding surface plasticity. I: Mathematical foundation and hypoplasticity. *Journal of Engineering Mechanics ASCE* **112**: 966–987.

6. Kolymbas, D. (1991). An outline of hypoplasticity. *Archive of Applied Mechanics* **61**: 143–151.
7. Habib, P., and Luong, M. P. (1978). Sols pulvurulents sous chargement cyclique, in *Materiaux and Structures Sous Chargement Cyclique*, Ass. Amicale des Ingenieurs Anciens Eléves de l'Ecole Nationale des Ponts et Chaussées (Palaiseau, 28–29), pp. 49–79.
8. Castro, G. (1969) Liquefaction of Sands. Ph.D. thesis, Harvard University, Harvard Soil Mech. Series no. 81.

SECTION **11.11**

# A Critical State Bounding Surface Model for Sands

MAJID T. MANZARI[1] and YANNIS F. DAFALIAS[2]

[1] *Civil and Environmental Engineering, The George Washington University, Washington, D.C., USA*
[2] *Department of Mechanics, National Technical University of Athens, 15773, Hellas, and Civil and Environmental Engineering, University of California, Davis, California, USA*

Contents



## 11.11.1 TRIAXIAL SPACE FORMULATION

The following constitutive model applies to sandy soils at different densities and pressures which do not cause crushing of the grains. It is a general purpose model for multiaxial, drained, undrained, monotonic, and cyclic loading conditions, within the general framework of critical state soil mechanics and bounding surface plasticity. The presentation is a direct derivative of the work by Manzari and Dafalias [1] and includes some additional expressions. The basic concepts and related equations of the sand plasticity model will first be presented in the classical triaxial space where $q = \sigma_1 - \sigma_3$, $p = (1/3)(\sigma_1 + 2\sigma_3)$, $\varepsilon_q = (2/3)(\varepsilon_1 - \varepsilon_3)$, and $\varepsilon_v = \varepsilon_1 + 2\varepsilon_3$, with $\sigma_i$ and $\varepsilon_i$ $(i = 1, 3)$ being the principal stress and strains ($\sigma_2 = \sigma_3$, $\varepsilon_2 = \varepsilon_3$).



*Handbook of Materials Behavior Models*. ISBN 0-12-443341-3.

#### 11.11.1.1 Basic Equations

The elastic relation will be assumed hypoelastic for simplicity, given in terms of the rates $\dot{q} = dq/dt$ and $\dot{p} = dp/dt$ as

$$\dot{\varepsilon}_q^e = \frac{\dot{q}}{3G} \quad \dot{\varepsilon}_v^e = \frac{\dot{p}}{K} \tag{1}$$

$$G = G_0\left(\frac{p}{p_{at}}\right)^a \quad K = K_0\left(\frac{p}{p_{at}}\right)^a \tag{2}$$

where $G$ and $K$ are the elastic shear and bulk moduli, respectively, $p_{at}$ is the atmospheric pressure, and the exponent $a$ is usually given a default value of $a = 0.5$.

The elastic range is represented by the shaded wedge shown in Figure 11.11.1 in the $q$, $p$ space, whose straight line boundaries Oc and Oe constitute the yield surface described analytically by

$$f = \eta - \alpha \mp m = 0 \tag{3}$$

where the stress ratio $\eta = q/p$ and the dimensionless (stress ratio type) quantities $\alpha$ and $m$ are shown in Figure 11.11.1. The $\alpha$-line is the bisector of the wedge angle, and 2 mp measures the wedge "opening." Equation 3 implies that upon constant $\eta$ loading no plastic loading occurs, which is approximately correct if the $p$ is not high enough to cause crushing of the sand grains and/or the sand sample is not very loose. Based on Eq. 3, plastic deformation occurs only when $\eta$ is on $f = 0$ and there is a change $d\eta = \dot{\eta}dt$ pointing outwards $f = 0$. In this case the plastic rate equations are given by

$$\dot{\varepsilon}_q^p = \frac{\dot{\eta}}{K_p} \quad \dot{\varepsilon}_v^p = D\left|\dot{\varepsilon}_q^p\right| \tag{4}$$

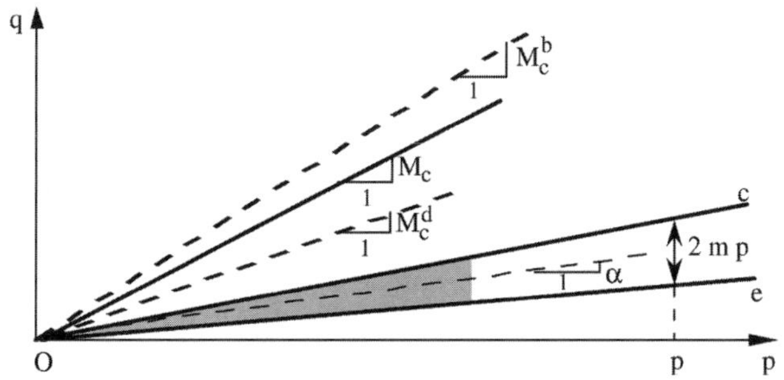


FIGURE 11.11.1 Schematic representation of the yield, critical, dilatancy, and bounding lines in $q$, $p$ space.

where $K_p$ and $D$ are the plastic modulus and dilatancy, respectively. Since $\dot{\varepsilon}_q = \dot{\varepsilon}^e_q + \dot{\varepsilon}^p_q$ and $\dot{\varepsilon}_v = \dot{\varepsilon}^e_v + \dot{\varepsilon}^p_v$, it is a trivial exercise to combine Eqs. 1 and 3 in order to express $\dot{\varepsilon}_q$ and $\dot{\varepsilon}_v$ in terms of $\dot{q}$ (or $\dot{\eta}$) and $\dot{p}$, and vice versa.

There remains the very important task of specifying $K_p$ and $D$. In reference to Figure 11.11.1, assume that the line shown with a slope $M^b_c$ represents a peak stress ratio for a given state. Henceforth, subscripts $c$ and $e$ imply association of a quantity with triaxial compression and extension, respectively. Such value of $M^b_c$ is a bound for $\eta$; hence, within the framework of bounding surface plasticity one can write a stress ratio "distance"-dependent plastic modulus expression such as

$$K_p = h\left(M^b_c - \eta\right) \tag{5}$$

in terms of a model parameter $h$. Similarly, the line shown with a slope $M^d_c$ in Figure 11.11.1, represents the phase transformation line, or for better naming, the dilation line. According to standard dilatancy theory, one can write

$$D = A\left(M^d_c - \eta\right) \tag{6}$$

with $A$ another model parameter. Hence, Eqs. 5 and 6 determine $K_p$ and $D$ via $M^b_c$, $M^d_c$, $h$ and $A$, and the model is complete.

## 11.11.1.2 Critical State

The critical state in soil mechanics is defined as a triplet of $q_c$, $p_c$, and void ratio $e_c$ values, at which unlimited plastic deviatoric strain occurs at zero volumetric strain rate. Such a critical state is defined simultaneously in the $q, p$ space by $\eta_c = q_c/p_c = M_c$, and in the $e-p$ space by $e_c = (e_c)_{ref} - \lambda ln(p_c/p_{ref})$, where $M_c$, $\lambda$, and $(e_c)_{ref}$ (for a chosen $p_{ref}$) are standard soil constants. The $M_c$ is related to friction angle, and its corresponding visualization in $q$, $p$ space is shown as a line of slope $M_c$ in Figure 11.11.1.

If left as is, the previous formulation will not meet the critical state requirements. For example, it follows from Eqs. 4, 5, and 6 that as $\eta$ approaches $M^b_c$, which may be assumed to be equal to $M_c$, $K_p \rightarrow 0$ and $\dot{\eta} \rightarrow 0$ while $\dot{\varepsilon}^p_q > 0$ and $\dot{\varepsilon}^p_v < 0$ since $M^b_c > M^d_c$. This implies unlimited dilation, contrary to physical expectation (negative volumetric strain rate means dilation). Furthermore, a fixed $M^b_c$ does not allow for the softening response in drained loading observed in many dense sand samples.

The remedy is to consider variable $M^b_c$ and $M^d_c$, such that at critical state $M^b_c = M^d_c = M_c$. If the so-called state parameter $\psi = e - e_c$ [2], is used as a measure of "distance" from critical state in the $e$, $p$ space ($e$ and $e_c$ refer to the same $p$), the idea put forth by Wood *et al.* [3] and supplemented by Manzari and Dafalias [1] in regards to the concept of $\alpha$ in Eq. 3 can be expressed by

$$M_c^b = \alpha_c^b + m = M_c - k_c^b \psi \tag{7}$$

where $k_c^b$ is a material constant and $\alpha_c^b$ is the "bound" for $\alpha$ in the same sense that $M_c^b$ is the bound for $\eta$. The second important modification refers to the variation of $M_c^d$ and was proposed in Reference [1] as

$$M_c^d = \alpha_c^d + m = M_c + k_c^d \psi \tag{8}$$

where $k_c^d$ is a material constant and $\alpha_c^d$ is a back-stress dilatancy ratio corresponding to the stress dilatancy ratio $M_c^d$. Observe that for $\psi < 0$ (denser than critical), $M_c^d < M_c < M_c^b$, while for $\psi > 0$ (looser than critical), $M_c^b < M_c < M_c^d$, reflecting standard properties of granular soils. For further reference, Eqs. 7 and 8 can be supplemented by

$$M_c = \alpha_c^c + m \tag{9}$$

which defines the critical back-stress ratio $\alpha_c^c$ in terms of $m$ and the critical stress ratio $M_c$.

Equations 1–8 provide a complete constitutive model in triaxial space. It only requires the specification of $M_e$, $k_e^b$, and $k_e^d$ in triaxial extension as well to describe reverse and cyclic loading. In such case, Eqs. 5 and 6 utilize the ensuing values of $M_e^b$ and $M_e^d$ in extension, following from Eqs. 7 and 8.

## 11.11.2 MULTIAXIAL STRESS SPACE GENERALIZATION

The multiaxial stress generalization of the model follows standard procedures [1]. Equation 1 becomes

$$\dot{\mathbf{e}}^e = \frac{\dot{\mathbf{s}}}{2G} \quad \dot{\varepsilon}_v^e = \frac{\dot{p}}{K} \tag{10}$$

where bold-face characters imply tensor quantities and $\mathbf{e}^e$ and $\mathbf{s}$ are the deviatoric elastic strain and stress tensors, respectively. Equation 2 remains as is. Equation 3 generalizes to

$$f = [(\mathbf{s} - p\boldsymbol{\alpha}):(\mathbf{s} - p\boldsymbol{\alpha})]^{1/2} - \sqrt{2/3}\, m\, p = 0 \tag{11}$$

where $\alpha$ of Eq. 3 generalizes to the back-stress-ratio tensor $\boldsymbol{\alpha}$ and : implies the trace of the product of two tensors. The plastic strain rate equations are expressed in the following in terms of $\dot{\mathbf{r}}$, where the deviatoric stress ratio tensor $\mathbf{r} = \mathbf{s}/p$ is the generalization of $\eta$.

After some algebra which involves the $\partial f/\partial \boldsymbol{\sigma} = \mathbf{n} - (1/3)(\mathbf{n} : \mathbf{r})\mathbf{I}$ with $\mathbf{I}$ the identity tensor and $\mathbf{n}$ given in following text, one has for the deviatoric and volumetric plastic strain rates, respectively,

$$\dot{\mathbf{e}}^p = <L> \mathbf{n} = <L> (2/3)^{-1/2}(\mathbf{r} - \boldsymbol{\alpha}) \quad \dot{\varepsilon}_v^p = <L> D \tag{12a}$$

$$L = \frac{1}{K_p} p\mathbf{n} : \dot{\mathbf{r}} = \frac{2G\mathbf{n} : \dot{\mathbf{e}} - K(\mathbf{n} : \mathbf{r})\dot{\varepsilon}_v}{K_p + 2G - KD(\mathbf{n} : \mathbf{r})} \tag{12b}$$

Equations 12 are the generalization of Eqs. 4 involving the plastic modulus $K_p$ and dilatancy $D$, with the added explicit feature of the loading index $L$ expressed either in terms of $\dot{\mathbf{r}}$ or in terms of $\dot{\mathbf{e}}$ and $\dot{\varepsilon}_v$. A combination of Eqs. 10 and 12 together with $\dot{\mathbf{e}} = \dot{\mathbf{e}}^e + \dot{\mathbf{e}}^p$, $\dot{\varepsilon}_v = \dot{\varepsilon}_v^e + \dot{\varepsilon}_v^p$ yields

$$\dot{\boldsymbol{\sigma}} = \mathbf{E}^{ep} : \dot{\boldsymbol{\varepsilon}} = \left[\mathbf{E}^e - \frac{(2G\mathbf{n} + KD\mathbf{I}) \otimes (2G\mathbf{n} - K(\mathbf{n} : \mathbf{r})\mathbf{I})}{K_p + 2G - KD(\mathbf{n} : \mathbf{r})}\right] : \dot{\boldsymbol{\varepsilon}} \tag{13}$$

for the total effective stress rate $\dot{\boldsymbol{\sigma}}$ in terms of the total strain rate $\dot{\boldsymbol{\varepsilon}}$ and the elastoplastic tangent stiffness moduli $\mathbf{E}^{ep}$. The latter is given explicitly by the bracketed quantity of the last member of Eq. 13, where $\otimes$ means tensor product and $\mathbf{E}^e$ is the well known isotropic elastic moduli tensor which in component form is given by $E^e_{ijkl} = 2G(\delta_{ik}\delta_{jl} - (1/3)\delta_{ij}\delta_{kl}) + K\delta_{ij}\delta_{kl}$.

The $K_p$ and $D$ in the multiaxial space will be obtained by generalization of corresponding triaxial concepts and equations. First, the bounding, dilatancy, and critical triaxial back-stress ratios $\alpha_c^b$, $\alpha_c^d$, and $\alpha_c^c$ in Eqs. 7, 8, and 9, correspondingly, are generalized as the bounding, dilatancy, and critical surfaces whose traces on the principal stress ratio $r_i = s_i/p$ $\pi$-plane are shown in Figure 11.11.2. The $\pi$-plane trace of the yield surface with its center $\boldsymbol{\alpha}$, Eq. 11, is also shown in Figure 11.11.2 as a circle. For a stress ratio $\mathbf{r}$ and

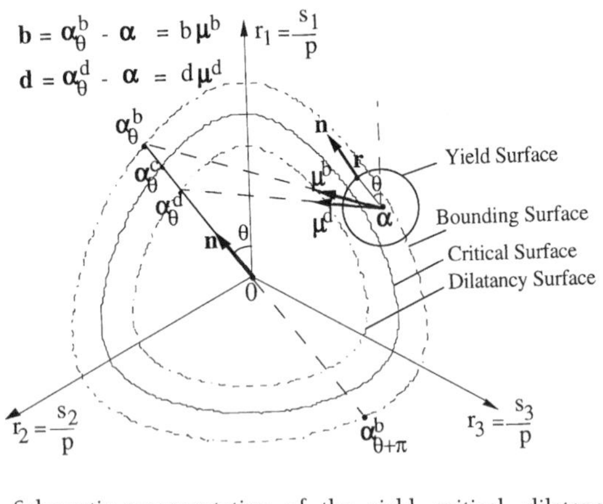


FIGURE 11.11.2 Schematic representation of the yield, critical, dilatancy, and bounding surfaces. (Reproduced with permission from Manzari, M. T., and Dafalias Y. F. (1997). A critical state two=surface plasticity model for stress. *Geotechnique* 47: 255–272.)

associated $\mathbf{n}$, the "image" back-stress ratio tensors $\alpha_\theta^b$, $\alpha_\theta^d$, and $\alpha_\theta^c$ are defined as the intersection of the $\mathbf{n}$ direction emanating from the origin with the foregoing three surfaces. Their scalar-valued norms are analytically given by

$$\alpha_\theta^b = g(\theta)M_c - g^b(\theta)k_c^b\psi - m \tag{14a}$$

$$\alpha_\theta^d = g(\theta)M_c + g^d(\theta)k_c^d\psi - m \tag{14b}$$

$$\alpha_\theta^c = g(\theta)M_c - m \tag{14c}$$

in terms of a third stress invariant, the Lode angle $\theta$ as shown in Figure 11.11.2, entering the interpolation functions $g$, $g^b$, and $g^d$ in order to account for the variation from triaxial compression to triaxial extension. Observe that for $\psi = 0$, $\alpha_\theta^b = \alpha_\theta^d = \alpha$, and the three surfaces collapse into the critical one. The corresponding image tensor quantities follow according to $\boldsymbol{\alpha}_\theta^a = \sqrt{2/3}\alpha_\theta^a\mathbf{n}$, with $a = b$, $d$, or $c$ (one must distinguish between the tensor $\boldsymbol{\alpha}_\theta^a$ and its norm $\alpha_\theta^a$). The kinematic hardening is given by $\dot{\boldsymbol{\alpha}} = <L> h \times (\boldsymbol{\alpha}_\theta^b - \boldsymbol{\alpha})$ in terms of a model parameter $h$, which, together with the consistency condition $\dot{f} = 0$ applied to Eq. 11, yields for $\dot{m} = 0$ the value of the plastic modulus as

$$K_p = ph\left(\boldsymbol{\alpha}_\theta^b - \boldsymbol{\alpha}\right) : \mathbf{n} \tag{15}$$

Similarly, the dilatancy $D$ is given by

$$D = A(\boldsymbol{\alpha}_\theta^d - \boldsymbol{\alpha}) : \mathbf{n} \tag{16}$$

Observe that a combination of Eqs. 3, 5–8 yields for the triaxial case that $K_p = h\left(\alpha_c^b - \alpha\right)$ and $D = A(\alpha_c^d - \alpha)$, hence, Eqs. 15 and 16 are their direct generalization. In applications, $h$ and $A$ may be constant or functions of the corresponding distances $b = (\boldsymbol{\alpha}_\theta^b - \boldsymbol{\alpha}) : \mathbf{n}$ and $d = (\boldsymbol{\alpha}_\theta^d - \boldsymbol{\alpha}) : \mathbf{n}$, respectively, Figure 11.11.2. In Reference [1] the expression $h = h_0|b|/(b_{ref} - |b|)$ was used in terms of a model constant $h_0$. The dependence of $K_p$, and by extension of $D$, on a distance between a stress-type quantity $\boldsymbol{\alpha}$ and its "image" $\boldsymbol{\alpha}_\theta^b$ on a surface is the classical constitutive feature of bounding surface plasticity.

## 11.11.3 IMPLEMENTATION AND MODEL CONSTANTS

The model is a usual bounding surface plasticity model, and its implementation follows standard procedures. The reader is referred to Manzari and Prachathananukit [4] for details of a fully implicit implementation. The model constants are summarized and divided in categories in Table 11.11.1, together with a set of typical values in parentheses employed in Reference [1].

TABLE 11.11.1 Model Constants

| Elastic | Critical state | State parameter | Hardening | Dilatancy |
|---|---|---|---|---|
| $G_0$ $(3.14 \times 10^4)$ | $M_c$ (1.14) | $k_c^b$ (3.97) | $h_0$ (1200) | $A$ (0.79) |
| $K_0$ $(3.14 \times 10^4)$ | $\lambda$ (0.025) | $k_c^d$ (4.20) | | |
| $a$ (0.6) | $(e_c)_{ref}$ (0.80) | | | |

The most peculiar to the model among the foregoing constants are the $k_c^b$ and $k_c^d$ (and their corresponding value $k_e^b$, $k_e^d$ in extension). The $k_c^b$ can be obtained from Eq. 7 and the experimentally observed values of the peak stress ratio $M_c^b$ and state parameter $\psi$. Similarly, the $k_c^d$ can be obtained from the observed value of $M_c^d$ when consolidation changes to dilation together with the corresponding value of $\psi$. These presuppose knowledge of the critical state line in $e$–$p$ space. For different $M_c^b$, $M_c^d$, and $\psi$'s, different $k_c^b$ and $k_c^d$ may be determined. It is hoped that these values do not differ a lot, and then an average value is the overall best choice. The $h_0$ and $A$ are obtained by trial and error (there are some direct methods also). All constants can be determined by standard triaxial experiments.

## ACKNOWLEDGEMENTS

M.T. Manzari would like to acknowledge partial support by the NSF grant CMS-9802287, and Y.F. Dafalias by the NSF grant CMS-9800330.

## REFERENCES

1. Manzari, M.T., and Dafalias, Y.F. (1997). A critical state two-surface plasticity model for sands. *Geotechnique* 47: 255–272.
2. Been, K., and Jefferies, M.G. (1985). A state parameter for sands. *Geotechnique* 35: 99–112.
3. Wood, D.M., Belkheir, K., and Liu, D.F. (1994). Strain softening and state parameter for sand modelling. *Geotechnique* 44: 335–339.
4. Manzari, M.T., and Prachathananukit, R. (2001). On integration of a cyclic soil plasticity model. Int. Journal for Numerical and Analytical Methods in *Geomechanics,* 25: 525–549.

SECTION **11.12**

# Lattice Model for Fracture Analysis of Brittle Disordered Materials like Concrete and Rock

J.G.M. van Mier

*Delft University of Technology, Faculty of Civil Engineering and Geo-Sciences, Delft, The Netherlands*

Contents



## 11.12.1 INTRODUCTION

Lattice-type models have a long history, but the development of useful models dates back to the end of the 1980s, when they were reinvented in statistical physics for simulating pattern growth in random media. Examples are fracture, dielectric breakdown of fuse networks, diffusion-limited aggregation, etc. The overviews written by Herrmann and Roux [2], Charmet *et al.* [1], Van Mier [8], and Krajcinovic and Van Mier [3] provide excellent introductions and examples of applications to this type of modeling. In this contribution we will deal with a lattice model for fracture analysis of concrete only.

*Handbook of Materials Behavior Models.* ISBN 0-12-443341-3.

## 11.12.2 BACKGROUND OF THE MODEL

In a lattice model the material is schematized as a network of linear elements, i.e., springs, truss, or beam elements that are used to describe dependency relations between the nodes in a lattice. The lattice is a regular or irregular construction in two or three dimensions. Examples are regular triangular lattices or triangular lattices with randomly varying element lengths. Triangular lattices are preferred since they yield more realistic estimates of Poisson's ratio [5, 7]. Earlier lattice models, for example, Herrmann and Roux, were based on square lattices, which result in a zero Poisson's ratio and — when fracture is considered — not very realistic crack patterns. A lattice construction of linear elements does not resemble the real microstructure or mesostructure of, for example, concrete or rock. The effects from material structure can be included in the model by superimposing an image of the relevant material structure on top of a lattice. Properties are then assigned to the lattice elements, depending on the specific location in the material structure. The image of the material structure can either be a computer-generated idealized structure [5] or a digital image from the real material [6]. Another method of introducing effects from material structure, i.e., randomness associated with variations of local material properties, is to assign lattice element properties according to a statistical distribution. For materials like concrete, the method of assigning properties by superimposing a "real" material structure seems the best method. An example of an overlay of a computer-generated structure on top of a regular triangular lattice is shown in Figure 11.12.1 The figure suggests that the lattice elements should be small enough to capture sufficiently small detail from the material structure. Indeed, the ratio between the length of a lattice element and the smallest material structural feature should be selected correctly.

The advantage of a lattice as "computational backbone" for numerical materials science is the fact that simple laws can be introduced to simulate fracture. These aspects will be treated in Section 11.12.4.

## 11.12.3 ELASTICITY PARAMETERS

The first step in an analysis is the determination of the lattice element size and the elastic properties of the global lattice. From this point on, only triangular lattices made with beam elements and a particle overlay will be considered. Three phases are generally distinguished in a lattice with particle overlay, namely, the aggregate and the matrix phases, and the interfacial transition zone. In Figure 11.12.1 b, c the lattice elements falling in these three phases

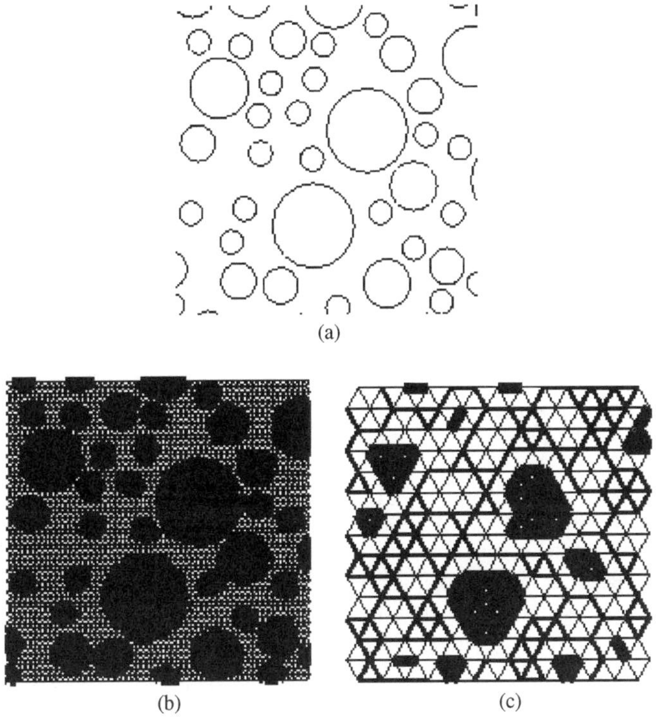


FIGURE 11.12.1 Example of an overlay of a computer-generated particle structure (a) on top of a lattice of different fineness: beam length $l = 0.5$ mm (b) and 2.0 mm (c), after Reference [9].

are given different line thicknesses. Each of these phases has own elastic properties. For convenience, all lattice elements are selected with the same cross-sectional properties, which makes the tuning of the elastic properties like the Young's modulus and the Poisson's ratio more straightforward. The first step is to set the length $l$ of the beam elements as a fraction of the smallest aggregate particle $d_{min}$. Experience learns that a ratio of $d_{min}/l$ between 3 and 5 gives a good balance between computational costs and obtained accuracy. Next the beam height is set. The value for $h/l$ depends on the overall Poisson's ratio of a lattice in relation to the Poisson's ratio of the relevant material. Schlangen and van Mier [7] derived a relation between $\nu$ and $h/l$ for two-dimensional regular triangular lattices as follows:

$$\nu = \frac{4}{3 + (h/l)^{\sqrt{3}}} - 1$$

After $l$ has been decided on the basis of the particle size, $h$ is defined depending on the required Poisson's ratio, which — for concrete — is usually in the order of 0.15 to 0.20. The global Young's modulus can now be set: ratios of $E_a/E_m$ and $E_b/E_m$ have to be defined. The subscripts *a, m,* and *b* stand for aggregate, matrix, and bond zones, respectively. The thickness of the beam elements is set to the thickness of the structure that is analyzed.

For a lattice with random beam length, Vervuurt [10] derived the relation between average *h/l* and the Poisson's ratio and Young's modulus from numerical analyses. In principle the procedure is identical to the one sketched previously. A closed-form solution does not exist for random triangular lattices.

## 11.12.4 FRACTURE ANALYSIS

Fracture can be modeled in a very simple and straightforward manner by removing in each load step the beam element with the highest (effective) stress-over-strength ratio. According to the beam distribution over the three material phases, different strength values are assigned to beams falling in the respective phases. A strength criterion is used because this simplifies the computational procedure and limits the duration of the computations. Other criteria might just as well be used. Generally, however, when iterative procedures must be adopted because nonlinear fracture laws are introduced, the computational effort will certainly grow out of bounds. The important part is the definition of effective stress in the beams. The simplest approach would be to use an effective stress based on the normal force *N* in each beam only, i.e.,

$$\sigma_{eff} = \frac{N}{A}$$

where $A = b \times h$ is the cross-sectional area of the beam. At this stage one might wonder why truss elements are not used. Removing lattice elements simulates fracture, and upon a certain amount of removals unstable regions might develop in the finite element mesh. This might impair the solution of the problem, i.e., specifically the inversion of the stiffness matrix.

A second possible effective stress is a combination of normal load and bending moment as proposed by Herrmann and Roux [2] following

$$\sigma_{eff} = \frac{N}{A} \pm \alpha \frac{(|M_i|, |M_j|)_{\max}}{W}$$

Here $M_i$ and $M_j$ are the bending moments in nodes *i* and *j* of the beam element, and $W = bh^2/6$ is the section modulus. $\alpha$ is a coefficient which regulates the amount of flexure that is taken into account. In fracture simulations of concrete this effective stress has proven to be quite effective, especially in cases when the external loading is tensile or a combination of tension and (global) shear. In case of global compression, this effective stress seems not very successful, and currently the search for a better candidate is under way. A Mohr-Coulomb-type effective stress might be a better option in that case. Recent analyses of a Brazilian splitting test revealed that the development of a shear cone in the parts of the specimen adjacent to the loading platens could be captured by means of a Mohr-Coulomb criterion. In that case, however, the vertical splitting crack did not appear, see Lilliu et al. [4].

## 11.12.5 FRACTURE PARAMETERS

The model requires a limited number of fracture parameters. They are the respective strengths of the beams falling in either of the material phases aggregate, matrix and bond. Next to that, when the second effective stress "law" is used (see previous section), the coefficient $\alpha$ must be specified. The determination of the coefficient $\alpha$ is not straightforward. For example, in order to simulate crack face bridging in concrete subjected to tension realistically, the value of $\alpha$ has to be selected close to zero (i.e., in the order of 0.005–0.010 [6]. For compressive failure, different values of $\alpha$ yield different results, and the ratio between the uniaxial compressive and uniaxial tensile strength is significantly affected. In that case, the highest ratio between the uniaxial compressive and uniaxial tensile strength is also obtained for a relatively low value of $\alpha$, but at the cost of an increased brittleness [7]. In van Mier [8] additional information on the development of inclined crack planes under uniaxial compression is shown. Table 11.12.1 contains examples of parameter settings used in many of the analyses carried out to date; see van Mier [8] for an overview.

## 11.12.6 COMPUTATIONAL PROCEDURE

The fracture simulation is carried out as follows. After the elastic properties have been determined, a unit test load is applied to the structure to be analyzed. For each beam element the effective stress is computed and divided by the strength of the phase in which it is situated. The beam with the highest stress-over-strength ratio is then removed from the mesh and a new linear elastic analysis (under the application of a unit test load) is carried out in order to decide which beam has to be removed next. After the first beam has been removed, the first point in the load-deformation

TABLE 11.12.1 Parameters for the Fracture Lattice Model (Values for Normal Strength Concrete Considered as a Three-Phase Model)

| | |
|---|---|
| Beam length $l$<br>Beam height $h$<br>Beam thickness $b$ | $l<d_{min}/3$ ($d_{min}$ is the smallest aggregate diameter)<br>set to meet the requirements for $\nu$ (see Section 11.12.3)<br>structure thickness $b$ |
| Young's moduli | $E_a/E_m = 75.000/25.000$ (MPa/MPa)<br>$E_b/E_m = 25.000/25.000$ (MPa/MPa) |
| Fracture strength | $f_{t,a}/f_{t,m}$=10/5 (MPa/MPa)<br>$f_{t,b}/f_{t,m} = 1.25/5$ (MPa/MPa) |
| Flexural parameter $\alpha$ | $\alpha$ =0.005 |

diagram is known, namely, by elongating the $P$-$\delta$ diagram until $P$ reflects the real fracture load of the considered lattice element. Thus ordinary scaling is applied. The computation is stopped after the last beam element has been removed and a complete fracture separates the structure into one or more pieces.

## REFERENCES

1. Charmet, J. C., Roux, S., and Guyon, E. (1990). *Disorder and Fracture*, New York: Plenum Press.
2. Herrmann, H. J., and Roux, S. (1990). *Patterns and Scaling for the Fracture of Disordered Media*, Elsevier Applied Science Publishers B.V. (North Holland).
3. Krajcinovic, D., and Van Mier, J. G. M. (2000). *Damage and Fracture of Disordered Materials*, CISM Courses and Lecture Notes # 410, Wien/New York: Springer.
4. Lilliu, G., Van Mier, J. G. M., and Van Vliet, M. R. A. (1999). Analysis of crack growth of the Brazilian test: Experiments and lattice analysis, in *Progress in Mechanical Behaviour of Materials, Proceedings ICM-8*, Ellyin, J., and Provan, J. W., eds., Victoria, Canada, May 16–21, Vol. I "Fatigue and Fracture," pp. 273–278.
5. Schlangen, E., and Van Mier, J. G. M. (1992). Experimental and numerical analysis of the micro-mechanisms of fracture of cement–based composites. *Cem. Conc. Comp.* 14(2): 105–118.
6. Schlangen, E. (1993). *Experimental and Numerical Analysis of Fracture Processes in Concrete*. Ph.D. thesis, Delft University of Technology, Delft, The Netherlands.
7. Schlangen, E., and Van Mier, J. G. M. (1994). Fracture simulations in concrete and rock using a random lattice, in *Computer Methods and Advances in Geomechanics*, pp. 1641–1646, Siriwardane, H. J., and Zaman, H.H., eds., Rotterdam: Balkema.
8. Van Mier, J. G. M. (1997). *Fracture Processes of Concrete: Assessment of Material Parameters for Fracture Models*, Boca Raton, FL: CRC Press.

9. Van Mier, J. G. M., and Van Vliet, M. R. A. (1999). Experimentation, numerical simulation and the role of engineering judgement in the fracture mechanics of concrete and concrete structures. *Constr. Build. Mater.* **13**: 3–14.
10. Vervuurt, A. (1997). *Interface Fracture in Concrete*. Ph.D. thesis, Delft University of Technology, Delft, The Netherlands.

# INDEX

## B

## D

## E



## F

## G

## H

# Q



# R

## S

## X



## Y



## Z